GENETIC ALGORITHMS AND GENETIC PROGRAMMING IN COMPUTATIONAL FINANCE

Contents

List of Figures

List of Tables

Preface

The applications of genetic algorithms and genetic programming to computation finance have been seen over the last decade in various journal publications, chapters in books, and magazine articles. Their relevance to computational finance is further strengthened when these tools are already deployed and used in many financial firms. Given the trend, these tools seem to deserve an independent place in computation finance. In fact, over the last few years, we have already recognized organizations of conferences' special sessions which were entirely devoted to financial applications of genetic algorithms and genetic programming. Nonetheless, a volume exclusively devoted to this subject has not been seen since the publication of *Genetic Algorithms and Investment Strategies* by Richard Bauer in 1994. Even in that book, genetic programming was not included. All the way back to 1994, financial applications of John Koza's genetic programming were indeed just getting underway.

In the year 2000, Allard Winterink, a then senior publishing editor of Kluwer Academic Publishers, brought the idea to me of publishing a special volume on this subject. Allard's enduring efforts to put forward this project and invite my involvement are very impressive. His enthusiastic support facilitates the launch of this project.

Ten chapters of the volume are based on a selection of papers presented at the *6th International Conference of the Society for Computational Economics on Computing in Economics and Finance* (**SCE'2000**), which was held at Universitat Pompeu Fabra, Barcelona, Catalonia, Spain on July 6-8, 2000. In addition to the conference papers, we also invited other leading scholars on the subject of this volume to contribute chapters. All proposed articles were then reviewed together with the conference papers and most of them were asked to revise. Five submissions which failed to meet the standard of the volume are therefore not included in this volume, which finally ended up with the 19 contributed chapters of this volume plus one introduction chapter by the editor. Many thanks is due to all the authors for cooperating with the production of the volume.

During the editing process, we saw the need for a simple menu-driven program so that beginners in this field can have no difficulty in implementing some procedures that have been suggested in the volume. Two research fellows of the *AI-ECON Research Center*, Tzu-Wen Kuo and Yun-Pyng Shieh, organized a research team and spent half a year writing a user-friendly program, called **Simple GP**. The current version of **Simple GP** enables users to implement a procedure like the *rolling forward scheme* used in Neely and Weller's chapter in the volume. The research team are currently still working hard to upgrade the program so that it can help users in the future to implement many other procedures mentioned in the volume. Including the chapter written for the software, there are a total of twenty-one chapters in this volume.

For instructors who plan to use the materials of this volume in their lectures, we also prepare some Powerpoint files for them to use. The Powerpoints are prepared by many research fellows at the *AI-ECON Research Center.* Among them, the editor is particularly grateful to Chueh-Yung Tsao, Ya-Chi Hwang, Chiao-Lin Cheng, and Chia-Lin Chang. Compiling the twenty chapters into a consistent format is not a trivial task, in particular when some chapters were originally not written in Latex. For this, I am grateful to Chung-Chin Tai and Bin-Tzong Chie. Their line-by-line, page-by-page, time-consuming task made them spend many holidays in their solemn office. My thanks also extend to Chung-Chih Liao, who read each chapter of the volume and carefully prepared the index of the volume based on his superb mastery of this field.

Finally, my biggest debt goes to my colleague, Shu George Wang, and my wife, Li-Chu, who helped me establish the *AI-ECON Research Center* and who has given me the greatest support for making the ideas of the center come true.

This book is dedicated to my parents, Kung-Tseng Chen and Hwa-Sheng Chen, and parents-in-law, Chih Sun and Chin-Ying Tseng, who give me the best family life in the world.

Chapter 1

GENETIC ALGORITHMS AND GENETIC PROGRAMMING IN COMPUTATIONAL FINANCE: AN OVERVIEW OF THE BOOK

Shu-Heng Chen

AI-ECON Research Center, Department of Economics, National Chengchi University

Taipei, Taiwan 11623

chchen@nccu.edu.tw

Abstract This chapter reviews some recent advancements in financial applications of genetic algorithms and genetic programming. We start with the more familiar applications, such as forecasting, trading, and portfolio management. We then trace the recent extensions to cash flow management, option pricing, volatility forecasting, and arbitrage. The direction then turns to agent-based computational finance, a bottom-up approach to the study of financial markets. The review also sheds light on a few technical aspects of GAs and GP, which may play a vital role in financial applications.

Keywords: Genetic Algorithms, Genetic Programming, Agent-based Computational Finance, Financial Engineering

Introduction

It has been exactly ten years since the first published application of genetic algorithms to computational finance (Bauer and Liepins (1992)). After a decade of development, it is now time to reflect upon how genetic algorithms (**GAs**) and genetic programming (**GP**) have contributed to computational finance. Even though, to date, there are only about 150 publications in this area, their application coverage is continuously increasing. The twenty-one chapters presented in this volume give us a general picture of the current state. In this volume you will see the application of genetic algorithms and genetic programming to a large domain in computational finance. In addition to the conventional applications to financial forecasting, trading strategies, trading system development, and portfolio management, there are novel applications to cash flow management, volatility modeling, option pricing, index options, and futures arbitrage.

The materials presented in this book are divided into two halves. The eleven chapters in the first half (Chapters 4- 14) discuss the modeling of financial optimization. The seven chapters (Chapters 15-21) in the second half focus on the modeling of financial markets. These two halves are, however, related as what microeconomics (individuals) is to macroeconomics (aggregation). The first half provides a blueprint for modeling individual *financial agents*, whereas their collective interacting behavior is dealt with in the second half. The connection of these divisions is fascinatingly described in Kwok Yip Szeto's and Markose's chapter in this volume. To give a background review of this research area, the volume starts with three introductory chapters (Chapters 1-3), which provide some basic materials, literature reviews, and computing practices.

1. Introductory Chapters

A great number of introductory materials to this research area is available. Bauer (1994a) is the oldest yet still the best textbook on introducing genetic algorithms to finance people, helping them to see the relevance of genetic algorithms to computational finance. Unfortunately, there are no equivalent textbooks on genetic programming. Nonetheless, Smith (1998) and Chen (1998a) provide a comprehensive review of the financial applications of genetic programming. Furthermore, in a more broader content, genetic algorithms and genetic programming are a branch of evolutionary computation and, even in a more broader sense, a branch of computational intelligence. Hence, Deboeck (1994) and Chen (2002a), while not exclusively devoted to GAs and GP, also include many overview articles. In particular, Chen and Kuo (2002) provides a bibliography covering almost 400 papers on evolutionary computation in economics and finance. Among the 400 papers, there are about 150 which are directly related to the subject of this volume. Given this rich resource, it is no longer necessary to duplicate too many introductory materials on this already sizeable volume. The two chapters presented in **Part I** of the volume do, however, make this volume self-contained.

Chapter 2, **Genetic Algorithms in Economics and Finance: Forecasting Stock Market Prices and Foreign Exchange**, by Adrian Drake and Robert Marks is a good start for those who have no background in genetic algorithms. In addition to a nice introduction to the basics of genetic algorithms, the chapter also reviews the *first half* of a decade's development of this area, from Bauer and Liepins (1992) to Pereira (1996). The review leads us through the earliest three financial domains to which genetic algorithms are applied, including financial status identification (classification), portfolio selection, and trading strategies.

A basic issue which may interest beginners is *implementation.* Implementation of genetic algorithms may not be difficult thanks to the many commercially available software packages. However, at this moment, very few packages on computation finance have a module of GP. While some GP software can be downloaded from websites, it is not written for finance people. To use it, one needs to know some programming language, and for those who are not equipped with programming skills, it may be a daunting task to apply GP. To help general finance people to overcome any technical obstacles and to exploit this novel tool, this volume provides a menu-driven program on GP.

Chapter 3, **Genetic Programming: A Tutorial with the Software Simple GP**, by Shu-Heng Chen, Tzu-Wen Kuo and Yun-Pyng Shieh is written to acquaint those beginners without a programming background with the six essential elements of genetic programming, i.e., the survival-of-the fittest principle, selection schemes, disruption avoidance, search intensity, primitives, and genetic operators. None of them are far-fetched for readers, because they can directly play and interact with each of these elements via the software's main menu, **Simple GP**. The software is demonstrated with a series of simulations to justify *a set of simple rules of thumb* in order to conduct an effective implementation of GP. A special feature of this chapter is that the authors describe the behavior of GP with the application of *production theory* from economics and *portfolio theory* from finance.

2. Main Application Domains

Financial *forecasting* and *trading* are the most active financial application domains of GAs and GP. Based on the bibliography prepared by Chen and Kuo (2002), there are about 40 publications for the former and 35 publications for the latter.[1] These two application domains are tightly connected, because one of the main purposes in making high-quality financial forecasting is to enhance the profitability of *trading.* For that purpose, one can first apply GAs or GP to evolve and build forecasting models, and then one can base trading decisions upon the resultant forecasts. Chapters 4 and 6 are applications of this style. Alternatively, one can also use GAs or GP to evolve trading decisions *directly.* Chapters 7 to 9 are cases in point.

2.1. Forecasting Financial Time Series

Forecasting financial time series can be generally described as follows. Given a time series data, $\{x_t\}$, we look for a mathematical function, $f(\ )$, such that

$$x_{t+1} = f(x_t, x_{t-1}, x_{t-2}, ...) + \epsilon_{t+1} = \hat{x}_{t+1} + \epsilon_{t+1}, \tag{1.1}$$

where the series $\{\epsilon_t\}$ is statistically independent or patternless. Term $\hat{x}_{t+1}$ is the forecast of x_{t+1}. A trading decision $g(\ )$ is the mapping,

$$g : \hat{x}_{t+1} \rightarrow \{buy, sell, hold\}. \tag{1.2}$$

Two issues can arise here. First, which series $\{x_t\}$ should interest us? Second, what is the effective characterization of the function $f(\ )$ which we are looking for? These two issues are not separate as they are evidently related through Equations 1.1 and 1.2. Ideally, the series $\{x_t\}$ should be very informative as far as market timing is concerned. Nonetheless, that information generally is hidden and cannot be effectively extracted without an appropriate choice of function $f(\ )$. In **Part II** of the volume, Chapter 4 will address the first issue, as the authors assert that activities on the internet may be an interesting series to forecast. Chapters 5 and 6 will show *two progresses* made for the second issue. The former is motivated by a *function approximation approach*, while the latter is based on the concept of *multi-stationarity*.

As already mentioned, one important issue in forecasting financial time series is the *set of variables* upon which a forecast is made. In the case of predicting stock prices, the variables range from the history of the stock price and trading volumes to technical indicators. However, attention has never been given to the information (communication) flows among market participants. Would it be a *signal* for us to predict the stock price if we see a case of *unusually* active communication among traders, or for that matter, a case of unusual quietness? In brief, would *communication density* help predict the stock price? This question has never been addressed in the literature partially because direct observations on communication density are not available. Interestingly, Chapter 4, **Genetic Programming and the Predictive Power of Internet Message Traffic**, by James Thomas and Katia Sycara, uses *the message board volume* data as a proxy for communication density, and addresses the question: *whether the message board volume data has predictive power.*

From yahoo.com and ragingbull.com, the authors collect *the volume of message board postings* on 68 stocks from Russell 1000. They then construct a time series of message traffic volume numbers, and use GP to search for trading signals. Their initial results are quite encouraging. A standard GP applied to the message traffic volume can generate trading rules for which the performance is superior to that of a buy-and-hold strategy in terms of excess returns, the excess Sharpe ratio, and the differential Sharpe ratio. Based on the bootstrap test, its dominance is statistically significant. The authors further show that message traffic volume provides genuinely new information which may not be revealed by returns and trading volumes.

The paper also deals with a few technical issues of GP. The authors place *proper representation* and *overfitting avoidance* as the keys to GP success, and consider tinkering with the parameters or using sophisticated search techniques of only secondary importance.

While genetic programming is extensively applied to financial forecasting, its mathematical or statistical foundation has not been rigorously laid. In particular, there are no rules to follow in the choice of *primitives*, and more often than not it is quite an arbitrary decision. Over the past few years, Hitoshi Iba has contributed some foundation works based on the *function approximation approach*. Chapter 5, **Genetic Programming of Polynomial Models for Financial Forecasting**, by him and Nikolay Nikolaev, shows a concrete application of the function approximation approach.

Based on *power series expansion*, or more precisely, the *Kolmogorov-Gabor polynomial*, the paper uses a set of *transfer polynomials* as the function set. This function set makes the GP system studied by the authors dramatically different from the standard GP, which uses basic arithmetic operations, e.g., +, -, ×, /, as the function set. The GP system with the transfer polynomials, called **STROGANOFF**, is then compared to the standard GP in forecasting financial time series.

In addition to the function set, another crucial issue addressed by the authors is the design of the *fitness function*, as the vanilla error function generally leads to overfitting solutions. Generally speaking, there are two approaches to tackling this issue. One is to add a *validation* step between the training and testing step. The other is to incorporate a parsimony and/or smoothness criterion into the fitness function, and that is what is done by the authors. They find that the elaborated fitness function enables one to find better forecasting solutions than the vanilla error function.

The authors' attention is also drawn to *data pre-processing*. Extremely noisy raw data may give GP a hard time here. The chapter considers three different transformations of raw series, and their empirical findings contribute to our understanding of the following issues. First, should one use raw data or *moving-average transformations*? Second, should one use price series or return series to make forecasts? The answers depend on the GP system which we apply. It is found that simple moving-average transformations may not help the GP of polynomials (**STROGANOFF**) to evolve profitable polynomial models from given price movements, though they are helpful to the standard GP system. On the other hand, GP of polynomials produces models that outperform those from standard GP on *return series*, and these polynomials yield the highest profits in all the experiments.[2]

Nikolaev and Iba's application of GP to financial forecasting is based on *universal approximation.* In contrast to the *global* modeling strategy, Chapter 6, **NXCS: A Hybrid Approach to Stock Indexes Forecasting**, by Giuliano Armano, Andrea Murru and Michele Marchesi proposes *local approximation* based on a novel technique on domain decomposition. Their chapter contributes to the volume in several different ways. Firstly, like the previous one, the chapter is motivated by a theoretical consideration, i.e., *multi-stationarity* in a financial time series, or to put it differently, *piecewise stationarity* or *quasi-stationarity.* The authors assert that *under the hypothesis that financial time series are multi-stationary, obtaining a single model that holds for different regimes can be extremely difficult.* Therefore, instead of identifying a global model, they attempt to identify different local models, known in the literature as the *guarded experts framework.*

Secondly, the authors give a magnificent presentation of the idea of *guarded experts*, tracing the origin of the concept. The literature review broadly covers a significant proportion of *computational intelligence* and *time series analysis*, from the early nearest neighborhood (classification and regression trees, decision trees, threshold autoregressive models, multivariate adaptive regression splines), to the most recent neural networks and extended classifier systems. The idea of guarded experts is shown to have long been pursued throughout the history of machine learning. It is composed of two parts, namely, *building the guards* (input domain decomposition) and *inviting the experts* (domain-specific models). Approaches vary in how guards and experts are established.

The system proposed by the authors, **NXCS**, is an *evolutionary system* whereby a population of NXCS experts, each characterized by an XCS (eXtended Classifier System) classifier and corresponding ANN (artificial neural network) predictor, is raised in a typical XCS-like environment. The XCS classifiers are then renewed and revised by the standard genetic algorithm. The system has been tested on COMIT, S&P500, and Nasdaq. In terms of the normalized Sharpe ratio, the results point to the good forecasting capability of the system, which repeatedly outperforms the "Buy and Hold" strategy.

This chapter is also the only application of an extended classifier system (**XCS**) in the volume. Unlike genetic algorithms and genetic programming, XCS may sound less familiar to readers with a background in finance, and the appendix and references given in the chapter may help interested readers to gain some familiarity with it.

2.2. Trading

Like financial forecasting, one of the most fundamental issues in the application of GAs or GP to financial trading is still *representation*, i.e., *how to effectively characterize a trading strategy which one is looking for.* Research on this issue is very much motivated by the format of existing trading strategies, and there are generally two approaches to this. The first approach, called the *decision tree* approach, was pioneered by Bauer and Liepins (1992), Bauer (1994a) and Bauer (1995). In this approach each trading strategy is represented by a decision tree. At the early stage, Bauer used the bit string to encode these decision tress, and generated and evolved them with genetic algorithms. However, since the expression power of the bit-string representation is very limited, it makes GAs very difficult to represent and to evolve decision trees with various shapes and sizes. Allen and Karjalainen (1999), Neely et al. (1997), Neely and Weller (1999) and many others started to use the parse trees of genetic programming to represent the decision trees of trading strategies, as GP enables users to explore a significantly larger space of trading strategies.[3]

The second approach, called the *combinatoric* approach, was first seen in Palmer et al. (1994).[4] The combinatoric approach treats each trading strategy as one realization of $\binom{n}{k}$ combinations, where $1 \leq k \leq n$, and n is the total number of trading rules. Using GAs, one can encode the *inclusion* or *exclusion* of a specific trading rule as a bit and the whole trading strategy as a chromosome. However, this approach can only *combine* the rules known to the users, but cannot *generate* anything that sounds *novel* to the users.

The three chapters presented in **Part III** of the volume provide some of the most recent advancements for both of the two representations. Chapter 7 proposes a *novel fitness function* to evolve financial decision trees, whereas Chapter 8 uses the *Backus-Norm Form* to address the semantic issue of financial decision trees. Chapter 9 deals with the novelty issue of the combinatoric approach via the *fuzzification* of the trading rules.

Chapter 7, **EDDIE for Financial Forecasting**, by Edward Tsang and Jim Li introduces a financial trading system developed at the University of Essex, called **EDDIE**, which is used to generate trading rules in the form of decision trees. To evaluate and evolve trading strategies, one has to notice that financial agents usually pursue more than just one goal. They care about profits, but they hate risks. They do not want to miss any golden opportunity, but they are afraid of catastrophe. Usually, these conflicting desires are compromised via a fitness measure which assigns different weights to different goals, and GAs or GP maximizes or minimizes these *unconstrained* fitness functions. In this chapter the authors introduce an alternative way to solve the conflicts based on a *constrained* fitness function, called **FGP-2**. The constraints, in their case, are the minimum and maximum percentage of investment recommendations,whereby FGP-2 takes two parameters from the users and then trades the rate of precision with the rate of missing opportunities. They test the return performance of FGP-2 against three artificial neural networks and a linear classifier on 10 stocks.

Chapter 8, **Forecasting Market Indices Using Evolutionary Automatic Programming**, by Michael O'Neill, Anthony Brabazon, and Conor Ryan introduces a non-standard application of GP which was seldom seen in previous financial applications. This non-standard application is based on the *Backus-Naur Form* (**BNF**) in computer theory.[5]

Financial users have long complained about the *semantic meaning* of the programs evolved from the standard GP. Chen 2002 included this issue as one of the four main issues in the use of genetic programming in economics and provided a lengthy discussion of it. Recent efforts made to cope with this issue subject the standard GP with *grammar*. Two pioneering applications of this kind are Bhattacharyya et al. (1998) and Nikolaev and Iba (2000), but Duffy and Engle-Warnick (2002) is the first one that explicitly referred to the *Backus-Naur Form*. What makes O'Neill et al's chapter further different from these early applications is their *representation*. Rather than representing the programs as *syntax trees*, *a linear genome representation* is used. Each trading rule is encoded as a *variable-length* binary string, containing in its condons (groups of 8 bits) the information to select production rules from **BNF** grammar.

Chapter 9, **Genetic Fuzzy Expert Trading System for NASDAQ Stock Market Timing**, by Sze Sing Lam, K. P. Lam and Hoi Shing Ng applies the conventional combinatoric approach to encode and evolve a trading expert system with a genetic algorithm. However, in their application, trading rules are not crispy, but *fuzzy*. The advantages of using fuzzy trading rules over the crispy ones are well discussed in Tay and Linn (2001). The fuzzy approach to modeling agents' behavior is intuitively sound, and its combined use with genetic algorithms or genetic programming should be a promising direction for further research.

Financial markets are complex in the sense that the underlying law of motion, if it exists, is *non-linear* and *time-variant.* The *non-linear* characteristic justifies the use of GAs or GP as the data mining toolkit, but that is not good enough. The *time-variant* characteristic requires the use of GAs or GP in an adaptive way, usually called *dynamic learning* or *adaptive learning.* The authors of this chapter also notice the significance of this issue and show the extra gains which one may have if GAs are combined with the use of an adaptive learning scheme.[6].

3. Miscellaneous Application Domains

Financial forecasting and trading take up 60% of the published applications of GAs and GP to financial engineering. The other 40% of them do not belong to any single application domain. Instead, they are unevenly distributed. The five chapters presented in **Part IV** of the book show these miscellaneous applications. They are *portfolio optimization* in Chapter 10, *cash flow management* in Chapter 11, *option pricing* in Chapter 12, *volatility modeling* in Chapter 13, and *arbitrage* in Chapter 14.

3.1. Portfolio Management

Portfolio management is not a new application domain, given that the first journal publication appeared in 1995 (Leinweber and Arnott, 1995).[7] *Portfolio optimization* is a very standard financial problem. The traditional approach to portfolio optimization is the mean variance framework, which is known as a *quadratic optimization* problem and can be solved analytically. However, financial reality may complicate both the *objective function* and *constraints* facing financial agents, which transforms the standard problem into one that is difficult to solve analytically. For example, Baglioni et al. (2000) showed how financial regulation can complicate the objective function and constraints in the case of a pension fund; Hiemstra (1996) exemplified how the short-run fluctuation in excess returns and volatility can modify the standard problem into a more difficult tactical asset allocation problem. They all suggested the use of GAs to tackle the optimization problem.

Chapter 10, **Portfolio Selection and Management Using a Hybrid Intelligent and Statistical System**, contributed by Juan Lazo, Marco Pacheco, and Marley Vellasco continues this line of research. They proposed a hybrid-system approach to portfolio selection and management. The distinguishing feature of this chapter is applying artificial neural nets to forecasting returns and the GARCH model to forecasting volatility. Based on the estimated returns and volatility, the authors use GAs to determine the optimal portfolios under different objectives, e.g., maximizing the Sharpe ratio and minimizing risks under a target return. The chapter is probably the first one that uses VaR (*Value at Risk*), one of the most popular risk measures, to evaluate the risk of GA-based portfolios.

3.2. Cash Flow Management

A subject related to portfolio management is *cash flow management*, a standard issue in corporate finance. While it should not be a surprise to see the application of GAs or GP to this area, any such work has never been done. Chapter 11, **Intelligent Cash Flow: Planning and Optimization Using Genetic Algorithms**, by Marco Pacheco, Marley Vellasco, Maíra F. de Noronha, and Carlos Henrique P. Lopes initiates this application to cash flow planning. Like many other financial issues, cash flow planning deals with an enormous amount of search space, which can be computationally demanding. As shown by the authors, cash flow planning for a period of 90 days can face 68^{90} possibilities if we consider 68 options of investment products for each day of a 90-day period.

Pacheco et al.'s paper also leads us to a very technical and important issue in GAs, namely, *epistasis*. Epistasis refers to *the lack of independence among bits* with respect to a fixed fitness function. A precise description of this phenomenon was given by Davidor (1991). With the presence of epistasis, high-performance schemata may point toward a poor area of the space: good low-order building blocks lead to poor higher-order building blocks. There are many studies regarding how the performance of GAs is affected by epistasis. These studies frequently center on the usefulness of operators such as crossover and mutation in solving epistatic problems. Nonetheless, the relevance of the epistatic issue to finance was not noted in financial applications of GAs until the appearance of their paper. To address the epistatic problem, they use the *partially matched crossover* (**PMX**) derived by Goldberg and Lingle (1985), which is a binary operator that combines ordering building blocks from above-average parents in a sensible way.

3.3. Option Pricing

Optional pricing is another new application domain. It was not until 1998 that the first journal article on this area was published. Jay White (1998) utilized *genetic adaptive neural networks* (**GANNs**) for pricing *interest rate futures call and put options*. In his application the option pricing formulae were *not* encoded *directly* by bit strings (chromosomes). Instead, they were represented by three-layer, feedforward, artificial neural networks. Genetic algorithms were then used to evolve and determine the weights of the neural nets. This representation is known as *indirect representation.* In indirect representation, a chromosome is not mapped directly to a *solution*; rather it is mapped to *a structure of a solution.* This indirect representation has become a common practice to enhance the expression power of genetic algorithms in many financial applications.

Option pricing formulae can also be directly represented by bit strings, though it is not that straightforward. Chen and Lee (1997) use a series expansion approach to represent a European call option formula, and truncated the infinite series to a finite one. The coefficients of the series were then encoded by bit strings, and evolved with genetic algorithms. However, since determining the size and shape of option pricing formulae is in general very difficult, the majority of recent studies have all adopted the parse-tree representation [Chen, Lee and Yeh (1999); Chidambaran et al. (2000); Keber (2000); Keber (2001)], as does the next chapter.

Chapter 12, **The Self-Evolving Logic of Financial Claims**, by Thomas Noe and Jonathan Wang demonstrates a standard application of genetic programming to option pricing. Using genetic programming, they first show that the Black-Sholes formula can be recovered from the simulated data. They then apply GP to S&P 500 futures options, finding that the performance of GP in pricing options is at least comparable to the performance of artificial neural nets in Hutchinson et al. (1994).[8]

The chapter also touches on a few technical issues in the financial application of GP. As in many other financial applications, the design of the fitness function is not always a trivial issue. To accurately measure the performance of a pricing formula from the range of *out-of-the-money* to the range of *in-the money*, the authors include in the fitness function both the *absolute error* and the *absolute percentage error.* The inclusion of the latter is particularly important when the option price is deep out of the money and the price is small. Another inclusion to the fitness function is a penalty based on the *program size.* The program size is measured by the number of nodes in the parse tree, the so-called *node complexity.* This modification is intended to bias the search toward functions with fewer nodes, which are simpler and therefore less prone to overfit the data.

An alternative approach taken to deal with overfitting is to add a *validation step* immediately after the training step. This approach does not impose a penalty to the program size directly, but uses a *selection period* to confirm that the expression power of GP has not been abused. In financial applications, the *selection* approach is first taken up by Allen and Karjalainen (1999) and Neely et al. (1997) and further used in Neely and Weller (1999) and Wang (2000). It has now become a standard procedure for GPers in financial data mining.[9]

3.4. Volatility

While the focus of financial forecasting is either price or return, for many financial decisions, these two factors are relatively less important than *fluctuation*, known as *volatility* in finance. A mathematical description of volatility, $\{v_t\}$, is given in Equations 1.3 to 1.5.

$$R_t = f_t(R_{t-1}, ..., R_{t-p}, \epsilon_{t-1}, ..., \epsilon_{t-q}) + \epsilon_t, \tag{1.3}$$

$$\epsilon_t = v_t \mu_t, \tag{1.4}$$

where R_t is the stock return, μ_t is a standard normal random variable and

$$v_t = h(R_{t-1}, ..., R_{t-r}, \epsilon_{t-1}, ..., \epsilon_{t-s}, h_{t-1}, ..., h_{t-m}). \tag{1.5}$$

We saw earlier the application of GP for modeling the function $f(\ )$ (Equation 1.1). However, few studies have extended the application of GP to modeling $h(\ )$.[10] Chen and Yeh 1997 and Chen (1998b) are the only such publications.[11] Based on the *jump process*, Chen and Yeh propose a non-parametric approach, called *adaptive genetic programming* (**AGP**), to model $h(\ )$.

Chapter 13, **Using a Genetic Program to Predict Exchange Rate Volatility**, by Christopher Neely and Paul Weller is an extension of the authors' early studies on forecasting exchange rates [Neely et al. (1997), Neely and Weller (1999)]. Those successful applications of genetic programming motivated the authors to advance from forecasting the conditional mean to forecasting the conditional variance. Obviously, one is curious as to whether genetic programming can forecast volatility better than the well-known *Generalized Autoregressive Conditionally Heteroskedastic* (**GARCH**) model. The authors conducted a series of careful experiments and tested the forecasting performance of GP against that of the GARCH model over different *time horizons*, using various *accuracy criteria*. Their results are mixed, with GP often outperforming the GARCH model on longer horizons and consistently returning lower mean absolute forecast errors. However, on short horizons the GARCH model outperforms GP in terms of the mean squared error.

Running GP involves many technical issues to which there are no general answers. Among the tricky ones are the determination of the function set, the fitness function, and the over-fitting avoidance strategy. In this chapter the authors provide two tests on these technical designs. The first test concerns the role that the function set plays. They consider two function sets, one with *primitive functions* and the other with primitive functions *plus* advanced functions. The second test concerns the benefits that one can gain by avoiding over-fitting. For this test, the authors incorporate into the fitness function a penalty depending on *node complexity*, as shown in the previous chapter. Interestingly, neither imposing a penalty for complexity nor expanding the set of data functions leads to any appreciable improvement in the performance of the genetic program. Adding a penalty function does not help in this case, probably because the program already has a *validation* step, which itself is a design for over-fitting avoidance.

3.5. Arbitrage

The last chapter of Part IV, **Evolutionary Decision Trees for Stock Index Options and Futures Arbitrage**, by Sheri Markose, Edward Tzang and Hakan Er applies genetic programming to *stock index options and futures arbitrage*; more precisely, the short **P-C-F** arbitrage. The ex post analysis of efficiency violations for short arbitrage positions shows on average that for all periods to maturity the profits from the arbitrage is substantial and statistically significant. It would be interesting to know whether GP can correctly identify and exploit profitable short arbitrage opportunities in a real time setting. The performance of **EDDIE-ARB**, a system developed by the authors, is compared with that of a naive strategy which executes an arbitrage trade whenever there is a contemporaneous profit signal. Due to the execution delay, the naive strategy faces execution price risk, as not every contemporaneous **P-C-F** profit signal will continue to be profitable. Therefore, with the assumed time delay in the execution of an arbitrage from an observed contemporaneous profit signal, an effective forecasting tool is needed to assess the success rate of such a strategy.

4. Agent-Based Computational Finance

Part V represents another active application domain of genetic algorithms and genetic programming in computational finance, namely *agent-based computational finance* (**ACF**). The term**ACF** implies a computational study of financial behavior and financial markets modeled as *evolving decentralized systems of autonomous agents.*[12] Introductory material can be found in LeBaron (2000). He also constructs a website

http://www.unet.brandeis.edu/ blebaron/acf

for this new area in finance. Interested readers are referred directly to this website.

A central element of **ACF** is the simulation of financial agents' evolution in financial markets. This is not just about a *single* financial agent. It is about a *population* of interacting (competing) agents. The previous three parts, usually known as *financial engineering*, only model an individual agent, but in ACF we need to "aggregate" these individual so that they evolve together or use *co-evolving*.

The purpose of GA and GP is to drive the evolution of a population. They can be applied towards evolving a population of financial strategies for an individual financial agent, as we see in financial engineering; so that they can be applied to a population of financial agents for a financial market, as we shall see in this part of ACF. While GA and GP are not the only tools used in ACF, they do play the most prominent role in the development of ACF. In fact, the earliest application domain of GA in computational finance is ACF.

As the publications of ACF pile up, the significance of it shall become gradually clearer for finance people. It is a promising approach for studying behavior finance, micro-structure, experimental finance, and psychological finance, while most of its current fruitful findings concentrate on the *financial econometrics.* The seven chapters in this part offer wide coverage of the recent progress in ACF, including the mathematical foundation of finance, modeling techniques of bounded rationality, price dynamics, market efficiency, trading mechanism, financial regulations, patterns of survival financial strategies or behavior, and methodological and philosophical issues. Below provides a quick grasp of them.

The foundation of mainstream financial economics or mathematical finance was built upon the *Walrasian general equilibrium analysis*, which is mainly concerned with the existence (or the non-existence) of equilibria associated with their characterizations. However, scant attention is drawn to the *market process* converging to these equilibria.[13] The usual argument that agents will *eventually learn* one of these equilibria (in particular, the Pareto superior one) is anything but well grounded.

The chapter by Thomas Riechmann, "**A Model of Bounded Rational Consumer Choice**", uses a standard general equilibrium model to show that even finding the *optimal consumption bundle* of three goods can be an extremely complicated issue for consumers. The model is very simple. It has 500 consumers. Each is endowed with the same utility function and the same budget constraint. These consumers interact with each other in an economy of three commodities, where the supply schedule of each commodity is exogenously fixed with the same elasticity. The prices are determined by equating supply with demand, which are not known to the consumers upon submission of their consumption plan. The trial-and-error process of consumers is driven by *genetic algorithms*. Under the circumstances, Riechmann examines two essential characterizations of consumer optimization. First, how well is the budget control done? Second, how good is the chosen consumption bundle?

Riechmann finds that the aggregate performance of consumers' choices crucially depends on how genetic algorithms are used to model the learning process of consumers. In particular, he evaluates the performance of the canonical GA and that with the addition of the *election operator* or the *elitist operator*. The last two operators basically prevent agents from rushing into any *new* idea without testing it first, making agents behave more prudently. While the economic meaning of these operators is intuitively sound, there is no guarantee that they will lead to desirable results.[14] The election operator tests the new ideas by estimating its preference, the so-called *potential fitness*. In some situations, this ex-ante fitness can be quite different from the realized one, and that may cause a problem.[15] In the author's words, for the case of flexible prices and low elasticity of supply, the election is far from leading to any kind of sensible consumer choice.

The performance with the elitist operator, called the *preselection operator* by the author on the other hand shows a degree of robustness. This result is interesting, because the elitist operator is usually neglected in agent-based computational economic models. Riechmann's description of the significance of the elitist operator reminds us that *memory* works in such a way as to enhance agents' learning capability.[16]

There is another technical novelty that should not go unnoticed, and that is the way Riechmann copes with the constraints in constrained optimization. Despite the many approaches that Michalewicz (1996) introduced to deal with this issue, few have been applied to financial engineering, not to mention agent-based computational finance. In this chapter the author gives a concrete example of how to use these techniques in an economic context by modifying fitness (utility) via a *penalty function*.

If boundedly-rational interacting heterogeneous agents cannot replicate the *equilibrium* in a simple general equilibrium model, then it would be no surprise if they cannot do the same thing in a more complicated financial models. The chapter by Shu-Heng Chen and Chung-Chih Liao, **Price Discovery in Agent-Based Computational Modeling of Artificial Stock Markets**, shows that indeed this is the case. They start with a standard asset pricing model with its homogenous rational expectations equilibrium price, and augment the standard asset pricing model with its *agent-based extension.* They then examine how well a population of financial agents can track the equilibrium price in the **AIE-ASM**, which is a variant of the Santa Fe Institute Artificial Stock Market. By simulating the artificial stock market with different dividend processes, interest rates, risk attitudes, and market sizes, they find that the market price is not an unbiased estimator of the equilibrium price. Except in a few extremely worse cases, the market price deviates from the equilibrium price moderately from minus four per cent to sixteen per cent.

The pricing errors are in fact not *patternless.* They are actually negatively related to *market sizes*: a thinner market size tends to have a larger pricing error, and a thicker market tends to have a smaller one. For the thickest market which they have simulated, the mean pricing error is only 2.17%. This figure suggests that the new classical simplification of a complex world may still provide a useful approximation if some conditions are met, such as, in this case, the market size.

At the end of the chapter, the authors sketch a research agenda for agent-based financial modeling. To make an agent-based computational approach a prolific tool for doing finance, they propose to include in the current artificial stock market a large variety of *financial products* and *financial agents.* Certainly, a trading mechanism should also be added to the list. In fact, the trading mechanism adopted in most agent-based stock markets either follows the *Walrasian tatonnement scheme* or the *rationing scheme.* Few studies have been done with a double auction.[17] The chapter by Shu-Heng Chen, Chung-Chin Tai, and Bin-Tzong Chie, **Individual Rationality as a Partial Impediment to Market Efficiency**, contributes to an agent-based version of the double auction market. To our best knowledge, this is the first paper to simulate the evolution of *bargaining strategies* within the context of an agent-based double auction market.[18]

The chapter re-visits a fundamental surprise in economics, i.e., the *inconsistency* between individual rationality and aggregate rationality. The usual way to put the surprise is that individual irrationality can lead to aggregate rationality. For example, in their computerized double auction market, Gode and Sunder (1993) show that a near 100% allocative efficiency can be generated from a group of zero-intelligence traders, who can only randomly bid or ask. Chen et al.'s chapter presents another way to see the surprise: individual rationality may also lead to aggregate irrationality. They considers two types of traders: smarts ones and mediocre ones. The smart traders are distinguished from the mediocre traders by a *privilege* which gives them the potential to learn sophisticated bargaining strategies with genetic programming. They then examine the allocative efficiency of 20 double auction markets composed exclusively of either smart traders or mediocre traders. Their simulation evidences that higher allocative efficiency is not achieved from a trading room of smart traders, but from one of mediocre traders. Financial regulations, from this paper, can be read as an annihilation to the evil-side of smartness.

In addition to asset pricing, another area to which agent-based simulations can be applied, is *portfolio theory*. This connection is particularly clear from Chapter 18, **A Numerical Study on Portfolio Optimization**, by Guido Caldarelli, Marina Piccioni, and Emanuela Sciubba. This chapter is motivated by an old debate in the theory of portfolio choice: *the normative appeal of logarithmic utility maximization* versus *the mean-variance approach*. The authors believe that an effective solution to the debate can come from the *evolution approach* which aims at studying long-run financial market outcomes as a result of a process akin to *natural selection*. The evolutionary approach to portfolio behavior was first taken up by Blume and Easley (1992), and was followed by Sandroni (2000). While both studies attempt to single out the key factors which determine the surviving portfolio rules, their restrictive assumptions make their analysis difficult to extend to the consideration of some interesting portfolio behavior, such as the CAPM rule.

The departure of Caldarelli et al's chapter provides more general insights into the debate based on numerical computation with less restrictive assumptions. They consider two types of traders, namely, *logarithmic traders* and *CAPM traders*. They then let these two groups of traders compete against each other in wealth. The dominance, survival, and extinction of these traders are then examined based on the asymptotic wealth shares of the traders. While the authors do not conduct their analysis with agent-based modeling, and hence do not use either genetic algorithms or genetic programming to evolve agents' portfolio behavior, it would be worthwhile to give it a try in the next step of the research.

One essential element of agent-based computational financial modeling is the *complex heterogeneity* of agents. While agents can be heterogeneous in many aspects, most agent-based computational financial models only address heterogeneity in *expectations*. Little attention has been drawn to other aspects of heterogeneity. From the previous chapter, we saw that in order to tackle the debate on the dominance of the MEL rules, it would be necessary to consider agents with different *preferences*. More generally, the artificial life of financial agents can be enriched if different *human characters* are taken into account. Chapter 19 by Kwok

Yip Szeto, **Adaptive Portfolio Managers in Stock Markets**, opens up an avenue for this direction.[19]

In Szeto's artificial stock market, the heterogeneity of portfolio managers is heterogeneous in two psychological measures, namely, a degree of *fear* and a degree of *greed.* These managers are otherwise homogeneous, including that they share the same forecasting rule. The forecasting rule is obtained by applying a genetic algorithm to extract patterns from financial time series. He finds a universal property in all his time series data: *greedy and confident investors are the winners.* The conclusion itself, while quite interesting, may not be the most important, because different enrichments of this rather simple model may lead to different results. What, however, really motivates us in this paper is the research opportunity of using agent-based financial modeling to address issues in psychological finance: e.g., *what types of personality determine a successful portfolio manager?*

While ACF brings us a great research opportunity, there is a known weakness: ACF is largely a computational model without immediate prospects for rigorous mathematical results. Since small changes may engender radically different results[20], a sensitivity analysis is required before one can ascertain a finding. Chapter 20, **Learning and Convergence to Pareto Optimality**, by Chris Birchenhall and Jie-Shin Lin provides perhaps the most extensive coverage of robustness checks ever seen in the **ACF** literature. A specific question is posed in this chapter. *How can we be sure that a certain kind of observed interesting behavior from an ACF model is attributed to real economic forces rather than to technical (genetic) parameters?* It is therefore up to them to check whether the observed behavior is *robust* to different designs of genetic operators.

Their work covers two different levels of GA designs: one is *genetic operators*, and the other is *architecture.* For the former, they consider different implementations of the four main GA operators, i.e., selection, crossover, mutation, and election. For the latter, they consider a *single-population GA* (population learning or social learning) vs. a *multi-population GA* (individual learning). They then apply all these different designs to re-run *the model of inflation* in Bullard and Duffy (1999). They find that Bullard and Duffy's results are sensitive to two main factors: the *election operator* and *architecture.* Their experimental results in fact lend support to some early findings, e.g., the significance of the election operator (Arifovic, 1994), and the different consequences of social learning and individual learning (Vriedn, 2000; Vriend, 2001; Yeh and Chen, 2001). What is particularly interesting is that *individual learning reduces the rate of convergence to the same belief.* This is certainly an important finding, because most studies on the convergence of GAs to Pareto optimality are based on the social learning version, e.g., Dawid (1996), Riechmann (1999), Riechmann (2001).

Since the result is sensitive to the design of GA, the immediate question is whether we can still attribute the economic behavior (in this case, the Pareto superior low inflation) to some economic forces. The answer depends on whether we have an economic-theoretic foundation to support a particular GA design; in their case, the use of the election operator and that of the social-learning architecture. Unfortunately, at this moment, such a foundation does not exist, and any specific GA design can be *ad hoc.*[21] To avoid *arbitrariness*, one can have the design determined *endogenously.* While this idea does not sound peculiar, it has never been tried in the context of ACF. This chapter is probably the first one to give it a try.

The authors propose an approach of *meta learning* (*open learning*), using an individual learning scheme to model agents' forecasts. The genetic operators for agents are neither homogeneous nor fixed. Instead, they evolve through *social learning.* Thus, some agents use tournament selection to evolve their forecasts, while others use a roulette-wheel selection to do the work. The same goes for other genetic operators. Over time, the market will determine which design is the best. The interesting findings from these experiments are two-fold. First, the authors find that the GA design used in Bullard and Duffy (1999) is one of the most popular survivors. Second, all runs converge and they converge to the Pareto superior low inflation equilibrium. While meta learning still has its limits, one cannot but acknowledge the novelty of this approach.

5. Concluding Chapter

After a decade's development of financial applications of GAs and GP, it is time to reflect what has been done and to examine what has been taken for granted. The volume starts with the question: *what is it?* We, however, are aware all the time of the more basic issue: *why is it so?* There are probably good reasons not to start with the fundamental issue, but there is no excuse to end this volume without touching on it. The concluding chapter, **The New Evolutionary Computational Paradigm of Complex Adaptive Systems: Challenges and Prospects for Economics and Finance**, by Sheri Markose presents a thought-provoking discussion of the issue: *were the financial applications of GAs and GP well anticipated?*

The author places the financial agents and financial markets in the context of modern *complex sciences*, and examines the research methodologies for them from the development of *computational theory* and *history of economic thoughts.* The message of the chapter is clear and strong: *dynamical system outcomes produced by algorithmic agents need not be computable.* Generally we have no way (*algorithm*) of inferring what would result from a system except by running its course. Whatever will be will come to us as *emergent properties.* Neoclassical economics fails to see this, but it was clearly identified at the provenance of the economics in the 18th century.

Given the nature of uncomputability, it would be more appropriate to treat financial agents, not as *neo-classical optimizing agents*, but as *adaptive agents* whose goals and means are changing over time. Part II to Part IV show how these adaptive agents can be built with genetic algorithms and genetic programming, whereas Part V demonstrates simulations of markets composed of these adaptive agents. In the last chapter, the author examines some theoretical issues concerning the complexity of minority games, double-auction markets and stock markets.

6. Concluding Remarks

What is the current state of financial applications of genetic algorithms and genetic programming? From a review of the 20 chapters distributed over the five parts of the volume, one can see the following observations. First, *the application coverage is continuously enlarging.* There is little doubt that new application domains will emerge in the next few years. In fact, apart from what have been said on the volume, the recent publication Noe (2000), which applies genetic algorithms to the study of *takeover* behavior, shows another novel application.

Second, to reflect credits of GAs and GP in financial applications, a *rigorous statistical analysis is imperative.* This point is well taken by the authors of the volume, as we see the involvement of various statistical procedures applied to the performance evaluation, such as Monte-Carlo simulation, bootstrap testing, and kernel estimation of error density. The statistical rigors also extend to experimental designs, as we see from Chapter 2 where the performance of GAs and GP sensitively depends on the design. Wang (2000), who casts doubt on the superior performance of GP to that of the buy-and-hold strategy, is another recent example to show such rigor.

Third, *several technical issues are still the main concerns of financial applications of GAs and GP.* The choice of *fitness function* which can correctly measure the quality of solutions is not trivial, and neither is the *representation* of solution candidates. The expression power of GP is striking, but should not be abused. It becomes clear that the canonical genetic algorithms and standard genetic programming would not be so productive for financial applications without further modifications, and chapters of this volume suggest ways to do so.

Notes

1. See Chen and Kuo (2002) pp. 425-426 for details.

2. These results can be compared to the main finding in Kaboudan (2000), which shows that one should use price series rather than return series to forecast the price.

3. There is, however, another approach to enhance the flexibility of Bauer's trading strategies. That is, to *parameterize trading rules*, and then encode them with bit strings. The GA is then used to evolve these strings. Examples can be found in Pereira (2002).

4. The combinatoric approach was also frequently seen is other application domains. See Farley and Jones (1994).

5. For those who are not familiar with the Bcakus-Naur Form, ? could be a useful reference.

6. For a systematic study of the significance of adaptive learning schemes in trading applications of GAs, one is referred to Chen and Lin (1997).

7. Bauer (1994b) applied genetic algorithms to solve a related problem, namely, selection of mutual funds.

8. Hutchinson et al. (1994) is the first journal publication of an application of artificial neural nets to option pricing.

9. For instance, also see Chapters 4 and 13 of the volume.

10. In addition to genetic programming, there are other non-parametric or semiparametric approaches to modeling the function $h(\ )$. The interested reader is referred to Chen (1998b).

11. Not in the context of a financial time series, but in the context of option pricing, Christian Keber has conducted a series of studies on *implied volatility* using genetic programming. See Keber (1999), Keber (2000) and Keber (2001).

12. Italics are borrowed from the Tesfatsion (2001) in the definition of *agent-based computational economics.*

13. While *computational general equilibrium models* do provide a *constructive* proof of the existence of the equilibrium, the construction itself is not a real market process. For a full discussion, see the Markose chapter in this book.

14. It is true that the use of the election operator is already a standard procedure in agent-based computational economics. It has also been shown in many cases that without the inclusion of the election operator, one can have quite disappointing results, but there are also cases where the election operator is preferred not to be used. For details, the interested reader is referred to Chen 2002.

15. Birchenhall (1995) is probably the first one who acknowledges this problem.

16. Economic studies comparing the performance of the two operators are limited.Novkovic 1998 is the only paper that shows some advantages of the elitist operator over the election operator.

17. See Chen (2001) for an account of this development.

18. Chen (2000) provided a literature review of some experimental and computerized double auction markets.

19. The first half of the chapter is devoted to a review of the author's early applications of genetic algorithms to financial time series prediction, including Fong and Szeto (2001), and Szeto and Luo (1999).

20. See Fogel et al. (2002) for two interesting examples.

21. See Chen (2001) and Chen 2002 for an in-depth discussion.

References

Allen, F. and R. Karjalainen (1999). "Using genetic algorithms to find technical trading rules," *Journal of Financial Economics*, 51(2), 245–271.

Arifovic, J. (1994). "Genetic Algorithm Learning and the Cobweb Model," *Journal of Economic Dynamics and Control*, 18(1), 3–28.

Baglioni, S., D. Sorbello, C. C. Pereira, and A. G. B. Tettamanzi (2000). "Evolutionary Multiperiod Asset Allocation," in Whitley D., Goldberg D., Cantú-Paz E., Spector L., Parmee I., Beyer H.-G. (eds.), *Proceedings of the Genetic and Evolutionary Computation Conference*, 597–604. Morgan Kaufmann.

Bauer, R. J. Jr. (1994a). *Genetic Algorithms and Investment Strategies.* New York: John Wiley & Sons.

Bauer, R. J. Jr. (1994b). "An Introduction to Genetic Algorithms: A Mutual Fund Screening Example," *Neurove$t Journal*, 2(4), 16–19.

Bauer, R. J. Jr. (1995). "Genetic Algorithms and the Management of Exchange Rate Risk," in Biethahn J., Nissen V. (eds.), *Evolutionary Algorithms in Management Applications.* 253–263, Heidelberg and New York: Springer.

Bauer, R. J. Jr. and G. E. Liepins (1992). "Genetic Algorithms and Computerized Trading Strategies," in O'leary D. E., Watkins R. R. (eds.), *Expert Systems in Finance.* North Holland.

Bhattacharyya, S., O. Pictet, and G. Zumbach (1998). "Representational Semantics for Genetic Programming Based Learning in High-Frequency Financial Data," in Koza J. R., Banzhaf W., Chellapilla K., Deb K., Dorigo M., Fogel D. B., Garzon M. H., Goldberg D. E., Iba H., Riolo R. (eds.), *Genetic Programming 1998: Proceedings of the Third Annual Conference*, 11–16. Morgan Kaufmann.

Birchenhall, C. R. (1995). "Modular Technical Change and Genetic Algorithms," *Computational Economics*, 8(3), 233–253.

Blume, E. and E. Easley (1992). "Evolution and Market Behavior," *Journal of Economic Theory*, 58, 9–40.

Bullard, J. and J. Duffy (1999). "Using Genetic Algorithms to Model the Evolution of Heterogeneous Beliefs," *Computational Economics*, 13(1), 41–60.

Chen, S.-H. (1998a). "Evolutionary Computation in Financial Engineering: A Roadmap to GAs and GP," *Financial Engineering News*, 2(4).

Chen, S.-H. (1998b). "Modeling Volatility with Genetic Programming: A First Report," *Neural Network Worlds*, 8(2), 181–190.

Chen, S.-H. (2000a). "Toward an Agent-based Computational Modeling of Bargaining Strategies in Double Auction Markets with Genetic

Programming," in K.S. Leung, L.-W. Chan, and H. Meng (eds.), *Intelligent Data Engineering and Automated Learning- IDEAL 2000: Data Mining, Financial Engineering, and Intelligent Agents*, Lecture Notes in Computer Sciences 1983. 517–531. Springer.

Chen, S.-H. (ed. 2002a). *Evolutionary Computation in Economics and Finance.* Physica-Verlag.

Chen, S.-H. (2002b). "Fundamental Issues in the Use of Genetic Programming in Agent-Based Computational Economics," in A. Namatame, T. Terano and K. Kurumatani (eds.), *Agent-based Approaches in Economic and Social Complex Systems*, 208–220. IOS Press.

Chen, S.-H. (2002c). "Evolutionary Computation in Economics and Finance: An Overview of the Book," in Chen, S.-H. (ed.), *Evolutionary Computation in Economics and Finance*, 1–26. Physica-Verlag.

Chen, S.-H. and T.-W. Kuo (2002). "Evolutionary Computation in Economics and Finance: A Bibliography," in S.-H Chen (ed.), *Evolutionary Computation in Economics and Finance*, 419–444. Physica-Verlag.

Chen, S.-H. and W.-C. Lee (1997). "Option Pricing with Genetic Algorithms: The Case of European-Style Options," in T. Back (ed.), *Proceedings of the Seventh International Conference on Genetic Algorithms*, 704–711, San Francisco, CA: Morgan Kaufmann Publishers.

Chen, S.-H. and W.-Y. Lin (1997). "Financial Data Mining with Adaptive Genetic Algorithms," in Philip T. (ed.), *Proceedings of the 10th International Conference on Computer Applications in Industry and Engineering*, 154–159

Chen, S.-H. and C.-H. Yeh (1997). "Using Genetic Programming to Model Volatility in Financial Time Series," in Koza J. R., Banzhaf W., Chellapilla K., Deb K., Dorigo M., Fogel D. B., Garzon M. H., Goldberg D. E., Iba H., Riolo R. (eds.), *Genetic Programming 1997: Proceedings of the Second Annual Conference.* 58–63. Morgan Kaufmann.

Chen, S.-H., W.-C. Lee, and C.-H. Yeh (1999). "Hedging Derivative Securities with Genetic Programming," *International Journal of Intelligent Systems in Accounting, Finance and Management*, 8(4), 237–251.

Chen, S.-H., W.-Y. Lin, and C.-Y. Tsao (1999). "Genetic Algorithms, Trading Strategies and Stochastic Processes: Some New Evidence from Monte Carlo Simulations," in Banzhaf W., Daida J., Eiben A. E., Garzon M. H., Honavar V., Jakiela M., Smith R. E. (eds.), *GECCO-99: Proceedings of the Genetic and Evolutionary Computation Conference.* 114–121. Morgan Kaufmann.

Chidambaran, N., C.-W. J. Lee, and J. Trigueros (2000). "Option Pricing via Genetic Programming," in Abu-Mostafa Y. S., LeBaron B., Lo

A. W., Weigend A. S. (eds.), *Computational Finance – Proceedings of the Sixth International Conference*, Cambridge, MA: MIT Press.

Cohen, D. I. A. (1991). *Introduction to Computer Theory.* Wiley.

Davidor, Y. (1991). "Epistasis Variance: A Viewpoint on GA Hardness," in Rawlins, G. J. E. (ed.) *Foundation of Genetic Algorithms*, 23–35. San Mateo: Morgan Kaufmann Publishers.

Dawid, H. (1996). *Adaptive Learning by Genetic Algorithms: Analytical Results and Applications to Economical Models*, Heidelberg and New York. Springer.

Deboeck, G. J. (ed. 1994). *Trading on the Edge: Neural, Genetic, and Fuzzy Systems for Chaotic Financial Markets.* John Wiley & Sons.

Duffy, J. and J. Engle-Warnick (2001). "Using Symbolic Regression to Infer Strategies from Experimental Data," in S.-H. Chen (ed.), *Evolutionary Computation in Economics and Finance.* Physica-Verlag.

Farley, A. M. and S. Jones (1994). "Using a Genetic Algorithm to Determine an Index of Leading Economic Indicators," *Computational Economics*, 7(3), 163–173.

Fogel, D. B., K. Chellapilla, and P. J. Angeline (2002). "Evolutionary Computation and Economic Models: Sensitivity and Unintended Consequences," in S.-H. Chen (ed.), *Evolutionary Computation in Economics and Finance.* Physical-Verlag.

Fong, A. L. Y. and K. Y. Szeto (2001b). "Rules Extraction in Short Memory Time Series Using Genetic Algorithms," *European Physics Journal B*, 20, 569–572.

Gode, D. K. and S. Sunders (1993). "Allocative Efficiency of Market with Zero-Intelligence Trader: Market as a Partial Substitute for Individual Rationality," *Journal of Political Economy*, 101(1), 119–137.

Goldberg, D. E. and R. Lingle (1985). "Alleles, Loci, and the Traveling Salesman Problem," *Proceedings of the First International Conference on Genetic Algorithms and Their Applications*, 154–159.

Hiemstra, Y. (1996). "Applying Neural Networks and Genetic Algorithms to Tactical Asset Allocation," *Neurove$t Journal*, 4(3), 8–15.

Hutchinson, J. M., A. W. Lo, and T. Poggio (1994). "A Nonparametric Approach to Pricing and Hedging Derivative Securities via Learning Networks," *Journal of Finance*, 851–889.

Jay White, A. (1998). "A Genetic Adaptive Neural Network Approach to Pricing Options: A Simulation Analysis," *Journal of Computational Intelligence in Finance*, 6(2), 13–23.

Kaboudan, M. A. (2000). "Genetic Programming Prediction of Stock Prices," *Computational Economics*, 16(3), 207–236.

Keber, C. (1999). "Genetically Derived Approximations for Determining the Implied Volatility," *OR Spektrum*, 205–238.

Keber, C. (2000). "Option Valuation with the Genetic Programming Approach," in Abu–Mostafa Y. S., LeBaron B., Lo A. W., Weigend A. S. (eds.), *Computational Finance – Proceedings of the Sixth International Conference*, 689–703, Cambridge, MA: MIT Press.

Keber, C. and M. G. Schuster (2001). "Evolutionary Computation and the Vega Risk of American Put Options," *IEEE Transactions on Neural Networks*, 12(4), 704–715.

LeBaron, B. (2000). "Agent Based Computational Finance: Suggested Reading and Early Research," *Journal of Economic Dynamics and Control*, 24, 679–702.

Leinweber, D. and R. Arnott (1995). "Quantitative and Computational Innovation in Investment Management," *Journal of Portfolio Management*, 21(2), 8–15.

Michalewicz, Z. (1996). *Genetic Algorithms + Data Structures = Evolution Programs.* Springer.

Neely, C. J. and P. A. Weller (1999), "Technical Trading Rules in the European Monetary System," *Journal of International Money and Finance*, 18(3), 429–458.

Neely, C. J., P. A. Weller, and R. Dittmar (1997). "Is Technical Analysis in the Foreign Exchange Market Profitable? A Genetic Programming Approach," *Journal of Financial and Quantitative Analysis*, 32(4), 405–426.

Nikolaev, N. I. and H. Iba (2000). "Inductive Genetic Programming of Polynomial Learning Networks," in Yao X. (ed.), *Proceedings of the IEEE Symposium on Combinations of Evolutionary Computation and Neural Networks*, 158–167. IEEE Press.

Noe, T. H. and L. Pi (2000). "Learning Dynamics, Genetic Algorithms, and Corporate Takeovers," *Journal of Economic Dynamics and Control*, 24(2), 189–217.

Novkovic, S. (1998). "A Genetic Algorithm Simulation of a Transition Economy: An Application to Insider-Privatization in Croatia," *Computational Economics*, 11(3), 221–243.

Palmer, R. G., W. B. Arthur, J. H. Holland, B. LeBaron, and P. Tayler (1994). "Artificial Economic Life: A Simple Model of a Stockmarket," *Physica D*, 75, 264–274.

Pereira, R. (1996). "Selecting Parameters for Technical Trading Rules Using Genetic Algorithms," *Journal of Applied Finance and Investment*, 1(3), July/August 27–34.

Pereira, R. (2002). "Forecasting Ability But No Profitability: An Empirical Evaluation of Genetic Algorithm-Optimized Technical Trading Rules," in S.-H. Chen (ed.), *Evolutionary Computation in Economics and Finance*, 287–309. Physica-Verlag.

Riechmann, T. (1999). "Learning and Behavioural Stability: An Economic Interpretation of Genetic Algorithms," *Journal of Evolutionary Economics*, 9(2), 225–242.

Riechmann, T. (2001). "Genetic Algorithm Learning and Economic Evolution," in Chen, S.-H (ed.), *Evolutionary Computation in Economics and Finance*, 45–59. Physica-Verlag.

Sandroni, A. (2000). "Do Markets Favor Agents Able to Make Accurate Prediction?" *Econometrica*, 68(6), 1303–1341.

Sciubba, E. (1999). "The Evolution of Portfolio Rules and the Capital Asset Pricing Model," DAE Working Paper No. 9909, University of Cambridge.

Smith, S. N. (1998). "Trading Applications of Genetic Programming," *Financial Engineering News*, 2(6).

Szeto, K. Y. and P. X. Luo (1999). "Self-Organizing Behavior in Genetic Algorithm for the Forecasting of Financial Time Series," *Proceeding of the International Conference on Forecasting Financial Markets*, FFM99, CD-Rom.

Tay, N. and S. Linn (2001). "Fuzzy Inductive Reasoning, Expectation Formation and the Behavior of Security Prices," *Journal of Economic Dynamics and Control*, 25, 321–361.

Tesfatsion, L. (2001). "Introduction to the Special Issue on Agent-Based Computational Economics," *Journal of Economic Dynamics and Control*, 25, 281–293.

Vriend, N. (2000). "An Illustration of the Essential Difference between Individual and Social Learning, and Its Consequence for Computational Analysis," *Journal of Economic Dynamics and Control*, 24(1), 1–19.

Vriend, N. (2001). "On Two Types of GA-Learning," in Chen, S.-H. (ed.) *Evolutionary Computation in Economics and Finance*, 233–243, Heidelberg: Physica-Verlag.

Wang, J. (2000). "Trading and Hedging in S&P 500 Spot and Futures Markets Using Genetic Programming," *Journal of Futures Markets*, 20(10), 911–942.

Yeh, C.-H and Chen S.-H. (2001). "Market Diversity and Market Efficiency: The Approach Based on Genetic Programming," *Journal of Artificial Simulation of Adaptive Behavior* (**AISB Journal**), 1(1).

I

INTRODUCTION

Chapter 2

GENETIC ALGORITHMS IN ECONOMICS AND FINANCE: FORECASTING STOCK MARKET PRICES AND FOREIGN EXCHANGE — A REVIEW

Adrian E. Drake
Sirius International Insurance Corporation
ABB Financial Services Group
Stockholm, Sweden
adrian.drake@se.abb.com

Robert E. Marks
Australian Graduate School of Management
Universities of New South Wales and Sydney
UNSW Sydney 2052 Australia
bobm@agsm.edu.au

Abstract After a brief overview of the history of the development and application of genetic algorithms and related simulation techniques, this chapter describes alternative implementations of the genetic algorithm, their strengths and weaknesses. Then follows an overview of published applications in finance, with particular focus on the papers of Bauer, Pereira, and Colin in foreign exchange trading. Many other rumored applications remain unpublished.

Keywords: Genetic Algorithms, Genetic Programming, Finance, Forecasting, Exchange Rates, Trading Rules

Introduction

The increased availability of computing power in the past two decades has been used to develop new techniques of forecasting. Today's computational capacity and the widespread availability of computers have enabled development of a new generation of intelligent computing techniques, such as expert systems, fuzzy logic, neural networks, and genetic algorithms. These "intelligent" computing techniques are concerned with performing actions that imitate human decision makers. They are flexible and so can adapt to new circumstances. They can learn from past experience. Moreover, they can find solutions to problems unsolvable by traditional means. Of the computing techniques mentioned above, genetic algorithms are the most powerful and yet the most simple innovation. They are showing very promising results in improving or replacing existing statistical methods.[1]

The genetic algorithm is a surprisingly simple problem-solving technique, inspired by biological evolution. It is based on an artificially simulated process of natural selection, or survival of the fittest, known as Darwinian evolution. Genetic algorithms have emerged as a powerful general-purpose search-and-optimization technique and have found applications in widespread areas.[2]

1. History of Genetic Algorithms

Darwin's natural selection is the inspiration for genetic algorithms. Computational problems often require searching through a large number of possible solutions. What can help in this case is the application of parallelism, where many possibilities are explored simultaneously, and an intelligent strategy for selecting the next set of sequences to evaluate is used. These and other helpful features can be observed in the search for constantly evolving fitness criteria in natural evolution.[3] Scientists have been impressed by the remarkable power of natural evolution for a long time, and had been trying to mimic it in procedures for computer programs. The first attempts to combine computer science and evolution were made in the late 1950s and early 1960s. They were at first relatively unsuccessful, mainly because they relied too much on the operator of mutation, the evolution process that was emphasized by biology in those days.[4]

By the mid-1960s, John Holland at the University of Michigan had added the operators of mating and cross-over to mutation and was able to create a programming technique that he believed could solve difficult problems by mimicking natural selection. Since then, he has been regarded as the founder of the field of generic algorithms. Research in this area has been extended by two scholars of Holland's, and by many others all over the world, who have created several variations of Holland's original genetic algorithms and applied them in many different areas (Nissan 1995).

Today, the term "evolutionary algorithms" is used as an umbrella term for problem-solving computer optimization techniques that are based on some mechanisms of natural evolution as key elements in their design and implementation.[5] The major evolutionary algorithms are genetic algorithms, genetic programming, evolution strategies, and evolutionary programming, all being evolution-based systems with a population of individuals that undergo some transformations, during which the individuals or their "genes" "fight" for survival.[6]

Genetic algorithms are the most popular form of evolutionary algorithms. We will examine them in detail below. Genetic programming is a modification of genetic algorithms for evolving computer programs, rather than simple strings (Koza 1993). Evolution strategies (Evolutionsstrategien) is a form of evolutionary algorithms that uses non-string representation. It was created in Germany and relies more on the operator of mutation. Evolutionary programming is very similar to evolution strategies but it was developed independently. It has no restriction on solution representation.[7] The boundaries between the different branches of evolutionary algorithms have been blurred somewhat in recent years. When people use the term "genetic algorithms" today, it is not always clear whether they relate to Holland's original conception or to one of the numerous modifications and further developments.[8]

Before we move to an description of how genetic algorithms function and how they can contribute in forecasting foreign exchange markets, we discuss a wide range of genetic algorithms applications, which shows how widely genetic algorithms are already being used.

2. Applications of Genetic Algorithms

Genetic algorithms have a number of characteristics that make them very suitable for use in applications. They can solve hard problems quickly and reliably. They are broadly competent algorithms that can solve problems with many hard-to-find optima. Since genetic algorithms use very little problem-specific information, they are easy to connect to existing application codes, making them easy to interface with existing simulations and models. Genetic algorithms are also easy to hybridize, enabling the input of problem-specific information.[9]

As the theory and practice of genetic algorithms were being developed, they were being used in a wide spectrum of real-world applications. The benefits of genetic algorithms in engineering can be illustrated by their application in the **design of turbine blades** for jet engines, as done by General Electric. The design of turbines involves more than 100 variables. The resulting search space contains more than 10^{387} points.[10] The use of genetic algorithms in this case rendered an improvement of 2 percent in efficiency, an immense gain in this field.[11]

The **design of communications networks** has also been aided by genetic algorithms. The genetic algorithm approach was found to perform well in this setting, cutting design time from two months to two days and saving up to 10 million dollars per design.[12] In addition, genetic algorithms have been applied to problems where the network topology was fixed. They have been found to perform well in this area as well, where the goal of the software was to carry the maximum possible amount of data with the minimum number of transmission lines and switches in an existing network topology.[13]

Genetic algorithms have been used in **controlling and managing gas pipelines** that consist of many branches carrying different amounts of gas. The solution cut costs, optimized energy use, minimized false leak alarms as well as creating a set of rules for responding properly to punctures in the pipelines.[14]

In distribution, companies have used genetic algorithms for what is commonly known as the classical **traveling salesman problem**. Manufacturing companies must make efficient use of production capacity, while shipping the most goods on the fewest trucks using the best routes. The new model, based on genetic algorithms, efficiently searches through almost limitless possibilities to find the schedule which makes the best use of company resources and also cuts delivery time.[15]

Another type of problem that has been addressed successfully with genetic algorithms is **scheduling**. This is the problem of finding the optimal sequence to execute a set of operations in a way that a certain set of constraints are not violated. Usually, one attempts to maximize the utilization of individuals or machines and to minimize the cost or time required for completing the entire process. Conflicts may arise when one individual or machine is scheduled for more than one operation at the same time, or when more resources than available are utilized. Some tasks may have priority over others.[16]

Likenesses of the faces of criminals from witnesses' descriptions are important for **criminal identification**. Artists' sketches have been used for many years, but nowadays this task can be done by a genetic algorithm application, developed by the psychology department of New Mexico State University. The system presents several computer-generated faces and then asks the witness to rate the faces in terms of which looks most like the criminal. It is much easier for a human mind to recognize similarity in a face than to recall and describe facial features. The system takes the witness' rating and creates new faces. The process continues until a likeness of the criminal's face evolves. The system has proven to be very successful in tests, and New Mexico State University has applied for a patent.[17]

In **computer chip manufacturing**, Texas Instruments has used genetic algorithms for finding a design on the smallest piece of silicon possible. The new chip took 18 percent less space by using a strategy of cross connections that no human had thought of.[18]

There are of course many more applications that could be added to this list and the number is growing every day.[19] Genetic algorithms have also been used in financial applications, such as credit scoring, bankruptcy prediction, database mining, and for currency trading, which we will discuss later.

We can see that genetic algorithms offer the potential of assisting a broad range of different applications. They imitate natural evolution. Hence, they are designed to cope with the most challenging tasks imaginable. After all, natural evolution has been fine-tuned during billions of years of evolution.

3. Description of Genetic Algorithms

3.1. Representation

Natural evolution operates on encodings of biological entities at the level of chromosomes, rather than on the entities themselves. Similarly, genetic algorithms operate on representations of solutions to problems. Since they work with encoded parameters of the optimization problem, the choice of a representation form has a large impact on the performance. There are different ways of encoding solutions, and probably no single best way for all problems. The performance of genetic algorithms depends on the choice of a suitable representation technique.[20]

Most genetic algorithms applications use Holland's fixed-length simple binary coding. This is historically the most widely used representation.[21] Each chromosome is comprised of zeroes and ones, with each bit representing a gene. A conceptually simpler technique would be the real-number representation, in which each variable being optimized is represented as a conventional floating-point number. Another approach might use decimal digits only and many other encoding techniques are possible. Researchers in genetic algorithms are divided on this issue. In the following, the fixed-length simple binary coding will be used to explain how the genetic algorithm works.[22]

The terminology used in the genetic algorithms literature is a mix of expressions originating in natural genetics and in computer science.

Many researchers in the area of genetic algorithms came from the biology domain, therefore the terminology of natural genetics seems to have prevailed. In order to obtain a better understanding, Table 2.1 summarizes the correspondence between natural and artificial terminology.[23]

Table 2.1. Comparison of Natural Genetics and Genetic Algorithms Terminology

Natural genetics	Genetic algorithms
chromosome	string
gene	bit, feature, character, or detector
allele	feature value, value of a gene
locus	position
genotype	structure, coded solution
phenotype	behaviour, parameter set, solution alternative, decoded structure, decoded solution, or a point in the solution space

3.2. Genetic Algorithms Step by Step

The typical genetic algorithm consists of a number of steps on its way to finding optimal solutions. Different authors describe different numbers of steps, but all include the characteristic procedures that a genetic algorithm runs through.[24] These will be covered in the following.

3.2.1 Initialization. The first step is the creation of an initial population of solutions, or chromosomes. The population of chromosomes is generally chosen at random, for example, by flicking a coin or by letting a computer generate random numbers. There are no hard rules for determining the size of the population. Generally, population size is chosen to be between 100 and 200. Larger populations guarantee greater diversity and may produce more robust solutions, but use more computer resources. The initial population must span a wide range of variable settings, with a high degree of diversity among solutions in order for later steps to work effectively.[25]

The population will converge through the application of the genetic operators. Convergence means that the initially diverse population of chromosomes that has been constructed by a random number generator tends to become more similar as it undergoes the genetic recombinations. The population may even become identical after many generations.[26] If explicit knowledge about the system that has to be optimized is at hand, it may be included in the initial population, for example, by restricting the initialization to a domain of interest.[27]

3.2.2 Fitness Evaluation. In the next step, the fitness of the population's individuals is evaluated. In biology, natural collection means that chromosomes that are more fit tend to produce more offspring than do those that are not as fit. Similarly, the goal of the genetic algorithm is to find the individual representing a particular solution to the problem which maximizes the objective function, so its fitness is the value of the objective function for a chromosome.[28] Genetic algorithms can of course also solve minimization problems.[29]

The fitness function (also called objective function or evaluation function) is used to map the individual's chromosomes or bit strings into a positive number, the individual's fitness. The genotype, the individual's bit string, has to be decoded for this purpose into the phenotype, which is the solution alternative. Once the genotype has been decoded, the calculation of the fitness is simple: we use the fitness function to calculate the phenotype's parameter values into a positive number, the fitness. The fitness function plays the role of the natural environment, rating solutions in terms of their fitness.[30]

3.2.3 Selection. In the third step, the genetic algorithm starts to reproduce. The individuals that are going to become parents of the next generation are selected from the initial population of chromosomes. This parent generation is the "mating pool" for the subsequent generation, the offspring.[31] Selection determines which individuals of the population will have all or some of their genetic material passed on to the next generation of individuals. The object of the selection method is to assign chromosomes with the largest fitness a higher probability of reproduction. There are many alternative selection procedures, a live-or-die turning point for the chromosomes.[32] The most common way is Holland's spinning of the weighted roulette wheel.

Roulette wheel selection gets its name from the fact that the algorithm works like a roulette game. Segments of the roulette wheel are allocated to population individuals in proportion to the individual's relative fitness score (see Figure 2.1).[33] Selection of parents is carried out through successive spins of the roulette wheel. The selection process is random, but fitter individuals will be more likely to be selected because of the higher probability of getting picked. This is a simple technique of rewarding fitter individuals, but because possible random fluctuations it is not ideal. With the relatively small population sizes typically used in typical genetic algorithms, an unfavorable series of spins of the

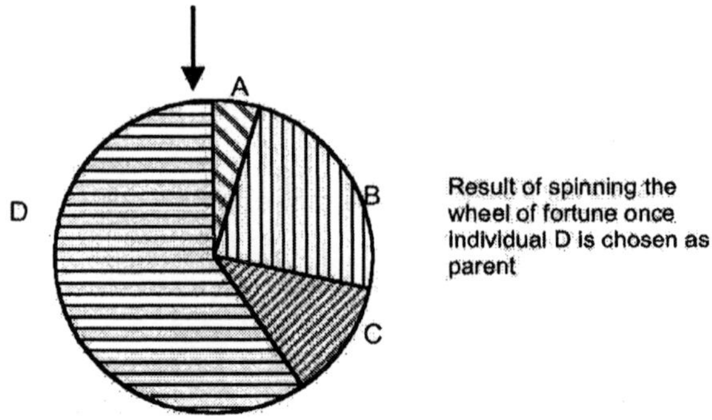

Figure 2.1. Roulette wheel selection: a larger segment implies larger fitness.

roulette wheel can select a disproportionately large number of unfit parent chromosomes.[34]

There are many other ways of accomplishing effective selection, such as stochastic remainder selection, elitism selection, linear fitness selection, genitor selection, ranking selection with roulette wheel, or tournament selection.[35]

Tournament selection can involve two or more individuals at a time. In the two-party tournament selection, for example, pairs of strings are randomly chosen from the initial population. Their fitness values are compared and a copy of the better performing individual becomes part of the mating pool. The tournament will be performed repeatedly until the mating pool is filled. That way, the worst performing string in the population will never be selected for inclusion in the mating pool.[36]

The choice of the selection method has a major impact on the balance between "exploitation" and "exploration".[37] Exploitation means taking advantage of information already obtained, while exploration refers to searching new areas. Holland illustrates this with the chess example.[38] After finding a good strategy for playing chess, one can of course solely concentrate on exploiting this strategy. But this behaviour inhibits the discovery of new, possibly superior, strategies. Improvements will only be discovered by trying new and more risky strategies, by exploring.

If we solely applied the selection procedure, only exploitation would occur. But one attribute of genetic algorithms that makes them superior to conventional problem-solving methods is their remarkable balance

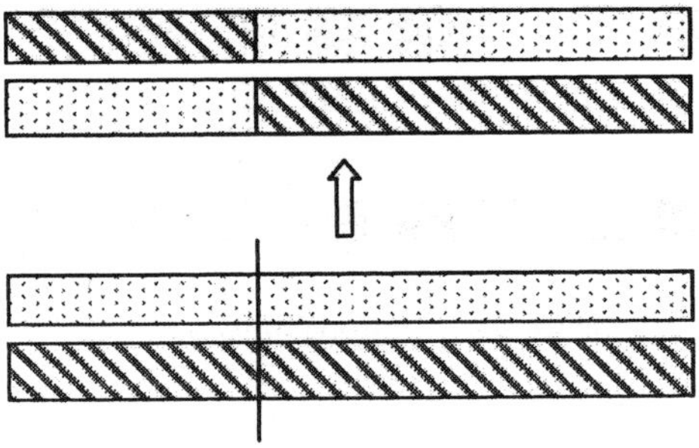

Figure 2.2. Crossover diagram.

between exploitation and exploration.[39] Exploration is being achieved by the genetic operators, to be examined in the next section.

3.2.4 Genetic Operators.

Crossover. The primary exploration operator in genetic algorithms is crossover, a version of artificial mating. If two strings with high fitness values are mated, exploring some combination of their genes may produce an offspring with even higher fitness. Crossover is a way of searching the range of possible solutions based on the information obtained from the existing solutions.

There are many ways in which crossover can be implemented, such as one-point crossover, two-point crossover, n-point crossover, or uniform crossover.[40] In the following, we will stay with the simplest form, Holland's one-point crossover technique.

Single-point crossover is the simplest form, yet it is highly effective. First, two potential parents are selected from the population. Then, a random position on the bit string is selected, the cross point or cut point. The bits to the left of the cross point on parent one are combined with the bits to the right of the cut point in parent 2 to form child 1. The opposite segments are combined to create child 2 (see Figure 2.2).[41]

Both children will tend to be different from either of their parents yet retain some features of both. If both parents had high fitness (which is likely because of the selection procedure), then there is a high chance that at least one of the children is as fit or as fitter than either parent. In this case, further selection procedures will favor the child's reproduction, otherwise they will favor the extinction of its genetic information, since it will not be selected as a parent for the next generation.

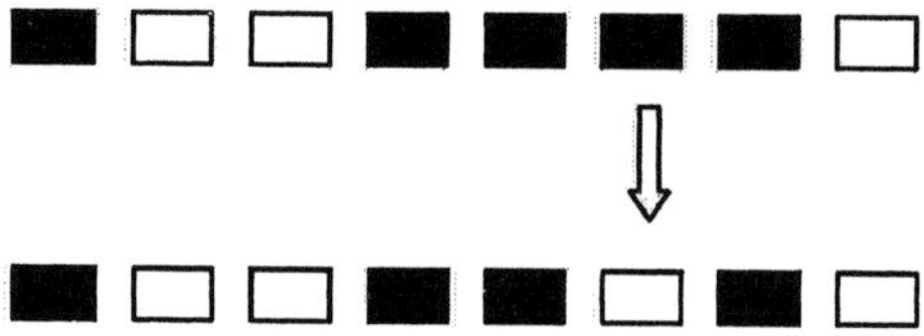

Figure 2.3. Mutation diagram.

In order to counter the possibility that most or all of the crosses produce children of lower fitness, a simulated coin is flipped before the crossover operation is performed. The coin is biased in a way that it lets crossover be performed more often than is required to pass the parents' genetic information into the next generation unchanged.

Mutation. If crossover is the main operator of genetic algorithms that efficiently searches the solution space, then mutation could be called the "background operator" that reintroduces lost alleles into the population.[42] Mutation occasionally injects a random alteration for one of the genes (see Figure 2.3).[43] Similar to mutation in nature, this function preserves diversity in the population. It provides innovation, possibly leading to exploration of a region of better solutions. Mutation is performed with low probability. Applied in conjunction with selection and crossover, mutation not only leads to an efficient search of the solution space but also provides an insurance against loss of needed diversity.[44]

Inversion. Another operator in Holland's genetic algorithms was the inversion operator. Inversion is a process that shifts the locus of one or more genes in a chromosome from one point to another. This does not change the meaning of the genotype, in that a genotype before and after inversion will still decode to the same phenotype. This is the same as in natural genetics, where the function of a gene is independent of its position in the chromosome.[45] In the genetic algorithm setting, inversion has not been found to be useful.[46]

3.2.5 Evaluate Offspring and Create a New Population.

Having applied the genetic operators, the new individuals must be evaluated in order to assess their fitness. Hence, the children undergo evaluation through the fitness function, similar to the process in Section 3.2.2 Once they have been assessed, the new, better offspring will replace those chromosomes of the population that perform the weakest, according to the fitness function.

3.2.6 Termination Criteria. The last step is the verification of a termination criterion. If the termination criterion is not satisfied, the genetic algorithm returns to Section 3.2.3 and continues with the selection and the genetic operators. The termination criterion is satisfied when either the maximum number of generations is achieved or when the genotypes (the structures) of the population of individuals converges. The maximum number of generations is set by the creator of the genetic algorithm before running it, which ensures that the genetic algorithm does not continue indefinitely. Convergence may occur in two ways: first, convergence of the genotype structure occurs when all bit positions in all strings are identical. In this case, crossover will have no further effect.[47] It is also possible that there is phenotypic convergence without genotypic convergence— identical behaviour with dispersant structures—this will happen when only certain, converged genes are tested.

3.3. Why do Genetic Algorithms Work?

We have visited the steps of a genetic algorithm and have seen that it is based on relatively simple procedures. But how can a limited number of individuals efficiently explore vast search spaces with trillions of different points? This question leads us to the Schema Theorem or Fundamental Theorem of Genetic Algorithms,[48] which states that each time an individual is processed, the genetic algorithm is sampling not simply one but many points in the search space.

With binary-digit-string genotypes, each position in the string is limited to the characters 0 and 1. We also need the "don't care" symbol (*), meaning that its value could be 0 or 1. The sample string 00**11 matches all of the string of the following set {000011, 000111, 001011, 001111}. Strings including the don't care symbol are called **schemata** or **similarity templates**. The success of genetic algorithms stems from the fact that they implicitly search for schemata with higher fitness. In a bit string of length five there are only $2^5 = 32$ individuals, but $3^5 = 243$ schemata. Every time a genetic algorithm processes a single individuum, it actually processes many schemata. Hence, a genetic algorithm that deals with a population of a few hundred or thousand strings in fact samples a vastly greater number of solutions. This **implicit parallelism**, combined with the fact that bad schemata are quickly eliminated, accounts for the genetic algorithm's central advantage over other problem-solving techniques and makes it a powerful search tool.[49]

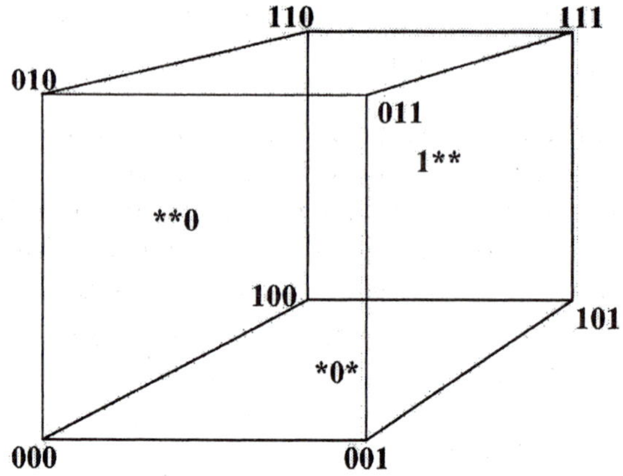

Figure 2.4. Visualization of schemata as hyperplanes in 3-dimensional space.

Although both crossover and mutation destroy existing schemata, they are necessary for building better ones. The degree of destruction is dependent on the order and the length of the schemata, where length is the distance between the first 0 or 1 and the last 0 or 1. For example, in 1***0*, the length is 4. Order refers to the number of fixed positions (0 or 1) in the schema. Highly fit, short, low-order schemata are called **building blocks**. They tend to stay intact, because when crossover chops up schemata as it recombines individuals, short schemata have a higher chance of surviving crossover intact. Low-order schemata process more individuals and are also more likely to stay intact.[50]

A good visualization of genetic algorithms and schemata is supplied by the geometric representation. The easiest representation is attained for strings and schemata of length 3 in a three-dimensional vector space (see Figure 2.4).[51] Points in the space represent string or schemata of order 3, lines are schemata of order 2, planes in the space are schemata of order 1. The whole space is covered by the schema of order 0: {***}.[52]

3.4. Summary

Genetic algorithms mimic natural reproduction and provide a simple yet powerful tool that can be easily transformed into a computer program. Further assessment uncovers the following benefits and drawbacks.

The **advantages** of genetic algorithms are that they can quickly and reliably solve problems that are hard to tackle by traditional means. Genetic algorithms can interface with existing applications, they are extensible, and they are easy to hybridize. Existing knowledge need not be discarded. Implicit parallelism makes genetic algorithms a very efficient optimization algorithm. Their greatest property is the ability to find approximate solutions to combinatorially explosive problems.[53]

There are, however, also a number of **disadvantages** cited in literature. For one, there is the fact that genetic algorithms may find only near-optimal solutions. There is the difficulty of verifying the optimality of a solution, especially when genetic algorithms have been applied on a problem that would be computationally impossible to solve when using conventional techniques.[54] Further restrictions are the difficulties of choosing a suitable representation technique, and making the right decision regarding the choice of the selection method and the genetic operator probabilities.[55]

4. Genetic Algorithms and Financial Applications

After this excursion into the technical world of the genetic algorithms, we return to financial markets. Before examining the technique's usefulness in forecasting foreign exchange markets, we take a closer look at the wide range of genetic algorithms applications in the financial domain, an area well-documented in the literature.

Genetic algorithms appear to be particularly well-suited for financial modelling applications because of three reasons:[56]

- They are payoff-driven. Payoffs can mean improvements in predictive power or returns over a benchmark. There is an excellent match between the problem-solving tool and the issues to be addressed.
- Genetic algorithms are inherently quantitative. They are well-suited to parameter optimization.
- Genetic algorithms allow a wide variety of extensions and constraints that cannot be provided in traditional methods. They are robust, revealing a remarkable balance between efficiency and efficacy. The genetic algorithm's use of a coding, the search from a population, its blindness to auxiliary information, and its randomized operators contribute to its robustness.[57]

Corporate distress and bankruptcy have been an area of research since the 1930s. The most common modelling technique in this area is one dating back to the 1960s: Multiple Discriminant Analysis. Genetic algorithms have been used to induce rules for **bankruptcy prediction** based on balance-sheet information. The genetic algorithm's task was to find rules formed on the basis of financial ratios which were indicating future bankruptcy. Genetic algorithms were able to come up with simple rules that consistently outperformed the Multiple Discriminant Analysis approach by more than 10%.[58]

Genetic algorithms have also been used for **credit scoring**. This is an universal business problem, having to assess the risk associated with any form of investment. Banks have been using models based on genetic algorithms, reporting considerable savings in some cases. Genetic algorithm approaches to this problem have been superior to others (such as neural nets) because of the transparency of the achieved results.[59]

Other applications of genetic algorithms in finance include database mining, training neural networks, financial spreadsheets, filtering insurance applications, investment portfolio selection and management,[60] as well as trading rules for stock or bond markets.[61] The latter application, creating rules for **trading in stock or bond markets**, is frequently illustrated in literature and has a very important characteristic: Even a small amount of performance improvement (through the use of genetic algorithms) is likely to yield significant payoffs! This of course also applies to trading rules for foreign exchange markets, which will be covered below.

Trading in stock markets can take advantage of the ability to predict time series containing market data. In one genetic-algorithm approach, 18 market indicators were included in the model. In this approach, the genetic algorithm's chromosomes consisted of a list of 36 real-valued numbers. Two of these numbers were associated with each of the 18 market indicators, one showing a low value of the indicator and one denoting a high value. Each chromosome was evaluated by determining the prediction system's historical performance when each indicator value had fallen between the low and the high values associated with it on the chromosome. After letting the genetic algorithm evolve the population, the result was a set of chromosomes that described combinations of parameter values correlated with high performance of the prediction system. The user of this system now only need wait until current indicator values fall into the ranges specified by one of these chromosomes. She then can perform a trade based on the predictions of the system.[62] The technique has been reported to perform well.[63]

A study of asset-allocation strategies was performed by Bauer and Liepins.[64] They examined strategies that involved switching between an investment in a Standard & Poor's 500 fund and a small-firm equity fund. The investor decides every month whether to be 100% in S&P 500 or 100% in small-firm stocks. He has a five-year investment horizon and wants to maximize terminal wealth. A particular switching strategy is represented as a 60-character bit string. The results from test runs for 12 five-year periods dating back to 1926 impressively demonstrated the genetic algorithm's ability to quickly find attractive problem solutions. The algorithm converged on average in approximately 50 generations. The average solution was within 0.3 percent of the optimal value.

An increase in complexity by incorporating transaction costs into the switching was easily accommodated by the genetic algorithm through changes in the evaluation function. It still performed reasonably well, showing average solutions that were within 3 percent of the optimal solution. The complexity was magnified even further by adding two more assets to the investment alternatives and assuming differential transaction costs. The genetic algorithm still performed quite well, despite the increase in complexity. The terminal wealth for the best performing individual was on average 94.5% of the optimal value.

Bauer and Liepins concluded that the power and flexibility of genetic algorithms enables searches of problem spaces in an extremely efficient manner. The algorithm converged quickly to optimal or near-optimal solutions, and variations of the basic problem could easily be accommodated through representation and by the evaluation function. Simplicity and flexibility are major advantages of genetic algorithms when programming. The authors suggest that genetic algorithms may be used in detecting and evaluating intra-day trading patterns.[65]

A major problem they found, however, was the representation issue. No generally accepted application code exists on this yet. This is a difficult but crucial decision that must be made before applying genetic algorithms to a particular problem. Another hindrance when applying the genetic algorithm to finance applications is the fact that tailoring it to the problem's specifications might produce better results. For example, one could hybridize the genetic algorithm with other algorithms currently used in the domain. As much as this promises better results, it is more of an art, requiring more knowledge and experience than when just using a genetic algorithm. Other problems are the task of coming up with a suitable evaluation function and the availability of **CPU** time. The latter problem of course is being eased by computing resources constantly becoming cheaper and more powerful.[66]

The potential benefits of using genetic alogrithms have been called mind-boggling, and Bauer and Liepins note that the genetic algorithm's efficiency suggests that they might be of use in real-time trading applications.[67] The more one wants to find out about current use of genetic algorithms in trading applications, however, the less people are willing to tell, referring to "proprietary knowledge". Does part of the answer to the question "If you're so smart, why aren't you rich?" lie herein?

5. Genetic Algorithms and Foreign Exchange

The literature in the area of genetic algorithms in forecasting on foreign-exchange market is viewed mainly from the perspective of speculators and traders. The genetic algorithm is used to come up with some trading rule that will generate profits. Most of the research is concentrated on finding technical trading rules. But we will first look at an approach assuming a fundamental perspective.

5.1. Fundamental Trading Strategies

In his paper on genetic algorithms and exchange rates, Bauer focused on fundamental strategies for trading currencies.[68] Fundamental strategies rely on economic relationships. The economic variables Bauer primarily chose to include in his model are: money supply, inflation, interest rates, and economic activity. He limited his study to the US dollar and the Deutchmark.

Bauer developed trading rules for the monthly allocation of assets either into U.S. dollars or Deutchmarks. He used a simple genetic algorithm with fixed-length binary-string representation and a population size of 100. One-point crossover was performed with a probability of 0.6, and mutation with a probability of 0.001.[69] Simple ranking selection was chosen and the genetic algorithm was limited to 300 generations. Each binary string represented a specific foreign-exchange monthly trading rule. Each month, the trading rule signalled whether to invest money in U.S. dollars or in Deutchmarks. The fitness measure was the U.S.-dollar value at the end of the five-year testing period. Transaction costs and interest were ignored in the calculations of the terminal value.

Bauer reported that the genetic algorithm found only near-optimal solutions in most cases, which he conquered by running multiple trials. The genetic algorithm found trading rules with substantially greater fitness levels than when using enumerative search. It also outperformed the buy-and-hold alternative. The performance of the rules in some intervals was mixed. None of the rules outperformed the buy-and-hold strategy during 1989-1990, an abnormal period, characterized by Germany expanding its money supply for financing the German reunification process.[70]

Hence, this application of a genetic algorithm for finding trading rules in the foreign exchange market based on fundamental economic variables is limited but shows promising results. Bauer recommends the following extensions and variations of his basic methodology for future research: the substitution or addition of other macroeconomic variables or even technical variables. Genetic-algorithm specifics such as representation, crossover probability, or the fitness function could also be altered in a sensitivity analysis.

5.2. Technical Trading Rules

Greater attention in the application of genetic algorithms in foreign exchange markets has been paid to the detection of trading rules based on technical data. Technical strategies rely on past price and trading-volume information. These variables have a far more short-term character and are hence more useful for the day-to-day trading and in real-time trading applications than are the macro-economic fundamentals.

Technical trading rules use historical price and volume data to predict the direction of future price movements. They then give either a buy or a sell signal.[71] Genetic algorithms have been applied differently in the area of technical trading rules. They have been used for optimizing conventional methods and for designing trading rules.

5.2.1 Optimizing Technical Trading Rules. Genetic algorithms can be applied in the search for the optimal parameter threshold values for technical trading models. The most elementary way of processing data is the **simple moving average** (see Figure 2.5).[72] It is simply calculated by taking an average of the prices of the last L days. The parameter L represents the length of the period of the moving average. If the actual price of a currency rises above the moving average, a buy signal is issued. When the price falls below the moving average, the model issues a sell signal.

This kind of a simple moving average can work well in trending markets. It ignores short-term changes and follows large moves. But since exchange rates usually show high volatility in a relatively narrow range, the simple moving average will lose money. To avoid this, the **double moving average** uses a combination of long-term and short-term moving averages (see Figure 2.6).[73] When the shorter moving average rises above the longer moving average from below, a buy signal is issued. The foreign currency is then held until the shorter moving average falls below the longer moving average, when it is sold.[74]

Another technique that has been used by technical traders is the **filter rule**: if the current investment position in a certain foreign currency is short, a buy signal is issued when its price rises by $x\%$ above the minimum local price (that is, the minimum observed in the historical series of prices since the last transaction). The foreign currency is then bought and held until the price falls $x\%$ below the maximum local price (that is, the maximum price observed since the last transaction), when the currency is sold.[75]

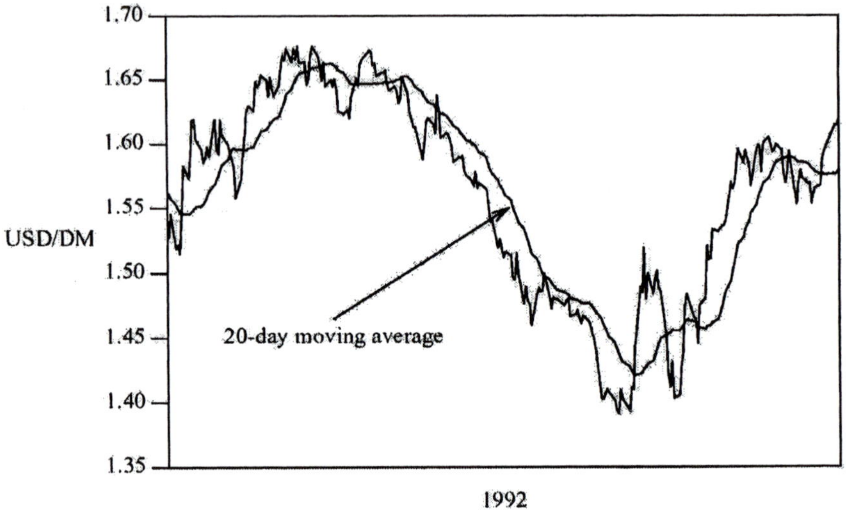

Figure 2.5. The simple moving average of the U.S.$/ DM exchange rate.

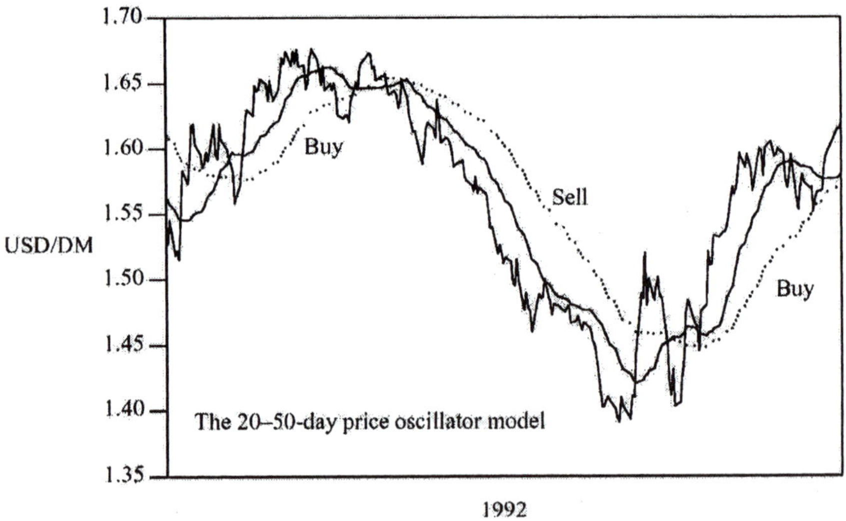

Figure 2.6. The double moving average of the U.S.$/ DM exchange rate.

A genetic algorithm can now be applied for searching for the optimal threshold values for the trading rule parameters (i.e., for the length L of the respective moving averages and for the percentage level x). Pereira has done this in a study for the U.S. dollar / Australian dollar exchange rate.[76] Each trading rule required slightly different genetic-algorithm features (i.e., population size). Still, all used fixed-length binary representation, the genitor ranking selection procedure, and the trading rule fitness was evaluated by the profit-maximization criterion. Crossover and mutation occurred with a probability of 0.6 and 0.001, respectively,[77] and the maximum number of generations was limited to 50 in all genetic algorithms.[78]

After letting the genetic algorithm find the optimal parameters for the three different rules and after applying these rules for trading U.S.$ / AUS$, the results showed profitable in-sample returns. The search for the complex double-moving-average parameter values was performed more than fifty times faster when using the genetic algorithm than when applying an exhaustive grid-search technique.

The out-of-sample profitability of the trading rules was much lower than in the rule-building period. Still, the trading rules rendered positive returns, with the filter rule being most profitable. But, the study did not consider risk, nor did it include any trading costs, such as transaction costs, taxes, or costs of information. The author concluded that more complex, statistically significant models must include risk and trading costs. A genetic algorithm is a suitable technique for these explosively complex trading systems.[79]

5.2.2 Genetic Programming for Rule Induction. After applying genetic algorithms for optimizing traditional prediction techniques, we will now examine whether they could also be used for the automatic induction of foreign-exchange trading models. We want to find a way of letting the computer develop its own strategies that can model the foreign exchange market's dynamics, based on a number of different inputs.

Suppose we have three different models:[80]

- if (10-day moving average > 0) then buy \$ else sell \$ (model 1)
- if (15-day moving average > 0) then buy \$ else sell \$ (model 2)
- if (25-day moving average > 0) then buy \$ else sell \$ (model 3)

We would like to create a genetic algorithm-based system that produces statements such as:

- if (model 1 says BUY) then BUY else SELL
- if (model 1 and model 2 say BUY) then BUY else SELL
- if (models 1 and 2 say BUY) OR (model 3 says BUY) then BUY else SELL

In this case it is unclear how to represent the statements in terms of a binary string, or how to perform crossover or mutation. In addition, each individual in the population can have a different length and complexity. This will handicap the crossover mechanism.

At this point, a new technique comes into play. **Genetic programming** is an important spin-off from genetic algorithms aiming at automatically evolving computer programs to solve problems.[81] It uses data structures of varying complexity that can grow or shrink through appropriately defined crossover and mutation operators.[82]

The representation now occurs by using a natural mapping between algebraic statements of the form shown above and a binary tree structure. The mutation operator now alters any one quantity in the set of data fields of a binary tree's node. The crossover operator selects two expressions with varying probability and it picks a crossover site in each. A copy of the first parent tree is made and the subtree consisting of the nodes below the crossover site in the second tree is implanted onto the first tree at the first tree's crossover site, replacing the original subtree.[83]

The trading system that has been created based on these specifications generated average returns of about 50% over the 7-year test period, or about 7% per year. Transaction costs were not taken under consideration, and the potential amounts that could have been made or lost by trading for the interest rate differentials between the currency pairs were ignored. Colin concludes that major diversification gains can be rendered by generating numerous non-correlated models (internal diversification), or by diversifying across markets.[84] The main disadvantage of a multiple-model approach is, however, the significant investment that has to be made in computer technology in order to be able to handle the large number of signals generated. Nevertheless, colin recommends extensions of the model including several markets or economic variables, such as using interest rates and commodity indices, etc.[85]

6. Summary

Genetic algorithms have been presented as robust general-purpose optimization techniques. These very efficient problem-solving techniques have found to be useful in many engineering applications as well as in the domain of finance. They can quickly solve difficult problems, they can easily be tailored to a specific problem, and they are easy to interface with existing models. They possess the ability to find approximate solutions to combinatorily explosive problems. This near-optimality has also been viewed as a weakness, though. Further obstacles are the difficulties of choosing the representation form suitable to a specific problem, as well as the right selection method and the genetic operator probabilities.

Because of the huge potential profits, the genetic algorithm is a very interesting tool for financial applications. In the area of foreign exchange markets, it has been applied in fundamental as well as in technical trading models. Genetic algorithms seem to provide very promising results. Research, however, is still only in its infancy. Early publications on tests in this area report that the models were able to generate profits, but the assumptions that have been made were restricting the models.

Notes

1. Feldman and Treleaven (1994), p. 195
2. Feldman and Treleaven (1994), p. 199
3. Mitchell (1996), pp. 4-5
4. Holland (1992a), p. 44
5. Heitkoetter and Beasley. (1996), p. 1
6. Michalewicz (1994), pp. 1-3
7. Biethahn and Nissen (1995), pp. 4-33
8. Mitchell (1996), p. 3
9. Goldberg (1994), pp. 114-115
10. There are 10^{81} elementary particles in the known universe.
11. Hughes (1990), p. 22
12. Goldberg (1994), p. 117
13. Davis and Coombs (1987), pp. 252-256
14. Holland (1992a), p. 49
15. Goldberg and Lingle Jr. (1985), pp. 154-159 and Grefenstette et al. (1985), pp. 160-168
16. Syswerda (1991), pp. 332-349
17. Caldwell and Johnston (1991), pp. 416-421
18. Kennedy (1996), p. 1
19. For an extensive overview of the wide area of genetic algorithms applications, refer to Biethahn and Nissen (1995), pp. 48-62 or Goldberg (1989), pp. 126-129
20. Davis (1991), p. 4
21. Davis (1994), p. 137
22. For an overview of the work on the representation problem, refer to Koza (1993), pp. 63-68
23. Goldberg (1989), p. 22
24. For an overview of the genetic algorithms steps, refer to Holland (1992a), p. 46, Goldberg (1989), pp. 15-18, Davis (1991), p. 5, Bauer (1994), pp. 11-17 and 55-71, Biethahn and Nissen (1995), pp. 7-13, and Fogel (1995), pp. 89-97.
25. Especially the crossover operator (see Section 3.2.4 below).
26. Bauer (1994), p. 293

27. Biethahn and Nissen (eds., 1995), p. 9

28. Davis (1994), p. 135

29. Bauer (1994), p. 49

30. Michalewicz (1994), pp. 17-20

31. Goldberg (1989), p. 15

32. Goldberg (1994), p. 113

33. Beithahn and Nissen(eds., 1995, p.10)

34. Mitchell (1996), p. 167

35. For an overview of various alternative selection techniques, refer to Goldberg and Deb (1991), pp. 69-93, Kingdon and Feldman (1995), pp. 96-100, Mitchell (1996), pp. 166-171, or Bäck (1996), pp. 163-195.

36. Goldberg (1994), p. 113

37. Kingdon and Feldman (1995), p. 96

38. Holland (1992a), p. 47

39. Holland (1992a), p. 47

40. Bauer (1994), pp. 94–96

41. Bauer (1994), p. 63

42. Biethahn and Nissen (1995), pp. 9-10

43. Bauer (1994), p. 65

44. Goldberg (1994), p. 114

45. Mitchell (1996), p. 159

46. Davis (1991), p. 21

47. Kingdon and Feldman (1995), p. 92

48. Goldberg (1989), p. 33

49. Holland (1992a), p. 46

50. Goldberg (1989), pp. 41-45

51. Koza (1993), p. 34

52. Goldberg (1989), pp. 53-54

53. Feldman and Treleaven (1994), p. 201

54. Bodnovich (1995), p. 1

55. Feldman and Treleaven (1994), p. 201

56. Leinweber and Arnott (1995), p. 12

57. Goldberg (1989), p. 10

58. Kingdon and Feldman (1995), pp. 109-110

59. Davis (1994), p. 143

60. Eddelbuttel (1996a), pp. 1-23 and Eddelbuttel (1996b), p. 21

61. For an overview of financial applications of genetic algorithms, refer to Biethahn and Nissen (1995), pp. 49-59.

62. Davis (1994), pp. 141-142

63. Goldberg (1994), p. 116

64. Bauer and Liepins (1992), pp. 89-100

65. Bauer and Liepins (1992), p. 97

66. Davis (1994), pp. 145-146

67. Bauer and Liepins (1992), p. 97

68. Bauer (1995), pp. 253-263

69. as recommended in Bauer (1994), pp. 62-65

70. Bauer (1995), p. 262

71. Pereira (1996), p. 28

72. Colin (1994), p. 152

73. Colin (1994), p. 154

74. Pereira (1996), p. 28

75. Pereira (1996), p. 28

76. Pereira (1996), pp. 27-34

77. As recommended in Bauer (1994), pp. 62-65

78. Good instructions for setting up a genetic algorithm for this kind of problems is given in Colin (1994), pp. 152-160.

79. Pereira (1996), pp. 33-34

80. Colin (1994), p. 160

81. Koza (1993), and Biethahn and Nissen (1995), p. 25

82. Colin (2000), p. 176

83. Koza (1992), pp. 225-226

84. Colin (2000), pp. 183-186

85. Colin (2000), p. 188

References

Bäck, T. (1996). *Evolutionary Algorithms in Theory and Practice, Evolutionary Strategies, Evolutionary Progarmming, Genetic Algorithms.* New York and Oxford: Oxford University Press.

Barnett, W. A., C. Chiarella, S. Keen, R. E. Marks, and H. Schnabl (eds., 2000). *Commerce, Complexity, and Evolution: Topics in Economics, Finance, Marketing, and Management: Proceedings of the Twelfth International Symposium in Economic Theory and Econometrics.* Cambridge: Cambridge University Press.

Bauer, R. J. Jr. (1994). *Genetic Algorithms and Investment Strategies.* New York: Wiley.

Bauer, R. J. Jr. (1995). "Genetic Algorithms and the Management of Exchange Rate Risk," in J. Biethahn and V. Nissen (eds.), *Evolutionary Algorithms in Management Applications*, 253–263.

Bauer, R. J. Jr. and G. E. Liepins (1992). "Genetic Algorithms and Computerized Trading Strategies," in O'Leary and Watkins (eds.), *Expert Systems in Finance*, 89–100.

Belew, R. K. and L. B. Booker (eds., 1991). *Proceedings of the Fourth International Conference on Genetic Algorithms.* San Mateo: Morgan Kaufmann.

Biethahn, J. and V. Nissen (eds., 1995). *Evolutionary Algorithms in Management Applications.* Berlin, Heidelberg, New York: Springer-Verlag.

Bodnovich, T. (1995). "Genetic Algorithms in Business and their Supportive Role in Decision Making," <http://business.kent.edu/~tbodnovi/ga.html>, November.

Caldwell, C. and V. S. Johnston (1991). "Tracking Criminal Suspect Through 'Face-space' with a Genetic Algorithm," in Belew and Booker (eds.), *Proceedings of the Fourth International Conference on Genetic Algorithms*, 416–421.

Colin, A. M. (1994). "Genetic Algorithms for Financial Modeling," in Deboeck (ed.), *Trading on the Edge: Neural, Genetic, and Fuzzy Systems for Chaotic Financial Markets*, 148–173.

Colin, A. M. (2000). "A Genetic-programming-based Approach to the Generation of Foreign-exchange Trading Models," in Barnett, Chiarella, Keen, Marks, and Schnabl (eds.), *Commerce, Complexity, and Evolution: Topics in Economics, Finance, Marketing, and Management: Proceedings of the Twelfth International Symposium in Economic Theory and Econometrics*, 173–189.

Davis, L. (ed., 1991). *Handbook of Genetic Algorithms.* New York: Van Nostrand Reinhold.

Davis, L. (1994). "Genetic Algorithms and Financial Applications," in Deboeck (ed.), *Trading on the Edge: Neural, Genetic, and Fuzzy Systems for Chaotic Financial Markets*, 133–147.

Davis, L. and S. Coombs (1987). "Genetic Algorithms and Communication Link Speed Design: Theoretical Considerations," in Grefenstette (ed.), *Genetic Algorithms and their Applications: Proceedings of the Second International Conference on Genetic Algorithms*, 252–256.

Deboeck, G. J. (ed., 1994). *Trading on the Edge: Neural, Genetic, and Fuzzy Systems for Chaotic Financial Markets.* New York: Wiley.

Eddelbüttel, D. (1996a). "A Genetic Algorithm for Passive Management: Creating Index Funds with Fewer Stocks," <http://rosebud.sps.queensu.ca/~edd/papers.html> 26 December.

Eddelbüttel, D. (1996b). "A Hybrid Genetic Algorithm for Passive Management," <http://rosebud.sps.queensu.ca/~edd/papers.html> 26 December.

Feldman, K. and P. Treleaven (1994). "Intelligent Systems in Finance," *Applied Mathematical Finance*, 1(2), December, 195–207.

Fogel, D. B. (1995). *Evolutionary Computation: Toward a New Philosophy of Machine Intelligence.* New York: IEEE Press.

Goldberg, D. E. (1989). *Genetic Algorithms in Search, Optimization, and Machine Learning*, Reading. Massachussetts: Addison-Wesley.

Goldberg, D. E. (1994). "Genetic and Evolutionary Algorithms Come of Age," *Communications of the ACM*, 37(3), March, 113–119.

Goldberg, D. E. and K. Deb (1991). "A Comparative Analysis of Selection Schemes Used in Genetic Algorithms," in Rawlins (ed.), *Foundations of Genetic Algorithms*, 69–93.

Goldberg, D. E. and R. Lingle Jr. (1985). "Alleles, Loci, and the Traveling Salesman Problem," in Grefenstette (ed.), *Proceedings of an International Conference on Genetic Algorithms and their Applications*, 154–159.

Grefenstette, J. J. (ed., 1985). *Proceedings of an International Conference on Genetic Algorithms and their Applications.* Hillsdale: Erlbaum.

Grefenstette, J. J. (ed., 1987). *Genetic Algorithms and their Applications: Proceedings of the Second International Conference on Genetic Algorithms.* Hillsdale: Lawrence Erlbaum.

Grefenstette, J. J., R. Gopal, B. Rosmaita, and D. V. Gucht (1985). "Genetic Algorithms for the Traveling Salesman Problem," in Grefenstette (ed.), *Proceedings of an International Conference on Genetic Algorithms and their Applications*, 160–168.

Hetketter, J. and D. Beasley (1996). "Q1: What Are Evolutionary Algorithms (EAs)?" <http://www.cis.ohio-state.edu/hypertext/faq/usenet/ai-faq/genetic/part2/faq-doc-1.html>

Holland, J. H. (1992a). "Genetic Algorithms," *Scientific American*, July, 44–50.

Holland, J. H. (1992b). *Adaptation in Natural and Artificial Systems*, 2nd ed. Cambridge: MIT Press.

Hughes, M. (1990). "Improving Products and Processes—Nature's Way," *Industrial Management & Data Systems*, 6, 22–25.

Kennedy, S. (1996). "Introduction to Genetic Algorithms," <http://www.axcelis.com:80/articles.html>

Kingdon, J. and K. Feldman (1995). "Genetic Algorithms and Applications to Finance," *Applied Mathematical Finance*, 2(2), 89–116.

Koza, J. R. (1992). "The Genetic Programming Paradigm: Genetically Breeding Populations of Computer Programs to Solve Problems," in

Soucek and IRIS Group, *Dynamic, Genetic, and Chaotic Programming: The Sixth Generation*, 203–321.

Koza, J. R. (1993). *Genetic Programming.* Cambridge: MIT Press.

Leinweber, D. J. and R. D. Arnott (1995). "Quantitative and Computational Innovation in Investment Management," *The Journal of Portfolio Management*, 21(2), Winter, 8–15.

Michalewicz, Z. (1994). *Genetic Algorithms + Data Structures = Evolution Programs*, 2nd ed. Berlin, Heidelberg, New York: Springer-Verlag.

Mitchell, M. (1996). *An Introduction to Genetic Algorithms.* Cambridge: MIT Press.

O'Leary, D. E. and P. R. Watkins (eds., 1992). *Expert Systems in Finance.* Amsterdam and New York: North-Holland.

Pereira, R. (1996). "Selecting Parameters for Technical Trading Rules Using Genetic Algorithms," *Journal of Applied Finance and Investment*, 1(3), July/August, 27–34.

Rawlins, G. J. E. (ed., 1991). *Foundations of Genetic Algorithms.* San Mateo: Morgan Kaufmann.

Soucek, B. and IRIS Group (1992). *Dynamic, Genetic, and Chaotic Programming: The Sixth Generation.* New York: John Wiley.

Syswerda, G. (1991). "Schedule Optimization Using Genetic Algorithms," in Davis (ed.), *Handbook of Genetic Algorithms*, 332–349.

Chapter 3

GENETIC PROGRAMMING: A TUTORIAL WITH THE SOFTWARE SIMPLE GP

Shu-Heng Chen
AI-ECON Research Center, Department of Economics, National Chengchi University Taipei, Taiwan 11623
chchen@nccu.edu.tw

Tzu-Wen Kuo
AI-ECON Research Center, Department of Economics, National Chengchi University Taipei, Taiwan 11623
kuo@aiecon.org

Yuh-Pyng Shieh
Department of Computer Science and Information Engineering, National Taiwan University, Taipei, Taiwan 106
arping@turing.csie.ntu.edu.tw

Abstract This chapter demonstrates a computer program for tutoring genetic programming (GP). The software, called **Simple GP,** is developed by the **AI-ECON Research Center** at National Chengchi University, Taiwan. Using this software, the instructor can help students without programming background to quickly grasp some essential elements of GP. Along with the demonstration of the software is a list of key issues regarding the effective design of the implementation of GP. Some of the issues are already well noticed and studied by financial users of GP, but some are not. While many of the issues do not have a clear-cut answer, the attached software can help beginners to tackle those issues on their own. Once they have a general grasp of how to implement GP effectively, many advanced materials prepared in this volume are there for further exploration.

Keywords: Simple GP, Symbolic Regression, Data Generating Mechanisms, Chaotic Dynamic Series, Production Function

1. Features of Simple GP

Simple GP is designed to perform a task known as *symbolic regression* in machine learning since most applications of genetic programming (GP) in computational finance can be simplified as issues of *symbolic regression*. Given its function, the software is ill-equipped to perform some more advanced applications such as *agent-based financial modeling*. Another limitation of the software is its exclusion of *data processing*, an important dimension of *financial data mining*. We assume that this issue deserves a separate article and may be too complicated for an introductory tutorial.[1] **Simple GP** also escapes another important issue, the *fitness function*. In practice, to avoid overfitting, the fitness measure often imposes penalties for complex models. In some cases, the fitness function is *normalized* to avoid premature convergence. These treatments are skipped in **Simple GP**.

By assuming away those complications, **Simple GP** focuses on the fundamentals of GP. By that, we mean the following six items.

(1) Survival of the Fittest Principle

(2) Selection Schemes

(3) Disruption Avoidance

(4) Search Intensity

(5) Primitives

(6) Genetic Operators

Figure 3.1 displays the main menu of **Simple GP** covering these six fundamentals. In the next section, we shall discuss what these fundamentals mean for GP. Our discussion is based on multiple-trials of **Simple GP** on the following four *data generating mechanisms* (DGMs).

$$x_t = 4x_{t-1}(1 - x_{t-1}), \quad x_0 \in [0,1] \tag{3.1}$$

$$x_t = 4x_{t-1}^3 - 3x_{t-1}, \quad x_0 \in [-1,1] \tag{3.2}$$

$$x_t = 8x_{t-1}^4 - 8x_{t-1}^2 + 1, \quad x_0 \in [-1,1] \tag{3.3}$$

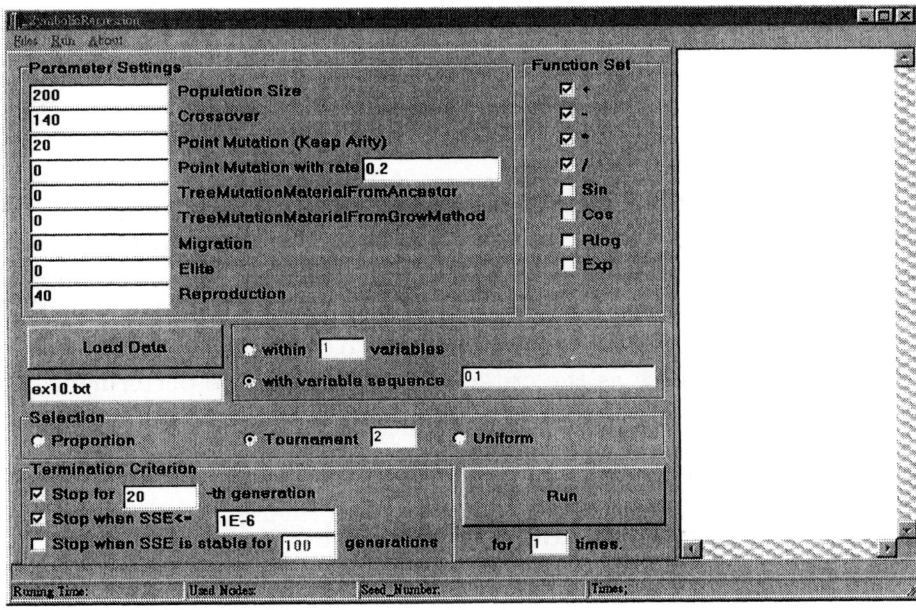

Figure 3.1. The Control Panel of Simple GP

$$x_t = 1 + 0.3x_{t-2} - 1.4x_{t-1}^2, \quad x_{-1}, x_0 \in [-1, 1] \tag{3.4}$$

These four equations can all generate *chaotic time series*, i.e., series which look random but are deterministic. The software **Chaos C++**, included as part of **Simple GP**, can generate data from these four DGMs. Call the generated series $\{x_t\}$. In each of our experiments, we present $\{x_t\}$ to **Simple GP** and test how well GP can find or approximate the underlying *data generating mechanism* (DGM). Based on the performance of GP under different settings, we ask what the six fundamentals mean for GP, and what constitutes an effective implementation of GP.

The fitness function used in **Simple GP** is the *Sum of Squared Errors* (SSE), defined as follows:

$$SSE = \sum_{t=1}^{T} (\hat{x}_t - x_t)^2, \tag{3.5}$$

where $\hat{x}_t$ is the GP prediction of x_t. In addition to SSE, **Simple GP** also outputs another frequently quoted *error statistic*, namely, the *Mean*

Absolute Percentage Error (MAPE),

$$MAPE = \frac{\sum_{t=1}^{T} \frac{|\hat{x}_t - x_t|}{|x_t|}}{T}. \quad (3.6)$$

Finally, since all our DGMs are deterministic, it is possible for GP to discover the true model. In this case, both the SSE and MAPE are zero, and hence cannot help us to see the difference of various implementations. Therefore, **Simple GP** also outputs a number N_g, i.e., the number of generations required for GP to find the DGM. Thus, N_g, SSE, and MAPE serve as the criteria for evaluating the performance of different designs of GP, which is to be discussed in the next section.

2. Fundamentals of GP

2.1. Survival of the Fittest Principle

The first element, perhaps the most important element, of genetic programming is the implementation of the *survival-of-the-fittest* principle. To examine the significance of this principle in GP, **Simple GP** provides two types of selection scheme: one that carries out the *survival-of-the-fittest* principle, and the other that does not. For the former, we have two selection schemes, namely, *proportionate selection* and *tournament selection*. For the latter, we have *uniform selection* (See "Selection" in Figure 3.1).

> *Remark 1:* While one may assume that GP without the survival of the fittest principle cannot work well, in some cases, the performance of uniform selection is, surprisingly, equal or even superior to that of *proportionate selection.*

Remark 1 provides an answer to a frequently asked question: *Is GP merely a repeatedly random guess or a blind search?* The answer is "*no*". In fact, GP generally outperforms random guessing, but in some special cases where a random search implemented via uniform selection can behave close to GP.

To see this, GP with uniform selection and its proportionate-selection counterpart are tested by DGM Equation (3.1). Figure 3.2 presents the histogram of N_g under 100 trials of each experiment. Both selection styles occasionally failed to discover the DGM within 200 periods (see the rightmost spike appearing in both the upper and lower diagram).

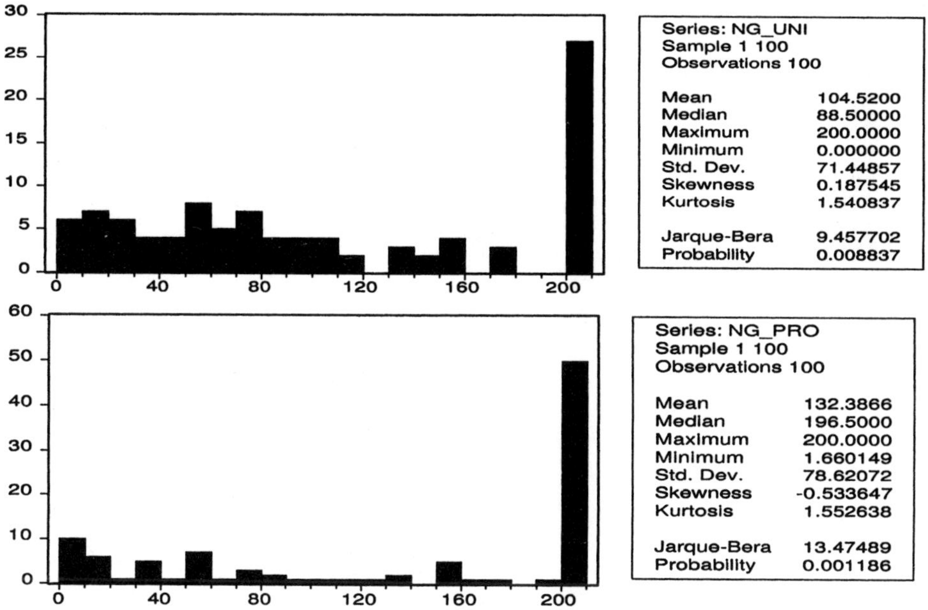

Figure 3.2. Comparison between Uniform Selection and Proportionate Selection in N_g (DGM = Equation 3.1, population size = 200, crossover = 140, mutation (tree, grow) = 40, reproduction = 20, number of generations = 200, function set = { +, -, × }, terminal set = { x_{t-1}, 0, 1, 2,...9 })

But, 50% (the median) of the trials with uniform selection located the true model within 88 generations, whereas it took 196 generations for proportionate selection to find it. However, uniform selection found the true DGM by pure luck (random guessing), and could not help GP improve the approximation. Therefore, if we compare the accuracy performance in terms of SSE, proportionate selection still performs better than uniform selection.

2.2. Selection Schemes

The survival of the fittest principle can be implemented via many different selection schemes, and they may not lead to the same results. In the literature, tournament selection was shown superior to proportionate selection on many occasions, as the following remark indicates.

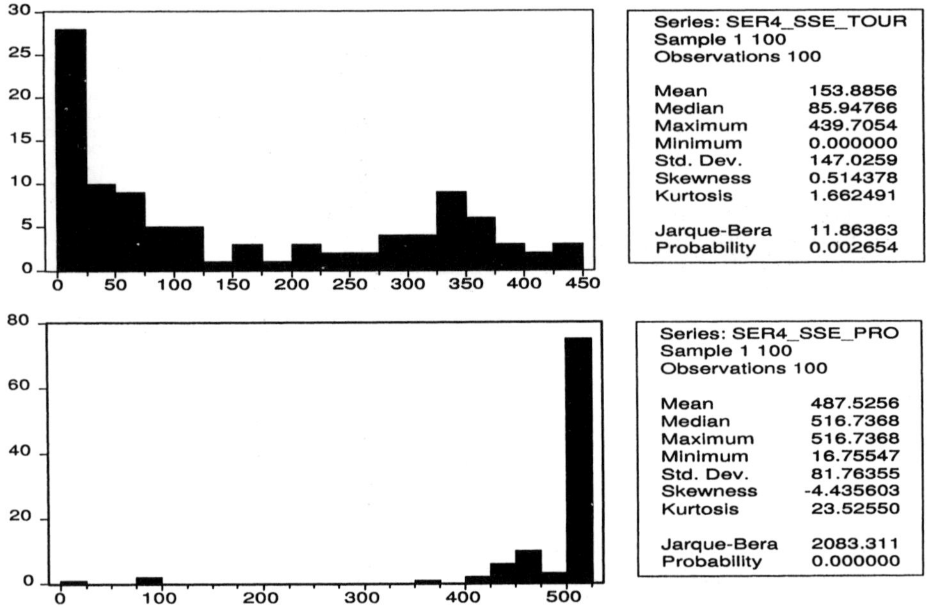

Figure 3.3. Comparison between Tournament Selection and Proportionate Selection in SSE (DGM = Equation 3.2, population size = 200, crossover = 140, mutation (tree, grow) = 40, reproduction = 20, number of generations = 500, function set = { +, -, ×, / }, terminal set = { x_{t-1}, 0, 1, 2,...9 })

> *Remark 2:* Other things being equal, tournament selection is preferred to proportionate selection.

Proportionate selection is weak in distinguishing well performing chromosomes when the range between best and worst is narrow. This may make it difficult to implement the survival-of-the-fittest principle.[2] In computational finance, the selection scheme has received far less attention than it deserves. Chidambaran, Lee, and Tigueros (2002) is the only exception known to us. They studied the performance of six selection schemes, and confirmed the remark given above. Furthermore, the study suggested that the *tournament size* used in tournament selection can play a role as well.

Simple GP supplies users with the choice of *tournament selection* and *proportionate selection.* It also allows users to try different tournament sizes (see Figure 3.1.) To exemplify the remark above, a series $\{x_t\}_{t=1}^{100}$ was generated from DGM Equation (3.2). As before, we ran 100

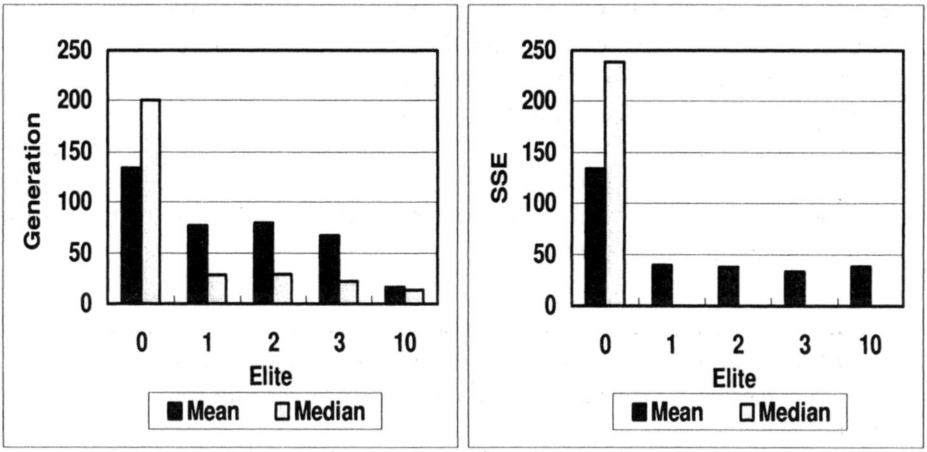

Figure 3.4. Effect of the Number of Elites on N_g and SSE (DGM = Equation 3.1, selection = tournament, population size = 300, crossover = 170, point mutation = 50, tree mutation (ancestor) = 30, elite = 0, 1, 2, 3, 10, number of generations = 200, function set = $\{ +, -, \times \}$, terminal set = $\{ x_{t-1}, 0, 1, 2,...9 \}$)

trials of **Simple GP** for each selection scheme, and the histogram of the 100 SSEs is drawn separately in Figure 3.3. The mean SSE for tournament selection (shown in the upper diagram) is 153; for proportionate selection (shown in the lower diagram), it is 487. The former performs significantly better than the latter.

2.3. Disruption Avoidance

The genetic operators *crossover* and *mutation* contribute to the generation of new rules (models, strategies). There is no guarantee that these new rules will make progress. Sometimes, they can perform even worse than the old ones. This phenomenon is called the *dark side of innovation* or simply *disruption. Disruption avoidance* is a design to prevent such things from happening.

In the literature, there are two popular approaches to avoiding disruption: one is the use of the *election operator*, and the other is the use of the *elitist operator*. For the election operator, disruption avoidance is carried out only at the end of the genetic operation step. It guarantees that offspring are at least no worse than parents, but does not warrant that the best individuals will be retained. For the elitist operator, is enforced right at the selection step. The elitist operator retains a number of the best individuals in each generation. Economic studies comparing the performance of these two versions of GAs are limited. Novkovic (1998) is the only paper that shows some advantages of the elitist operator over the election operator.

Simple GP offers the choice of the elitist operator (see "Elite" on the left panel of the main menu, Figure 3.1). By filling in the box next to "Elite", one can assign the number of best individuals to be retained to the next generation. The following remark gives an evaluation of the function of the elitist operator.

> *Remark 3:* The elitist operator can enhance the performance of GP. However, its effect quickly weakens when the number of elites increases.

As an illustration of the remark, Figure 3.4 presents the experiment of using different numbers of elites, from 0, 1, 2, 3, to 10. The left half of Figure 3.4 shows the mean and median of N_g taken over 100 trials given a number of elites, and the right half shows the mean and median of the SSE. From Figure 3.4, we can see that the performance of GP improves dramatically when the elitist operator is imposed (see the change from "elite=0" to "elite=1"). However, when we further increase the number of elites, the contribution becomes marginal. Therefore, it would be good enough to include only a few elites in the elitist operator.

2.4. Search Intensity

One key factor that determines the performance of GP is *how hard we are searching*, or simply *search intensity*. There are several different ways to measure the search intensity of GP. The two key determinants are the *number of generations* and *population size*. Consider these two determinant as input to a GP factory, and let the output of the GP factory O be a performance measure, then one can describe their relation in a familiar *production function* framework,

$$O = f(Pop, Gen), \tag{3.7}$$

where Pop refers to the population size, and Gen the number of generations. One amazing thing about Equation (3.7) is that it satisfies those properties usually assumed in *production theory* (Varian 1992), as summarized in the following remark.

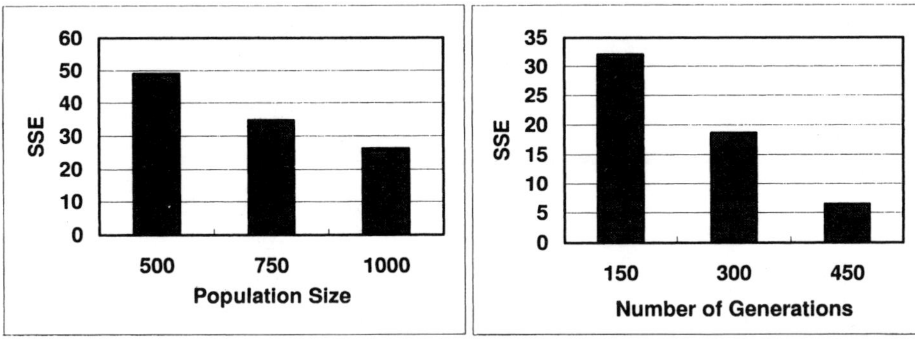

Figure 3.5. Effects of the Population Size and the Number of Generations on SSE (DGM = Equation 3.2, population size = 750 (right), crossover = 60% of Pop, point mutation = 15% of Pop, tree mutation (grow) = 5% of Pop, elite = 1, number of generations = 100 (left), function set = { +, -, ×, / }, terminal set = { x_{t-1}, 0, 1, 2,...9 }, selection = tournament)

Remark 4: Equation (3.7) has the following properties:

$$(i) \quad \frac{\Delta O}{\Delta Pop} > 0, \quad \frac{\Delta O}{\Delta Gen} > 0 \tag{3.8}$$

$$(ii) \quad \frac{\Delta^2 O}{\Delta Pop^2} < 0, \quad \frac{\Delta^2 O}{\Delta Gen^2} < 0 \tag{3.9}$$

$$(iii) \quad \frac{\Delta^2 O}{(\Delta Pop)(\Delta Gen)} > 0, \quad \frac{\Delta^2 O}{(\Delta Gen)(\Delta Pop)} > 0 \tag{3.10}$$

The first property (3.8) says that the *marginal productivity* of *Pop* and *Gen* is positive.[3] If one increases the population size or lengthens the evolution time (number of generations), the performance of GP is likely to improve. The second property says that the marginal productivity of *Pop* (*Gen*) diminishes as one further increases *Pop* (*Gen*) without simultaneously altering other parameters. This property is also known as *diminishing marginal productivity* in economics.

To illustrate these two properties, a series $\{x_t\}$ is generated from *DGM* Equation (3.2), and we ran 100 trials of **Simple GP** on this series. The left diagram of Figure 3.5 shows the mean SSE of the 100 trials associated with population sizes 500, 750 and 1000, and the right dia-

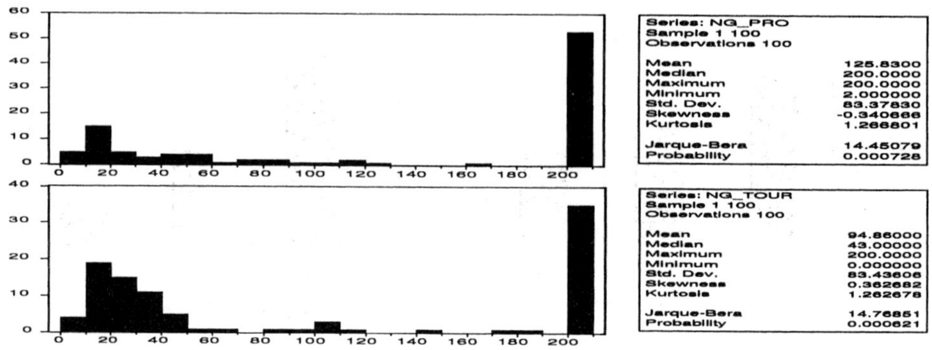

Figure 3.6. An Illustration of the Diminishing Marginal Productivity of Search Intensity (DGM = Equation 3.1, population size = 200, crossover = 100, point mutation = 50, number of generations = 200, function set = { +, -, × }, elite = 1, terminal set = { x_{t-1}, 0, 1, 2,...9 }), selection = proportionate (upper panel), tournament (lower panel))

gram shows the mean SSE associated with Gens 150, 300, and 450. The first property seems to be self-evident from Figure 3.5: GP performance improves with search intensity. However, as Figure 3.5 also shows, the improvement scale shrinks when either *Pop* increases from 750 to 1000, or *Gen* increases from 300 to 450.

Diminishing marginal productivity may have an important implication for the practice of financial data mining: when *Pop* or *Gen* comes to a scale, a further increase in the number of generations only waste a lot of computational resources without enhancing the performance. Figure 3.6 is a case in point. Here, we have a series generated from DGM Equation (3.1). A hundred trials of **Simple GP** were conducted, each lasting for 200 generations. Figure 3.6 is the histogram of N_g of these 100 trials. The upper diagram corresponds to proportionate selection, and the lower diagram corresponds to tournament selection.

The point of showing these two diagrams is their "*J*" distribution rather than the uniform distribution. The *J*-shaped distribution indicates that if GP can successfully discover the underlying DGM, it can be done in a limited period of generations, say within fifty generations. This is reflected on the cluster on the left tail of the histogram. If GP fails to discover the underlying DGM in the initial 50 generations, extending the evolution time, say by another 150 generations, may still help little. This corresponds to the spike appearing at the right end of the *J*-shaped histogram.

Table 3.1. The Effect of the Pop-Gen Combination on GP Performance

	Exp. 1	Exp. 2	Exp. 3	Exp. 4	Exp. 5
Gen	1000	200	100	50	10
Population	10	50	100	200	1000
Crossover	4	20	40	80	400
Point Mutation	1	5	10	20	100
Elite	1	5	10	20	100
Reproduction	4	20	40	80	400
Mean (SSE)	50	96	97	99	215

DGM = Equation 3.1, K= $POP \times GEN$ =10,000, crossover = 40%, point mutation = 10%, elite = 10%, reproduction = 40%, function set = { +, -, ×, / }, terminal set = { x_{t-1}, 0, 1, 2,...9 }, selection = tournament

Property (3.9) points out that the design of GP which focuses *only* on *Pop* or *Gen* may lead to disappointing results. That "larger population size did not improve much" and "evolving longer did not lead to a significant difference" are typical results of the negligence of the next property, Property (3.10), which says that the marginal productivity of *Pop* (*Gen*) increases with *Gen* (*Pop*). In economic parlance, the two production factors *Pop* and *Gen* are *complementary* to each other. Hence, an effective design of GP should take both *Pop* and *Gen* into account, otherwise it would be a waste of the factor intensively used.

Expanding *Pop* and *Gen* simultaneously can be, however, computationally demanding. Therefore, a more relevant question concerning designs should be: *given a budget constraint K, where*

$$K = Pop \times Gen, \tag{3.11}$$

what is the most productive combination of Pop and Gen?[4] In other words, what we are looking for is a combination (Pop^*, Gen^*) such that, $Pop^* \times Gen^* = K$, and

$$f(Pop^*, Gen^*) \geq f(Pop, Gen), \forall (Pop, Gen) \in (3.11). \tag{3.12}$$

Remark 5: The corner solutions $(1, K)$ and $(K, 1)$ are both inferior to the interior solution. However, it is not entirely clear on the optimum level of the quotient,

$$q^* = \frac{Gen^*}{Pop^*}. \tag{3.13}$$

There is no point in having a corner design because $(1, K)$ loses diversity, and $(K, 1)$ does not have a chance of evolving. However, there is no clear-cut answer for q^*. We have some limited experiments showing that q^* depends on the selection scheme, the DGM and K. Using the terms from production theory, Equation (3.7) can be a highly irregular isoquant from which the implied expansion line is non-linear and not unique.[5] In the following, we shall show some results from our limited experiments.

Table 3.1 reports five different experiments of **Simple GP**. They differ in the combination of *Pop* and *Gen*, but share the same K, which is 10,000. The leftmost column lists the set-up of the experiment with the smallest population size but the longest evolution time, whereas the rightmost lists the set-up of the experiment with the largest population size but the shortest evolution span. For each experiment, we ran 20 trials, and the $\{x_t\}$ was generated by DGM Equation (3.1). The mean SSEs of the 20 trials are given in the last row. Among the five combinations, the best one is $(10, 1000)$, or $q = 10$, which shows that *Pop* is surprisingly small.[6]

To be more careful about the result obtained above, we carried out another two experiments based on DGM Equation (3.4) by increasing K up to 150,000. The two experiments differ in the combination of (Pop, Gen), which is (500,300) for the first one, and (250, 600) for the second. Each experiment was tried 50 trails. Notice that q is less than 1 for the first experiment but greater than 1 for the second. A histogram of the 50 SSEs is shown in Figure (3.7). While these two diagrams have close mean SSEs (20.74 for the first, and 20.11 for the second), their medians are quite different (10.88 for the first, and 7.23 for the second). The result confirms our earlier finding that q^* is bigger than 1, even though the difference is not as noticeable as the previous case.

2.5. Primitives

Primitives concern the choice of the function set and the terminal set. This choice is crucial because the *algorithmic complexity* (*minimal*

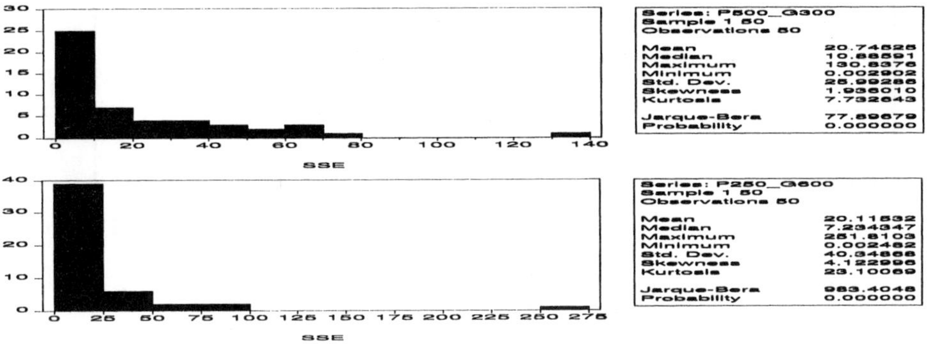

Figure 3.7. The Effect of the Pop-Gen Combination on GP Performance (DGM = Equation 3.4, K= $POP \times GEN$ =150,000, crossover = 40%, point mutation = 10% 10%(0.5), 2%, 2%, migration = 2%, elite = 1, reproduction = the remaining, function set = { +, -, ×, / }, terminal set = { x_{t-1}, 0, 1, 2,...9 }, selection = tournament)

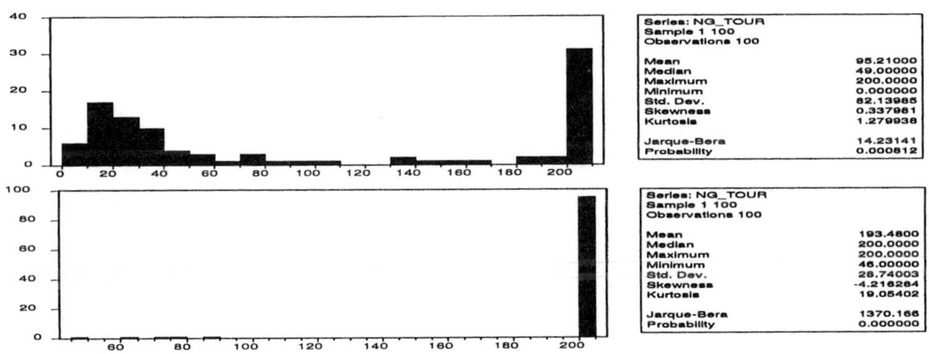

Figure 3.8. The Effect of the Algorithmic Complexity on the Discovery Probability (DGM = Equation 3.1 (upper diagram), Equation 3.2 (lower diagram), population size = 200, crossover = 140, point mutation = 40, reproduction = 19, elite = 1, number of generations = 200, function set = { +, -, × }, terminal set = { X_{t-1}, 0, 1, 2,...9 }), selection = tournament

description length) of any DGM depends on the choice of the function set and terminal set. Algorithmic complexity is relevant because the chance of discovering a specific DGM in a finite number of generations is a decreasing function of its algorithmic complexity.

> *Remark 6:* Let $\mathcal{T}$ be the terminal set and $\mathcal{F}$ the function set, and denote their cardinality by $\mid \mathcal{T} \mid$ and $\mid \mathcal{F} \mid$, then
>
> $$Prob(g^* \in G_n) = Prob[\mathcal{L}(g^* \mid \mathcal{F} \bigcup \mathcal{T})], \tag{3.14}$$
>
> and
>
> $$\frac{\Delta Prob(g^* \in G_n)}{\Delta \mathcal{L}} \mid_{\mathcal{F} \bigcup \mathcal{T}} \leq 0, \tag{3.15}$$
>
> where g^* is a targeted DGM, G_n the population at the nth generation, and $\mathcal{L}(g^* \mid \mathcal{F} \bigcup \mathcal{T})$ refers to the algorithmic complexity of g^* given $\mathcal{F}$ and $\mathcal{T}$.

Simple GP allows users to make their own choice of the function set and the terminal set (see Figure 3.1). We shall exemplify the remark based on Chen and Yeh (1997). Let $\mathcal{F} = \{+, -, \times, /\}$, and $\mathcal{T} = \{X_{t-1}, 0, 1, 2, ..., 9\}$. The $\mathcal{L}$ of DGM Equations (3.1) and (3.2) are 7 and 11 respectively. Thus, in a finite number of generations, the probability of discovering (3.2) is lower than discovering (3.1). To show how much lower it is, two experiments were conducted. In the first experiment $\{x_t\}$ was generated from Equation (3.1), and in the second experiment it was generated from Equation (3.2). For each experiment, we ran 100 trials of **Simple GP**. This enables us to get an estimate of $Prob(g^* \in G_{200})$.

Figure 3.8 gives the histogram of the N_g of the two experiments. The figure shows that the chance of discovering (3.1) within 200 generations is about 83%, whereas there is only 5% chance of discovering (3.2) with the same search intensity. However, if we add the variable "X_{t-1}^3" to the terminal set, then the $\mathcal{L}$ of DGM (3.2) becomes shorter–7, to be exact, which is the same as that of DGM (3.1). To see the effect of adding the terminal X_{t-1}^3, we conducted another two experiments. In both experiments, $\{x_t\}$ were generated from DGM (3.2). The first experiment did not include X_{t-1}^3 in $\mathcal{T}$, but the second did. 100 trials were run for each experiment, and the histogram of their N_g is shown in Figure 3.9.

For the experiment without including X_{t-1}^3, the distribution is almost degenerated at the point "200"; $Prob(g^* \in G_{200})$ is only 4%. But, after the inclusion of X_{t-1}^3, the spike on the right melts down, the N_gs are redistributed to the left. $Prob(g^* \in G_{200})$ increases up to 84%, which is about the same as the chance of finding (3.1) in the previous experiment (also see Figure 3.8).

Remark 6 has important implications for the design of GP. Users are advised to be more careful in the choice of their function set and terminal set. As seen above, simply adding one key variable can boost the finding

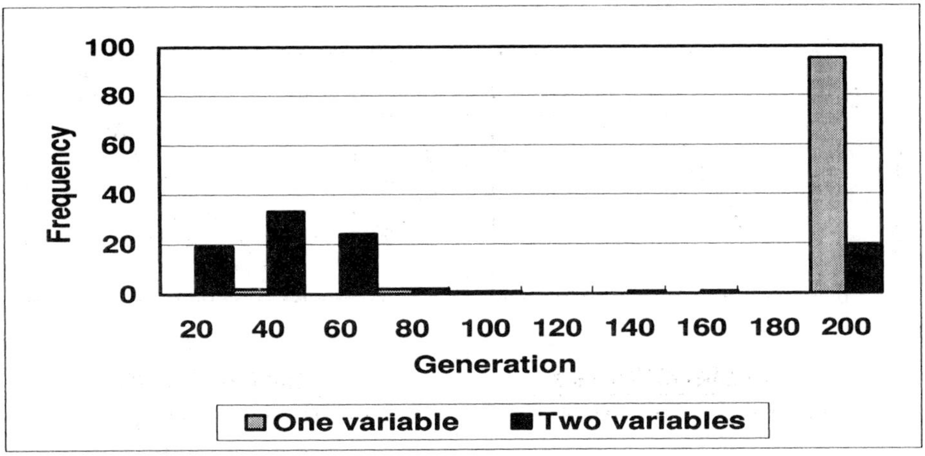

Figure 3.9. The Effect of the Terminal Sets on the Discovery Probability (DGM = Equation 3.2, population size = 200, crossover = 140, mutation (tree, grow)= 40, reproduction = 19, elite = 1, number of generations = 200, function set = { +, -, ×, / }, terminal set = { X_{t-1}, (X^3_{t-1}), 0, 1, 2,...9 }, selection = tournament)

probability from nil to 80%. A particular application of *Remark 6* is to incorporate some advanced functions or terminals (*benchmark*) into $\mathcal{F}$ and $\mathcal{T}$. Chidambaran, Lee, and Tigueros (2002), for instance, included in their function set the Black-Scholes model as an existing subroutine in their study of option pricing formula.

However, *Remark 6* does not tell us how to find the functions and terminals to economize the description length of g^*. For example, how do we know *ex ante* that X^3_{t-1} is a length-reducing terminal? If not, we may also run the risk of including some irrelevant terminals. In the following series of experiments, we shall give a concrete example to illustrate the consequences of including irrelevant terminals.

The $\{x_t\}$ used in this series of experiments were generated from DGM Equation (3.1). The experiments in this series were identical in all aspects except the terminal set. For the first experiment, we only included X_{t-1}, in addition to the integers. For the second experiment, we added a random series $\{Z_{1,t}\}$, where $Z_{1,t}$ is iid (identically independently) uniformly distributed ($Z_{1,t} \sim U[0,1]$), and for the third experiment, we added one more random series, $\{Z_{2,t}\}$, of the same kind. 100 trials were conducted for each of the experiments, and the summary statistics for the mean and median of the 100 N_gs and SSEs are shown in Figure 3.10. Here, we see two things. First, increasing the number of irrelevant

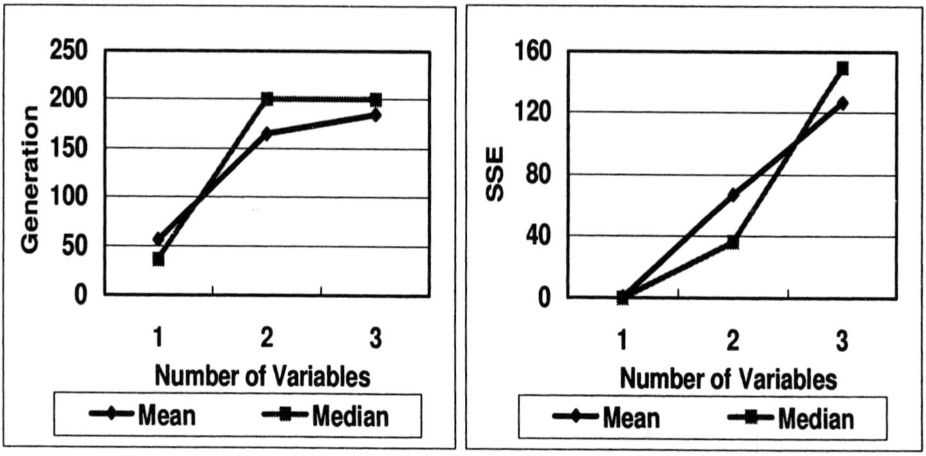

Figure 3.10. Consequences of Including Irrelevant Terminals (DGM = Equation 3.1, population size = 200, crossover = 140, mutation (tree, grow) = 40, reproduction = 19, elite = 1, number of generations = 200, function set = { +, -, × }, terminal set = { X_{t-1}, $(Z_{1,t}, U_{2,t})$, 0, 1, 2,...9 }, selection = tournament)

terminals has an adverse effect on the probability of finding g^*. In our cases, $Prob(g^* \in G_{200})$ started with 0.9, then dropped down to 0.32, when Z_1 was included, and further down to 0.12, when Z_2 was also included. Second, it also deteriorates the approximation capability of GP. This is particularly clear from the upward sloping curve of the mean or median SSE shown in the right half of Figure 3.10.[7]

2.6. Genetic Operators

In the main menu of **Simple GP**, what follows the "population size" is the *genetic portfolio.* The genetic portfolio indicates how offspring shall be generated by different genetic operators. We use the word "portfolio" because it behaves like a *financial portfolio.* Here, the total investment is the population size. Given the total investment, a financial portfolio shall be distributed into various financial assets. Instead of financial assets, what is offered in GP is a list of genetic operators. Basically, there are three kinds: *reproduction*, *crossover* and *mutation.*

Reproduction simply disseminates the promising solutions among the population. This doing is virtually *riskless* (non-disturbing), but it cannot generate novel solutions. The innovative work is mainly conducted by crossover and mutation, which is somewhat *risky* (disturbing). The immediate question facing the users of GP is *how to distribute the in-*

vestment (population size) into riskless and risky assets.

> *Remark 7:* The reproduction operator plays nothing more than a marginal role in generating offspring. Reducing the use of reproduction but increasing the use of risky genetic operators can boost GP performance.

Remark 7 basically points out that the riskless genetic operator is less productive than the risky ones. It is hard to say exactly how distribution should be made between riskless and risky genetic operators, but an example may help. Here, we conducted a series of seven experiments. Each experiment differed in the distribution between reproduction and mutation. Out of 200 seats, a total of 99 were assigned to the two operators. The allocation to mutation started from 0, then to 15, 30, 45, 60, 75, and 90, and the remaining of the 99 seats were left to reproduction. Both selection schemes were investigated. The series $\{x_t\}$ was generated from (3.2). We ran 50 trails for each experiment, and the median of the SSE of the 50 trials is depicted in Figure 3.11, which roughly shows a down-sloping curve (though less clear for proportionate selection). When we released more seats from reproduction to mutation, the performance of GP did improve. A portfolio with a high percentage of mutation and a low percentage of reproduction performed much better than one with a low percentage of mutation and a high percentage of reproduction.

The second question concerns the division among risky assets. In the literature, there are two diametrically opposing assertions regarding risky genetic portfolios. The first one contends that what can be done by mutation can also be done by crossover, but not vice versa. The second one claims that mutation is irreplaceable. Although the two assertions have different implications for the portfolio, few financial applications noticed this mooting point.[8] The first one suggests zero investment in mutation, while the second one recommends a degree of diversity. Let Q_c and Q_m be the percentage of crossover and of mutation in the risky genetic portfolio respectively, where $0 \leq Q_c, Q_m \leq 1$ and $Q_c + Q_m = 1$.

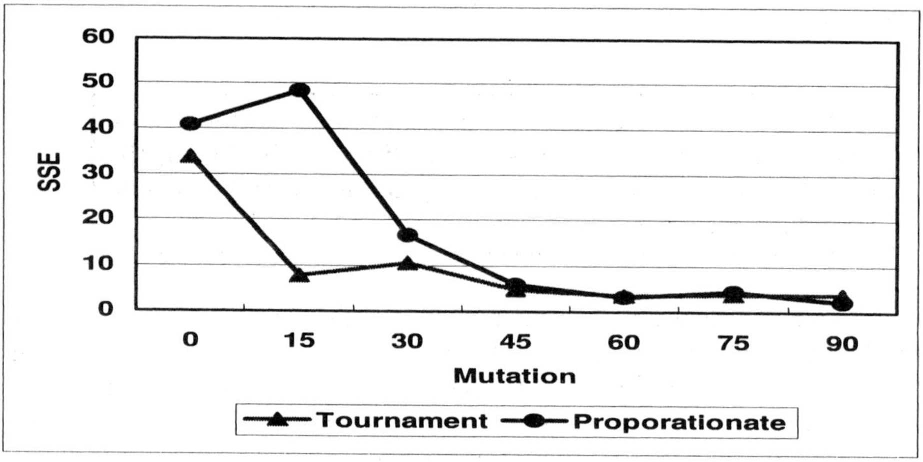

Figure 3.11. Distribution between Riskless and Risky Genetic Operators (DGM = Equation 3.1, population size = 200, crossover = 100, elite = 1, point mutation = {0, 15, 30, 45, 60, 75, 90 }, reproduction = the remaining, number of generations = 100, function set = { +, -, ×, / }, terminal set = { X_{t-1}, 0, 1, 2,...9 }, selection = proportionate and tournament)

> *Remark 8:* Let (Q_c^*, Q_m^*) be the optimum risky genetic portfolio. It is expected that
>
> $$Q_c^* > 0, \quad and \quad Q_m^* > 0. \tag{3.16}$$

The remark above excludes "all or nothing" from the optimal portfolio list. Its standpoint is closer to the second assertion. Experiments related to *Remark 8* are given as follows. The sixteen experiments we conducted differed in the risky genetic portfolio. The total number of seats allocated to crossover and mutation was 150. Among the 150, crossover took 0, 10, 20, ..., 130, 140, and 150 in the experiments, and mutation took the rest. $\{x_t\}$ was generated by (3.1). 100 trials were conducted for each experiment. Their mean SSEs are given in Figure 3.12. From Figure 3.12, we can see that GP performance is robust to different risky genetic portfolios as long as a degree of diversity is maintained. The change of the SSE is not significant and shows no pattern along the movement from portfolio (10, 140) to portfolio (130, 20). In contrast, GP delivered a poor performance when there was no diversity. This is clearly reflected from the corner portfolios, (0, 150) and (150, 0).

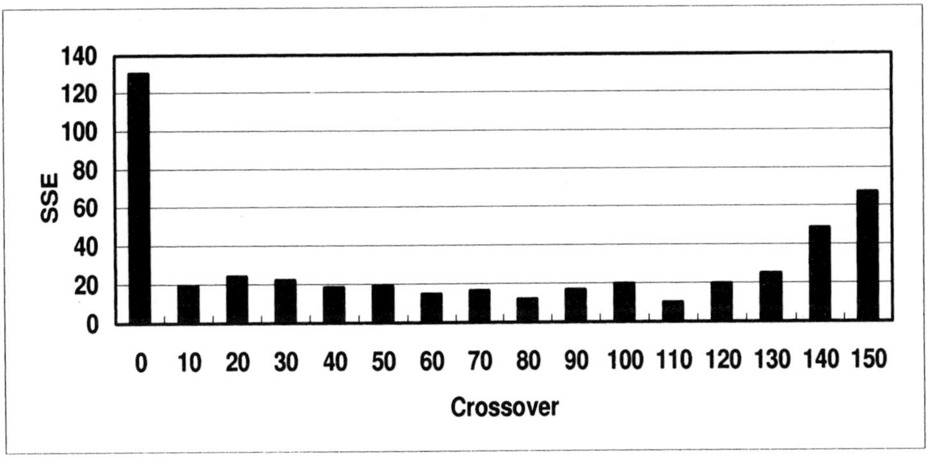

Figure 3.12. The Effect of Risky Genetic Operator on GP Performance (DGM = Equation 3.1, population size = 200, crossover = { 0, 10, 20, 30, ..., 150 }, point mutation = {150, 140, 130, ..., 0 }, reproduction = the remaining , elite = 1, number of generations = 100, function set = { +, -, × }, terminal set = { X_{t-1}, 0, 1, 2,...9 }, selection = tournament)

While *Remark 8* shows the necessity of including mutation as a part of portfolio, there are several different technical implementations of mutation. Are they equally effective? **Simple GP** offers two basic styles of of mutation, namely, *point mutation* and *tree mutation.* As the name suggests, point mutation only replaces the *node* chosen, whereas tree mutation would replace the whole *sub-tree* originating from the the node chosen.

> *Remark 9:* Both point mutation and tree mutation can are important for the implementation of the mutation operator. It is advisable to include both styles in the mutation portfolio. For a complex function, tree mutation is particularly important.

Remark 9 indicates something very similar to *Remark 8*, i.e., not taking the *all-or-nothing* portfolio. Empirical evidence suggests that these mutation styles have different ways of creation. One may expect that tree mutation will have a larger disturbing effect than point mutation and that it is good at exploration but not exploitation. Therefore, a balance between exploration and exploitation can be achieved by the use of both mutation styles.

Table 3.2. The Performance of Point Mutation and Tree Mutation

	Number of Successes		Ng	SSE	MAPE
Tournament	Point Mutation	12	189.22	76.66	1.21
	Tree Mutation	2	197.14	71.95	1.31
Proporationate	Point Mutation	5	195.34	354.43	1.40
	Tree Mutation	2	198.15	430.63	1.67

DGM = Equation 3.2, population size = 500, crossover = 350, tree mutation (grow) = 100 or point mutation (0.2) = 100, reproduction = 50, number of generations = 200, function set = { +, -, ×, / }, terminal set = { x_{t-1}, 0, 1, 2,...9 }, selection = tournament or proporationate

We conducted two experiments related to *Remark 9*: one using only point mutation, and the other using only tree mutation. The series $\{x_t\}$ was generated from (3.2). We ran 100 trials with both styles of selection, and the summary statistics are detailed in Table 3.2. Point mutation turned out to have a better chance of discovering the DGM than tree point mutation (see the column "Number of Successes"). Tree mutation beat point mutation in SSE under tournament selection, while point mutation outperformed tree mutation under proportionate selection.

In another series of experiments, we tried a more difficult DGM, namely, Equation (3.4). From the criterion of the minimum description length ($\mathcal{L}$), this function is more complex than Equation (3.2). As a result, the chance of discovering it is even smaller. It would be interesting to see how well point mutation and tree mutation can approximate the function when perfect discovery is extremely difficult. As before, we conducted two experiments. They were identical in all aspects, except the number of seats reserved for the two mutation styles. The first experiment had 10 seats for point mutation, and 50 seats for tree mutation; seat allocation in the second experiment was just the opposite. We ran 50 trials of the experiments, and the summary statistics are given in Table 3.3.

Table 3.3 shows that when perfect discovery is difficult, tree mutation can explore a large space of function and hence provides a better approximation of g^*. This is shown by its lower SSE and MAPE for both selection styles. The results support the remark above: since we do know how complex the DGM we are facing, it would be useful to have both styles of mutation in our genetic operator.

Table 3.3. Point Mutation and Tree Mutation in Exploration

	Proporationate		Tournament	
(Point, Tree)	**SSE**	**MAPE**	**SSE**	**MAPE**
(50, 10)	41.8608	0.4875	11.7384	0.24442
(10, 50)	10.8130	0.1965	9.7295	0.2265

DGM = Equation 3.4, population size = 250, crossover = 40%, mutation = 10%, 10%(0.5), 2%, 2% or mutation = 2%, 2%, 10%, 10%(0.5), migration = 2%, elite = 1, reproduction = the remaining, number of generations = 300, function set = { +, -, ×, / }, terminal set = { x_{t-1}, 0, 1, 2,...9 }, selection = tournament, proportionate.

3. Concluding Remarks

In this chapter, we use the software **Simple GP** to conduct quite a few experiments, each with multiple trials. Upon these experiments, nine remarks are established. While these remarks should not be read as *theorems* given the difficulty in mathematical proofs, their plausibility is justified, to some extent, by the related experimental results. Those who have doubts about these remarks can use **Simple GP** to run their own trials. Based on these nine remarks, a general conclusion is given as follows.

General Conclusion: *The driving force of GP is the survival-of-the-fittest principle. Selection plays an important role in the performance of GP. Tournament selection is in general preferred to proportionate selection. The use of the elitist operator to avoid disturbance can be another performance-boosting design, but there is no need to keep too many elites. Search intensity should be considered in a more general framework. Within a framework of production function theory, neither the population size nor the number of generations should be increased alone. They should be increased simultaneously by following an expansion line. It is advisable to enlarge the set of primitives by adding length-reducing functions and terminals, but one should beware of the adverse effect of including irrelevant primitives. Finally, portfolio theory applies to the distribution of genetic operators. Just as all-or-nothing is a bad strategy, a uniform portfolio is a poor design. A higher proportion of risky (disturbing) genetic operators to riskless (non-disturbing) genetic operators is recommended. Different risky genetic operators perform different kinds of search. While some are exploitation-oriented, others are exploration-oriented. It is important to include all of them in a risky portfolio.*

The conclusion above summarizes our fundamental understandings of the six essential aspects of GP. There is little doubt that each of these aspects can be further pursued. In fact, a lot has been done in the literature. For example, in addition to the proportionate and tournament selection schemes, many other schemes have been tested in computational finance.[9] Also, there is a general tendency to make primitives and genetic portfolios adaptive. We shall leave these advancements to **Financial GP**, a more advanced version of **Simple GP**.[10]

Acknowledgments

An earlier version of **Simple GP** was distributed to the students of the authors' class on Economic Forecasting. The authors are grateful to the following students for their valuable feedback, Kai-Ann Chan, Chia-Chih Chen, and I Wang .

Notes

1. As a result, different transformations of data series are beyond the scope of this chapter. But, one can find related materials in many chapters of the book. See Nikolaev and Iba's, Neely and Weller's chapter in this volume.
2. Of course, fitness can be normalized in such a way so as to avoid weak selection. For an example of fitness normalization, we refer the reader to Loraschi and Tettamanzi (1996).
3. Marginal productivity is usually defined by the *partial derivative*. However, since neither *Pop* nor *Gen* is continuous, marginal productivity has to be roughly defined in a discrete fashion. Some mathematical rigor is lost for convenience.
4. A similar question was raised by Lettau (1997) in a GA context. The trade-off he considered is *Gen* and the number of test cases, rather than *Pop*. He showed that a combination of a small number of test cases and a large *Gen* was the best.
5. For those who are not familiar with the isoqunat and the expansion line, any textbook on microeconomics may help.
6. This result is in accord with Lettau (1997).
7. Chen (2002) discussed the adverse effect of including irrelevant functions and terminals. Using **Simple GP**, one can re-estimate the probability of finding Equation (3.1) or (3.2) by including $sin, cos, exp, \log$ into $\mathcal{F}$.
8. The only one known to us is Tsang and Lajbcygier (2002).
9. See the chapters written by Drake and Marks, and Birchenhall and Lin in this volume. Also, see Chidambaran, Lee, and Tigueros (2002).
10. Users who are interested in **Financial GP** can contact the first author of the chapter via e-mail.

References

Chen, S.-H. (2002). "Fundamental Issues in the Use of Genetic Programming in Agent-Based Computational Economics," in A. Namatame and H. SATO (eds.), *Agent-based Approaches in Economic and Social Complex Systems*. IOS Press.

Chen, S.-H. and C.-H. Yeh (1997). "Toward a Computable Approach to the Efficient Market Hypothesis: An Application of Genetic Programming," *Journal of Economic Dynamics and Control*, 21(6), 1043–1063.

Chidambaran, N., C.-W. J. Lee, and J. Trigueros (2002). "Option Pricing via Genetic Programming," in S.-H. Chen (ed.), *Evolutionary Computation in Economics and Finance*, 383–398. Physica-Verlag.

Lettau, M. (1997). "Explaining the Facts with Adaptive Agents: the Case of Mutual Fund Flows," *Journal of Economic Dynamics and Control*, 21(7), 1117–1147.

Loraschi, A. and A. Tettamanzi (1996). "An Evolutionary Algorithm for Portfolio Selection within a Downside Framework," in C. Dunis (Ed.), *Forecasting Financial Markets: Exchange Rates, Interest Rates and Asset Management*, 275–285. John Wiley & Sons.

Novkovic, S. (1998). "A Genetic Algorithm Simulation of a Transition Economy: An Application to Insider-Privatization in Croatia," *Computational Economics*, 11(3), 221–243.

Tsang, R. and P. Lajbcygier (2002). "Optimization of Technical Trading Strategy with Split Search Gen Abu-Metic Algorithms," in S.-H. Chen (ed.), *Evolutionary Computation in Economics and Finance*, 333–358. Physica-Verlag.

Varian, H. R. (1992). *Microeconomic Analysis*, 3rd ed. Norton.

II

FORECASTING

Chapter 4

GP AND THE PREDICTIVE POWER OF INTERNET MESSAGE TRAFFIC

James D. Thomas
Department of Computer Science
Graduate School of Industrial Administration
Carnegie Mellon University
jthomas@cs.cmu.edu

Katia Sycara
Robotics Institute
Carnegie Mellon University
katia@ri.cmu.edu

Abstract This paper investigates the predictive power of the volume of messages produced on internet stock-related measure boards. We introduce a specialized GP learner and demonstrate that it produces trading rules that outperform appropriate buy and hold strategy benchmarks in measures of risk adjusted returns. We compare the results to those attained by using other relevant variables, lags of price and volume, and find that the the message board volume produces cleraly superior results. We experiment with alternative representations for the GP trading rule learner. Finally, we find a potential regime shift in the market reaction to the message volume data, and speculate about future trends.

Keywords: Genetic Programming, Computational Finance, Internet Message Boards

Introduction

There has long been a strong interest in applying computational intelligence to financial data. Such attempts have traditionally been concerned with forecasting the future based on past price data. However, recently new sources of data has become available due to the increased communication and community building afforded by the growth of the Internet. This paper uses genetic algorithms to examine the relevance of one new source of information—the volume of message board postings of stock specific message boards from the following two sources:

- Yahoo! (http://messages.yahoo.com/yahoo/index.html)
- Ragingbull (now Lycos finance) (http://ragingbull.lycos.com)

The key question this paper addresses is if the measures of message volume can be used as an effective predictor of stock movements. To this end we build a specialized GP learner that builds trading rules based on this message volume data. We have performed preliminary explorations on smaller versions of this data set (see Thomas and Sycara 2000); this paper extends those techniques to a larger datasets, generating more robust conclusions. Since our interests lie in both finance and AI, we are concerned both with uncovering possible market inefficiency, and understanding how to adapt AI algorithms to exploit it. Specifically, this paper addresses the following questions:

- Can we demonstrate that the message board volume data has predictive power; ie, can a reasonable GP approach produce statistically significant out-of-sample excess returns?
- Is the message board volume data contributing information that other traditional numerical data (price, volume, etc) are not?

The format of this paper is as follows: In section 1, we describe the data we use—the volume of postings on Internet message boards, the relevant financial task, and the specifics of the GP trading rule learner. Next, in section 2, we discuss the empirical results of the basic algorithm, as well as running the algorithm on other sources of data to ensure message volume really is adding novel information. In section 2.5, we discuss the possibility of a regime change toward the end of the test period, and alter the algorithm to explore this possibility. Finally, section 3 summarizes and poses new challenges.

1. Task and Data

Our basic methodology is to take variables that potentially have predictive power over financial data, apply a GP learner algorithms to build trading rules based on that data, and then see if those trading rules produce statistically significant excess returns. In order to understand our approach, this section presents the following topics:

- The specific data set we used—Internet message board volume statistics—and the preprocessing steps we applied to it.
- The assumptions and measures of success of the basic financial trading rule framework used to generate returns.
- The design of genetic programming learner used.
- The specific representation of the trading rules.

1.1. The Data

Due to limitations of data collection, we limited our universe of stocks were those that appeared on the Russell 1000 (a list of the 1000 largest US equities by market capitalization, updated yearly) index for both 1999 and 2000, and who had price data dating back to Jan 1, 1998, on the Yahoo quote server. This left us with 688 stocks. The time universe was January 1, 1998 to December 31, 2001. For purposes of tractability, and since most of these stocks failed to generate significant message traffic at all, we limited ourselves to the top 10% by message traffic volume over the time period, leaving us with 68 stocks. We then randomly split this set of stocks in half; one half is used as a design set, used to build the algorithm; the other half is used as a holdout test set to verify the results (discussed in sections 2.4,1.2.2).

For market data, we downloaded split-adjusted prices (prices that are backwards adjusted to account for the effect of stock splits) and trading volume off of the Yahoo quote server for each stock. We used those price figures to compute excess returns. We realize that this ignores dividends and renders the excess return figures inexact; however, since most of the bulletin board with high discussion are technology companies who pay no dividends, we feel that this is an acceptable compromise.

For the message traffic data itself, we collected posts off of both the Yahoo and Ragingbull bulletin boards for every stock in the stock universe from the period Jan 1, 1998 to December 31, 2001. Both Yahoo and Ragingbull keep a full archive of their posts, making the complete record publicly available). We handled these counts of message board volume in the following ways:

- To get our data, we are simply counting the number of posts made to a stock specific board in the relevant time period.
- Only posts made while markets were closed were counted—the intuition behind this is that information contained in posts made during market open should be factored quickly into the prices.

- The daily count of messages was normalized by a factor determined by the day of the week, so that the expected number of posts on each day of the week was the same. For example, Sunday nights have far lower message board volume than Thursday nights, and we didn't want this day of the week bias to interfere.

- For multi-day periods when the markets were closed (weekends or holidays), message counts for the appropriate non-market days were averaged.

- We added the message traffic volume from Ragingbull and Yahoo together to get a single message count.

This gave us a time series of message traffic volume numbers, corresponding to a time series of daily returns. The genetic programming algorithm described below takes those message traffic volume numbers and converts them into a trading signal.

To give an idea of what traffic levels are like, Table 4.1 presents the mean daily message volume for Yahoo and Ragingbull. Presented are the top 10% (by message volume), the average over all stocks, and then a breakdown by quartiles. Observing the data, it's clear that the distribution is very skewed—nearly half of all posted messages are posted about the top 10% in the distribution. Yahoo produces far more posts than Ragingbull—the bottom three quartiles for Ragingbull produce less than one message per day. The standard deviation, skew and kurtosis for the mean daily message traffic over the entire sample of stocks at Yahoo are 39.91, 5.00 and 28.00; for Ragingbull, 11.07, 9.01 and 101.87.

Table 4.1. Mean Daily Message Volume

	Top 10%	All Stocks	Quart. 1	Quart. 2	Quart. 3	Quart. 4
Yahoo	103.98	17.55	51.87	4.87	1.58	.49
Ragingbull	20.55	5.28	9.15	.25	.08	.03

1.2. Trading Rules Framework

Our basic task is to learn trading rules over a universe of stocks that perform better than merely buying and holding the universe of stocks. For each stock, we make a basic decision: long, or short. If we decide to short a stock, we take a corresponding long position in the broader market (proxied by the Russell 1000 index).

The following formula describes specifically how we calculate returns based on a trading strategy. Assume the following definitions (for convenience of calculation, using log returns is standard in finance):

- Let $r_{Strategy}$ be daily log return our strategy produces.
- Let $x(t)$ be our trading signal: 1 for "long," 0 for "short."
- Let $r_{stock}(t)$ be the daily log return on the stock at time t.
- Let $r_{Russell1000}(t)$ be the daily log return on the Russell 1000 at time t
- Let $tcost$ be the one-way log transaction cost.
- Let $r_{shortrate}$ be the rate we pay short.

Then the formula for daily log returns is as follows:

- if $x(t) = 1, r_{Strategy}(t) = r_{stock}(t) - |x(t) - x(t-1)| \cdot tcost$
- if $x(t) = 0, r_{Strategy}(t) = r_{Russell1000}(t) - r_{stock}(t) - |x(t) - x(t-1)| \cdot tcost - r_{shortrate}$

This provides our fitness measure: returns. It is important to note that we are not trying to maximize prediction accuracy, rather we are trying to maximize the total returns produced by the strategy. The two tasks are similar, but not identical. Our task is to examine market efficiency, and for that purpose excess returns are economically meaningful in a way that prediction accuracy is not. For example, we could have high prediction accuracy, but if the days we classified incorrectly had significantly larger returns than the days we did classify correctly, the trading strategy could produce negative returns.

1.2.1 Measures of Success. Given the returns produced by this strategy (generated by a GP learner to be described below), how do we know they are meaningful? There are three issues to be addressed when formulating a comparison strategy. The first is finding a proper benchmark, the second is measuring performance relative to the benchmark, and the third is evaluating statistical significance.

The proper benchmark here is the buy and hold strategy over the appropriate stocks. If our trading strategy—which consists entirely of holding the stock, or shorting the stock and holding the market—can produce risk adjusted excess returns while accounting for reasonable transaction costs, then this is a strong argument that the algorithm is picking up a meaningful pattern in the data that argues for a market inefficiency.

The second issue is measuring of performance relative to the benchmark. The obvious candidate is excess returns—by how many percentage points did we beat the benchmark algorithm? However, risk is crucial in finance; excess returns could be the result of additional risk (in fact this is often the core of counter arguments against work attempting to establish market inefficiencies).

Thus, in addition to presenting results for excess returns, we also present a measure of risk adjusted returns—annualized Sharpe ratios. The Sharpe Ratio is a standard formula for calculating risk adjusted returns—it is return in excess of some benchmark (usually a risk free rate such as treasury bills) divided by the standard deviation of returns. We measure the Sharpe ratio over daily returns, but annualize the measure. We present two particular Sharpe ratio measures: the first is the excess Sharpe ratio, the Sharpe ratio of the trading strategy minus the Sharpe ratio of the buy and hold strategy, where both Sharpe ratios are computed against the an assumed risk free rate of 5%. We also compute the Sharpe ratio of the trading strategy against a benchmark of the buy-and-hold strategy (instead of against risk free returns). The two Sharpe ratios measure slightly different things: the first is the proper way to compares the risk/return profiles of different strategies against a common benchmark; the second Sharpe ratio measures the risk/return profile of the trading strategy, using the buy-and-hold strategy as the benchmark.

As far as measuring the statistical significance of these excess returns, the traditional approach is bootstrap statistics (used, for example, by Brock et al in their exploration of moving average rules, Brock et al. 1992). Bootstrap hypothesis testing (for an excellent introduction, see Efron and Tibshrani 1993) works as follows:

- Define the null hypothesis.
- Generate a number of datasets by the null hypothesis.
- Run the algorithm on these bootstrap datasets.
- Compare what proportion of the bootstrap datasets produce results exceeding that of the real dataset; this is the appropriate p-value.

In our case, our null hypothesis is that the message volume statistics associated with a trading day has no predictive power. So, to generate our bootstrap datasets we simply scrambled the message volume numbers associated with each daily return, thus eliminating any possible predictive power, while maintaining an identical statistical distribution.

1.2.2 A Note about Testing. Given the slippery nature of financial data—we are competing against the hypothesis that it's essentially all noise—it is essential to guard against overfitting. Financial data is qualitatively different from the data usually handled be machine learning, and requires special care.

Overfitting (confusingly, called "data mining" in the economics literature) is an enormous concern in finance. The canonical worry is that by trying a large enough number of trading rules, some will be found to be effective by mere chance. There is a fair amount of work examining the relationship between the size of the universe of rules being considered (see, for example, Sullivan et al. 1998) and the testing of statistical significance.

In a sense, the current approach is a strong guard against this kind of overfitting—instead of selecting a few rules from a large universe of possible rules, the AI approach selects a *method of generating rules* from a smaller universe of ways to generate rules. However, this opens the approach to criticisms that we are overfitting over the universe of possible methods of generating rules (i.e., the universe of algorithms). There is an unavoidable conflict here: one of the core elements of AI research is process of iterative algorithm tweaking and performance measurement, but this carries the danger the tweaking will narrow the scope of the algorithm so much that it will fail to generalize to novel data. This danger looms particularly large in financial data, given the levels of noise present. To guard against this, we establish the following policy:

- Hold out a final testing set of data. This data will not be touched until the algorithm design process is complete.
- Split the remaining data into training and testing sets.
- Perform algorithm design on only this data—develop the algorithm by examining performance on the test set, guided by sound a priori principles from finance and artificial intelligence.
- Then, only when the algorithm has been settled, verify the conclusions based on the "holdout" set.

As discussed above, for this paper our holdout set consists of a split of our stock universe—a randomly generated selection of half of the stocks. Experimental results over the holdout set can be found below in section 2.4.

1.3. Genetic Programming

Now that we understand how to calculate returns according to our trading strategy and evaluate their significance, the real meat of the problem is how to learn an appropriate trading strategy to produce said excess returns. We turn to the genetic programming approach pioneered in Koza (1992). We follow the following basic algorithm:

- Split data into training, validation, and testing set.
- Generate a random population of trading rules.
- Run the following algorithm for n generations.
  - Evaluate the fitness of the entire population.
  - Perform selection and create a new population.
  - Mutate the surviving population.
- After this training phase is over, take the final population, and select the trading rule with the highest fitness on the validation set.
- Evaluate this individual's fitness on the testing set.

The algorithm is kept simple—no crossover, for example. Extensive search is not the crucial issue—proper representation and fighting overfitting are the keys to success. Informally, we have found that tinkering with parameters or using sophisticated techniques to make the search more efficient has little effect on performance, and has uncertain effects on the algorithm's ability to generalize.

We used the following parameters. Population size of 20, 10 generations, and binary deterministic tournament selection. For fitness, we use the aggregate returns produced by the trading rule. Note that we are not attempting to maximize prediction accuracy—we care about returns. And finally, the training and validation sets are always a 50/50 split of the available training data.

Financial data is slippery; although there hasn't been much detailed research along these lines, it is conventional wisdom that financial time series are rife with regime changes. Thus, we want to avoid applying trading rules to a data in test set temporally distant from the training set. We re-learn periodically, applying the algorithm to test sets 3 months large at a time. Specifically, we start the algorithm using six months (January-June, 1998) for the training/validation set (split 50/50), and test on the next three months of data (July-September,

1998). Then, we add that test set to the training/validation set (again, splitting it 50/50), re-run the algorithm, and evaluate the results on a test set consisting of the next three months of data. We continue this until we run through the entire data set.

1.4. Rulespace & Representation

Of course, using genetic programming techniques to learn trading rules depends crucially on the representations of possible trading rules . As we will explore in detail in the experiments section, this representation is crucial.

Past work in applying genetic programming to learn trading rules have drawn inspiration from technical analysis (for a good introduction, see Pring 1991), a quasi-scientific methodology for financial forecasting. While technical analysis has long had a small cultish following among practitioners, economists have only recently began to investigate it rigorously—see Brock et al. (1992) for a good example of applying technical analysis trading rules. Allen and Karjalainen (1999) took the idea of using a GP learner to learn technical analysis trading rules with some success. Others have extended this approach, particularly Neely's work (in Neely et al. 1997) and our own work (Thomas and Sycara 2000) with small sample sets of message volume scores.

These learners function by comparing ratios of summary statistics and issuing appropriate "buy/hold" or "sell/stay out" signals accordingly (although the GP frameworks used in the work cited above do possess more complexity than this).

Crucially, they have been unbiased with respect to being in the asset or not, being "in" or "out" of the appropriate asset with roughly equal probability. For some of this work—for example, that of exchange rate behavior—this lack of bias is well-motivated. However, it carries the implicit assumption that every day is equally easy for the learner to predict. In contrast, we will make only occasional predictions, based on message volume statistics, of a downward movement in a stock. Our strategy will be to stay long in a stock unless we get such a "trouble" signal; if so, then we will short the stock and go long in the market for a brief period.

The leaf nodes of our trading nodes will take the following form: if the current message traffic volume is greater than a threshold, we get out of the stock, and stay out for a certain period of time. The intuition behind this is that we do not always want to commit to making a prediction—we only care about spikes in message volume traffic (it's up to the GP to learn what those spikes should look like)—and if we're not making a prediction, we want to be in the stock. Formally, the leaf nodes of our trading rules have the following format:

- if $v_t > ma_{t,n} + k \cdot \sigma_{t,n}$, sell the stock short and go long in the market for m days.

Where:

- v_t is the message volume from market close of day $t-1$ until market open on day t.
- $ma_{t,n}$ is the n day moving average over message volume up to day t
- $\sigma_{t,n}$ is the n day standard deviation of the message volume up to day t
- m is the length of time we stay out of the stock once we've detected an event.

This means that we are effectively searching over the following parameters, with the following ranges:

- n, mean and standard deviation window size: 10 to 100 trading days, in increments of 10 trading days
- k, event threshold, measured in standard deviations: 3 to 6 standard deviations above the mean, in increments of 0.5.
- m, stay out period: One to ten trading days.

Thus, each terminal takes a time series of message volume data and outputs a corresponding series of in and out of the market signals signals. Now that web have defined our terminals, let us define our function set. Each function takes a time series of out of market and in market signals and outputs a corresponding time series of short and out of market signals.

- AND: day by day logical AND: For each day, if both time series are "out of market," issue an "out of market" signal, otherwise issue an "in market" signal.
- OR: day by day logical OR: For each day, if either time series is "out of market," issue an "out of market" signal, otherwise issue an "in market" signal.

In order to keep the trading rules relatively simple, we limit the maximum number of nodes to 10.

2. Experiments

This section presents experimental results. We are specifically concerned with addressing the following questions:

- Can we demonstrate that the message board volume data has predictive power; i.e., can a reasonable GP approach produce statistically significant out-of-sample excess returns?
- Is the message board volume data contributing information that other traditional numerical data (price, volume, etc) are not?

We take the following approach: first, we present the results of the standard approach that has been described above. Then, we vary key factors in the standard approach—data sources, parameters of the learning algorithm—re-test, to understand what features are crucial to performance. Finally, we test the standard approach on a final hold out set. Except where noted, all results presented are averaged over 30 runs of the algorithm.

2.1. The Standard Approach

The results for the standard approach are presented in Table 4.2. We include p-values from bootstrap hypothesis testing, briefly described above in section 1.2.1, with 200 bootstrap datasets generated under the null hypothesis that the message volume scores contained no predictive power over returns by scrambling the message volume score time series. The results are averaged over thirty trials.

Table 4.2. Results from Standard Approach

Approach:	GP Learner	Buy and Hold	Difference	Bootstrap p
Returns	164.56%	126.21%	38.34%	.015
Daily Std Dev	2.18%	2.45%	−.27%	.010
Sharpe Ratio	1.7455	1.1598	.5857	.005
Differential Sharpe	N/A	N/A	1.6076	.030

The key figures are the differences between the GP learner (referred to as the standard approach) and the buy and hold approach. While both approaches produce large returns, the standard approach produces an average excess return of 38.34% over the entire test set, with lower daily standard deviation of returns. This produces a difference in annualized Sharpe ratios of .5857. And, most interesting, the annualized differential Sharpe ratio (Sharpe ratio of the standard approach using the buy and

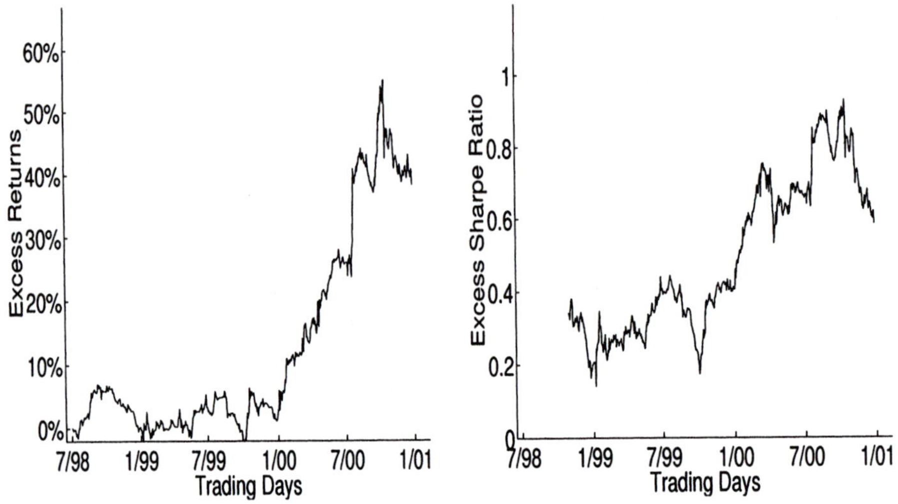

Figure 4.1. Results of GP trading strategy learner.

hold strategy as a benchmark) is an extremely impressive 1.6076, which indicates that the excess returns over the benchmark strategy have a favorable risk/return profile. Furthermore, all these results have bootstrap p-values of less than the traditional significance level of .05. These results are unambiguously positive.

Figure 4.1, presents the results for excess returns and excess Sharpe ratios plotted over time. The left plot contains the cumulative excess returns, and the right plot the excess Sharpe ratios. In addition, we do not plot the first three months of Sharpe ratios—calculating daily standard deviations over such a small sample of returns produces wildly variable results; this does not materially effect the results in any way.

These plots add some detail to the figures above. First, note that for the first year and a half, excess returns seem to be rising, although there is an occasional dip back to zero. It is important to note that at this same time, the excess Sharpe ratio is strongly positive—over .2. When we are out of the stock, we are invested in a much less risky asset (the market), our trading strategy is inherently less risky than the buy-and-hold strategy (of course, this is only true because the stocks with high message traffic volume tend to be more volatile than the market). So even though the trading strategy doesn't produce strong excess returns during the initial period, it would still be preferable on a risk/return basis.

But, in the beginning of 2000, the excess returns rise steadily and sharply for about seven months, peaking in October and declining slightly to end the year.

2.2. Other Possible Predictive Variables: Lagged Trading Volume and Lagged Returns

One possible explanation for the algorithm's success is that message volume is simply echoing information already present in other key variables more traditionally exploited by trading systems—the one day lags of trading volume or returns. There is some correlation between message traffic volume and these variables; average correlation between lagged trading volume and message traffic is a strong .5194, while for returns and message traffic it is -.1017. This suggests that while lagged returns are unlikely to contain the same information as the message volume, the high correlation between message volume and trading volume suggests the possibility that message volume is simply echoing trading volume.

One simple way to test this is to re-run the algorithm, substituting in lagged trading volume or lagged returns for the message volume time series. If the message trading volume is merely echoing data already found in trading volume or return data, then the algorithm should be able to perform equally well with lagged trading volume or lagged returns. On the other hand, performance using the other variables is below that of the algorithm using message volume, that is strong evidence that message volume is providing genuinely new information.

To test this, we re-ran the algorithm, inserting one-day lagged trading volume or returns for message volume. The end of period results are presented in Table 4.3, averaged over thirty trials. We present excess returns (over the buy-and-hold strategy), excess Sharpe ratio (over the buy-and-hold strategy), and Sharpe ratio using the buy-and-hold strategy as a benchmark.

Table 4.3. Message Volume vs. Trading Volume

Predictor Variable:	Excess Returns	Excess Sharpe	Differential Sharpe
Message Volume	38.34%	.5857	1.6076
Trading Volume	1.27%	.1519	.0438
Returns	3.13%	−.0088	−.3569

The results are clear: performance is far superior when using message volume than when using lagged trading volume or returns, in all three categories. Not surprisingly, lagged trading volume fares better than lagged returns. Using a 2-tailed T test we found that the differences

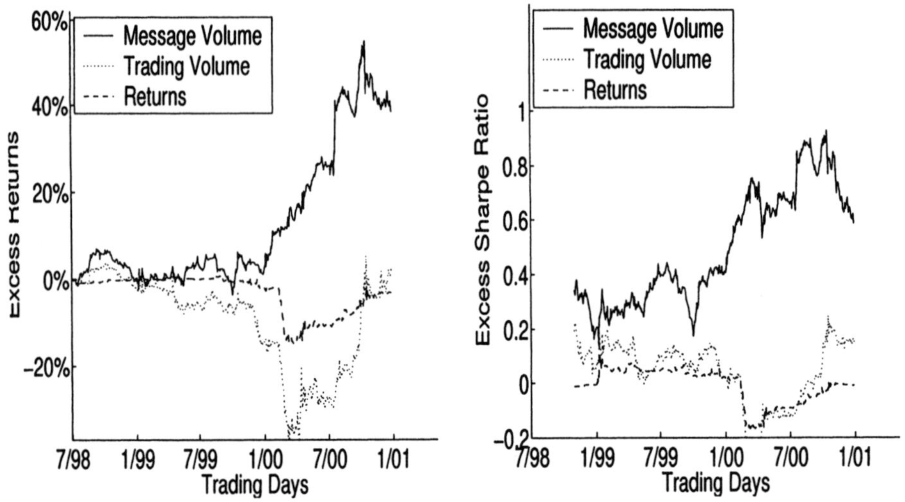

Figure 4.2. Message volume vs. lagged trading volume and returns as a predictor variable.

between the message volume results and the lagged trading volume and lagged returns results were all statistically significant, with p-values less than .001 in all cases.

This strongly suggests that message volume does contain novel information. Of course, it is possible that message volume and trading volume would be equally predictive inputs to some other algorithm or representational scheme—but exhaustively exploring this possibility is beyond the scope of this paper. What is important is that the information being extracted out of message volume signal is not present in the lagged trading volume or the lagged returns.

Plots presenting behavior over time are plotted below in Figure 4.2. The left graph plots excess returns, the right graph excess Sharpe ratio. Notice that on the excess return graph, the message volume approach is not clearly superior to the lagged returns approach until the final six months; however, superiority in excess Sharpe ratio occurs across the board. Note also that the shape of the cumulative excess returns and average Sharpe ratio produced by the algorithm using lagged trading volume is qualitatively very different than that of the algorithm using message volume; this further reinforces the idea that message volume contains information that lagged trading volume does not.

2.3. The Importance of Representation: Changing the Nature of Trading Rules

This section explores the effect of varying the representation built into the genetic programming algorithm. We have found that constraining the representation searched by the GP algorithm is far more important to good performance than tinkering with the individual GP parameters themselves. In order to demonstrate this, out next experiment dramatically changes the underlying representation for our trading rule. As discussed in section 1.4, our standard approach uses an event methodology: the trading rule stays in a stock unless it finds a rare event. It is instructive to compare this with the more technical analysis inspired approach that depends on comparing moving averages, that can be found in Allen and Karjalainen (1999) on the S&P 500 and later extended by Neely et al. (1997) to foreign exchange, as well as our own previous papers on building trading rules from message board data (Thomas and Sycara 1999; Thomas and Sycara 2000). Here, we present an extremely simplified version of this. We replace the leaf nodes of our GP learner with the following rule:

- if $ma_{t,n} > ma_{t,m}$, sell the stock short and go long in the market.

Where $ma_{t,n}$ is the n day moving average of the message volume. Here, although the tree structure can still vary, the parameters in the leaf node that can vary are limited to n and m, the moving average window sizes.

We keep the algorithm identical in all other respects—in preprocessing, algorithm parameters, and in the logical operations used to form trees out of leaf nodes. Intuitively, these kinds of trading rules contain one key difference: instead of looking for a rare event and pulling out of a stock, this kind of trading rule is neutral with regards to being in or out of a stock. The results are presented in Table 4.4.

Table 4.4. Event Approach vs. Moving Average Approach

Trading Rule Scheme:	Excess Returns	Excess Sharpe	Differential Sharpe
Event approach	38.34%	.5857	1.6076
Moving average approach	−46.39%	.4177	−.4751

The moving average approach performs abysmally, with large negative excess returns and differential Sharpe ratio. The excess Sharpe ratio is large—surprisingly large at .4177. Even though the excess returns are negative, the excess Sharpe ratio can be positive if the volatility of

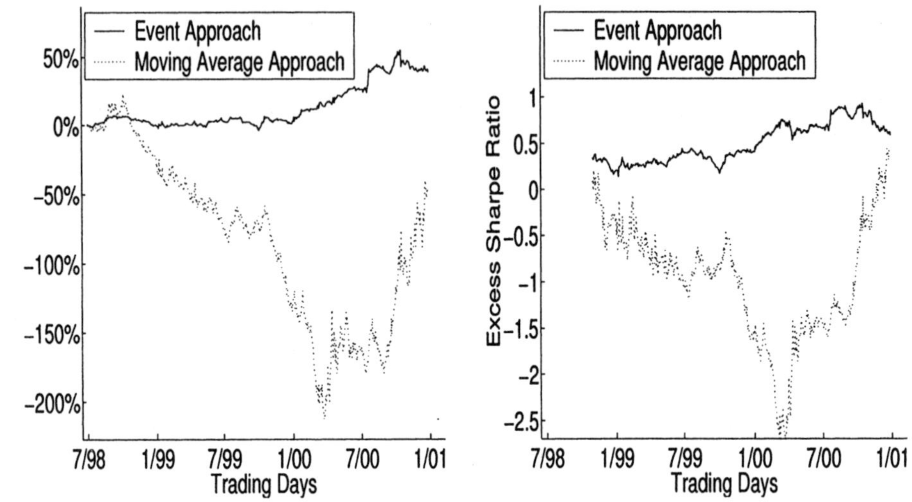

Figure 4.3. Event approach vs. moving average approach.

the moving average approach is very low—this indicates a low risk, low return profile.

These results are reinforced by the behavior over time plotted in Figure 4.3. For both excess returns and excess Sharpe ratio, performance of the moving average approach rapidly and consistently decays. There is a huge spike upward after March of 2000, roughly mirroring the performance of the standard approach, but the end of period returns are still strongly negative.

This fact is striking: that a reasonable representation, a variation of one that has been used successfully in the past, performs so poorly using the *exact same parameters* for the genetic programming component points to the overwhelming importance of representation. Of course, one could also argue that this is evidence we have tuned our representation specifically to this dataset; this is a concern we address below, in section 2.4.

2.4. Tests on Holdout Data

This section discusses tests performed on holdout data set as extra verification of the significance of these results. Our holdout data set consisted of a random selection of half of out stock universe. We re-ran the standard approach on this data; results are presented in Table 4.5.

At first glance, these results look event more impressive, with strong excess returns, excess Sharpe ratio, and differential Sharpe ratio. How-

Table 4.5. Standard Approach on Holdout Set

Approach:	GP Learner	Buy and Hold	Difference	Bootstrap p
Returns	149.01%	89.60%	59.41%	.045
Daily Std Dev	2.59%	3.03%	−.44%	< .005
Sharpe Ratio	1.3235	.6353	.6882	.030
Differential Sharpe	N/A	N/A	1.7548	.110

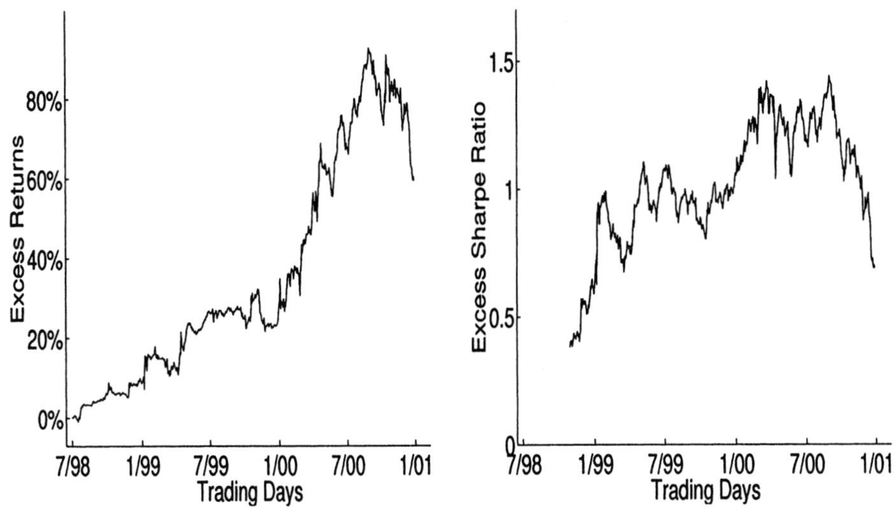

Figure 4.4. Results of standard approach on holdout test data set.

ever, the p-values are higher than in the test set—the excess returns and excess Sharpe ratio are still statistically significant by the bootstrap hypothesis testing; while the differential Sharpe ratio is not. Still, the fact that the algorithm performs similarly on the holdout set is strong confirmation that we were not simply overfitting our algorithm to the test set.

The performance over time plots presented below in Figure 4.4 reinforce this. The return and excess Sharpe ratio plots correspond strongly to the forms found in 4.1. The rise in excess returns from July of 1998 to January 2000 is stronger, with less volatility; the explosive growth in returns after January of 2000 is present; and the decline after October of 2000 is even more apparent. The excess Sharpe ratio results follow accordingly.

2.5. Regime Changes and Reversing Polarity

The decline in excess returns on both the test set and the holdout data set from October of 2000 to the end of the time period shown by the plots in Figures 4.1 and 4.4 is troubling. Will it continue? If we were investment managers, would we have much faith in the results of the algorithm going forward? Perhaps, but it would certainly provoke concern and exploration. Is there a regime change at work?

There is a strong assumption built into our algorithm—that message volume spikes portend trouble for a stock. If indeed a regime change has happened, that assumption might no longer be true. One way to explore the possibility is to change that assumption about how message volume is important: instead of looking for spikes in message volume, we look for slumps in message volume. We literally "reverse polarity" in the algorithm.

Specifically, we make the following changes. We replace the leaf nodes of our trading rules with the following, simply swapping the greater than sign for a less than sign:

- if $v_t < ma_{t,n} + k\sigma_{t,n}$, sell the stock short and go long in the market.

In addition, we change the range of minimum event thresholds from 3 to 6, to -1.5 to -3, and search in increments of .25 (we do this because the distribution of message volume traffic is skewed; there are many message volume spikes 3 standard deviations above from the mean, but almost literally no days message volume 3 standard deviations below the mean). Again, these changes reflect a shift in the underlying assumption in the algorithm from looking for high message volume to looking for low message volume.

We keep everything else the same, and re-run the algorithm with a test set starting on November of 2000. Of course, fundamentally changing the ground rules of our algorithm in the middle of our test set is hardly fair; so we present results that extend past our test set—to the end of March 2001—in Table 4.6.

These results are very promising, especially given the relatively brief (8 months) time period. Both on the test set and the holdout set, the trading strategy consistently and statistically significantly beats the buy and hold strategy. The results are more impressive on the holdout set than the test set, but statistically significant all around. Note that even though the results on the test set don't produce positive returns, we are competing with the buy and hold strategy, and it is the difference in returns and Sharpe ratios that is important both conceptually and for statistical testing purposes.

Table 4.6. "Reversed Polarity" Algorithm

Approach:	GP Learner	Buy and Hold	Difference	Bootstrap p
	Performance on Test Set			
Returns	−1.03%	−28.59%	27.56%	.020
Daily Std Dev	1.64%	3.07%	−1.43%	< .005
Sharpe Ratio	-.2390	-.9817	.7427	.020
Differential Sharpe	N/A	N/A	1.2640	.235
	Performance on Holdout Set			
Returns	21.15%	−35.94%	57.09%	< .005
Daily Std Dev	2.10%	3.88%	−1.78%	< .005
Sharpe Ratio	.8151	-.9549	1.7701	< .005
Differential Sharpe	N/A	N/A	2.1855	.025

If one examines the returns for the buy and hold strategy, they are negative both for the test set and the holdout set. The fact that the buy and hold returns were strongly positive in the first part of the data, and negative after the possible regime change is highly suspicious: a possible hypothesis is that reaction to message volume traffic numbers depends on the behavior of these stocks relative to the larger market. Sadly, this provocative suggestion must wait for more data to accumulate.

These results are reinforced by behavior over time plots presented in Figure 4.5. For space reasons, we present the both the results for the test set and the holdout set on the same graphs; excess returns on the left, excess Sharpe ratios on the right. Important to note is that the excess returns continue to accumulate into 2001, although the result is definitely more pronounced for the holdout set, it does seem that the regime change is "sticking". Second, examine the Sharpe ratio graph—the Sharpe ratios level off at the beginning of 2001, and even decline a little bit for the test set. Note that since the Sharpe ratio is an average Sharpe ratio, holding steady is a good result—it means that the level of excess return is continuing to accumulate—so while the drop off in the test set results is a little discouraging, the results on the holdout set is exactly what we expect to see, if the algorithm is working.

Although these results certainly are promising, they must be interpreted carefully: fundamentally changing an assumption underlying a machine learning algorithm in the middle of a test set is clearly problematic from an AI point of view. While we are cautious to read too much into these results—clearly, the acid test must wait for time to pass and further data to accumulate—in our defense we make three points:

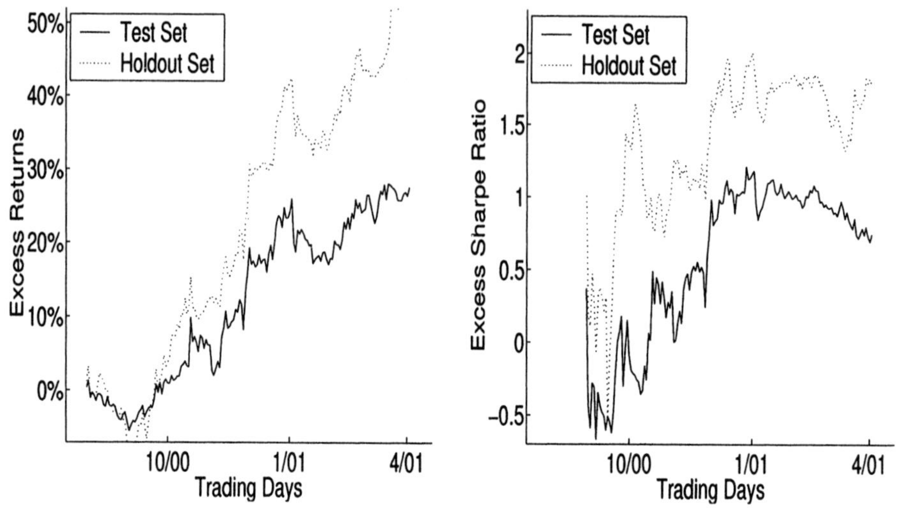

Figure 4.5. "Reversed Polarity" algorithm on test set and holdout set.

- In finance, the data is far to slippery to completely trust any machine learning algorithm; they should be used as decision support tools rather than automatic decision makers. As such, if a human practitioner suspects trouble or a regime change, they would immediately re-investigate, and perhaps adjust the algorithm.

- Our change was well motivated: given that we suspected a regime change, it is a perfectly legitimate to hypothesize the market had changed the way it reacted to message volume numbers, and adjust the algorithm accordingly. The change was simple and small.

- A key to future research in the application of machine learning to finance is understanding how to detect regime changes. While this is by its nature a difficult problem, we plan to explore it in future work.

From the economics point of view, on the other hand, these results are extremely intriguing: if you take these results at face value the data really does suggest that the market literally reversed the way it reacted to message volume scores in late 2000. This deserves further exploration, first to further determine if this shift is real and lasting, and to find possible explanations or market correlates.

3. Summary and Conclusions

We have described a GP learner tailored to learning financial trading rules based on the volume of message traffic on Internet chat boards, and demonstrated that it produces statistically significant excess returns both on our test set and a further holdout test set.

We investigated the possibility that this information could be found in other financially relevant variables—lagged trading volume and returns—found that in this framework, message volume numbers contain unique information. We demonstrated that a moving average framework that had worked in the past failed on this data, suggesting that our choice of an event-based framework was important to the success of the algorithm.

In addition, we suspected a regime change in the data, and hypothesized that the market's reaction to message volume numbers had shifted, and adjusted the GP learner accordingly. This change produced strong results over the post-regime change period, although this fact needs to be interpreted carefully as data for proper testing is limited.

While these results are extremely promising, we see two key areas requiring followup research. First, we are working with the volume of messages on Internet chat boards—while the mere count of messages is clearly informative, there is far, far more data on the message boards in the form of the text itself. Utilizing the information in this text—perhaps by integrating it with text classification methodologies (as tried by Thomas and Sycara 2000)—is key to augmenting these results. Second, as underscored by the results of this paper, understanding regime changes in financial data is crucial.

References

Allen, F. and R, Karjalainen (1999). "Using Genetic Algorithms to Find Technical Trading Rules," *Journal of Financial Economics*, 51(2), 245–271.

Brock, W., J. Lakonishok, and B. LeBaron (1992). "Simple Technical Trading Rules and the Stochastic Properties of Stock Returns," *Journal of Finance*, 47(5), 1731–1764.

Efron, B. and R. J. Tibshrani (1993). *An Introduction to the Bootstrap.* Chapman & Hall.

Koza, J. R. (1992). *Genetic Programming: On the Programming of Computers by Means of Natural Selection.* MIT Press.

Neely, C., P. Weller, and R. Dittmar (1997). "Is Technical Analysis in the Foreign Exchange Market Profitable? A Genetic Programming Approach," *Technical Report.* Federal Reserve Bank of St. Louis.

Pring, M. J. (1991). *Technical Analysis Explained.* McGraw Hill.

Sullivan, R., A. Timmermann, and H. White (1998). "Data-snooping, Technical Trading Rule Performance, and the Bootstrap," *Journal of Finance*, 54, 1647–1692.

Thomas, J. D. and K. Sycara (1999). "The Importance of Simplicity and Validation in Genetic Programming for Data Mining in Financial Data," in *Proceedings of the Joint GECCO-99 and AAAI-99 Workshop on Data Mining with Evolutionary Algorithms: Research Directions.*

Thomas, J. D. and K. Sycara (2000). "Integrating Genetic Algorithms and Text Learning for Financial Prediction," in *Proceedings of the GECCO-2000 Workshop on Data Mining with Evolutionary Algorithms.*

Chapter 5

GENETIC PROGRAMMING OF POLYNOMIAL MODELS FOR FINANCIAL FORECASTING

Nikolay Y. Nikolaev
Dept. of Math. and Computing Sciences
Goldsmiths College
University of London
London SE14 6NW
United Kingdom
n.nikolaev@gold.ac.uk

Hitoshi Iba
Dept. of Inf. and Comm. Engineering
School of Engineering
The University of Tokyo
7-3-1 Hongo, Bunkyo-ku
Tokyo 113-8656
Japan
iba@miv.t.u-tokyo.ac.jp

Abstract This paper addresses the problem of finding trends in financial data series using genetic programming (GP). A GP system STROGANOFF that searches for polynomial autoregressive models is presented. The system is specialized for time series processing with elaborations in two aspects: 1) preprocessing the given series using data transformations and embedding; and, 2) design of a fitness function for efficient search control that favours accurate, parsimonious, and predictive models. STROGANOFF is related to a traditional GP system which manipulates functional expressions. Both GP systems are examined on a Nikkei225 series from the Tokyo Stock Exchange. Using statistical and economical measures we show that STROGANOFF outperforms traditional GP, and it can evolve profitable polynomials.

Keywords: Genetic Programming, Polynomial Models, Overfitting Avoidance

Introduction

Economic forecasting is a challenging task that has been approached recently by Genetic Programming (GP) (Koza 1992). Various applications of GP to financial data analysis and prediction have already been reported (Colin 1994; Koza 1995; Bhattacharyya et al. 1998; Chen and Yeh 1998; Chen and Ni 1998; Allen and Karjalainen 1999; Chen and Lu 1999; Iba and Sasaki 1999; Lee 1999; Santini and Tettamanzi 2001; Nikolaev and Iba 2001; Zumbach et al. 2001, etc.). Such a financial problem instance is for example learning of the trend in market price series with the goal to predict future prices.

The most serious difficulties faced when attempting to solve such financial problems are (Refenes et al. 1996; Zapranis and Refenes 1999): what kind of model to use, whether to take standard linear models or non-linear models like polynomials, neural networks etc.; which input variables should enter as factors the selected model; how to identify the model structure; how to estimate the model parameters; how to organize search for the most promising models; how to balance the statistical bias, statistical variance and model complexity in order to make the search navigation less sensitive to the intrinsic series noise.

GP can be used to overcome these difficulties in financial data processing. It provides an evolutionary search paradigm that can be easily tuned to search for econometric models. Having selected a concrete model, it offers several advantages: 1) GP performs efficient stochastic exploration of the search space with a population of models that progressively adapt to the given data; 2) GP finds automatically the dependencies among the input variables, and thus it identifies which variables should enter the model; 3) GP discovers the model structure adequate to the data, that is it tailors the structure to the data; and, 4) GP allows to employ discontinuous objective (fitness) functions for search control that can help to achieve good generalization.

We propose a GP system for learning polynomial autoregressive (PAR) models from data. The system is specialized for time series modeling with emphasis on two key aspects: 1) preprocessing the given series by specific transformation techniques for extracting the significant information from the raw observables, followed by identification of delayed variables; and, 2) design of a fitness function for efficient evolutionary search navigation. The idea is to make such a fitness function that avoids overfitting with the data and locates well performing models by balancing the statistical bias with the statistical variance. Well performing are considered such models that exhibit good accuracy on the training data, high predictability on unseen data, and parsimonious structure.

A GP system STROGANOFF (Iba et al. 1994; Nikolaev and Iba 2001) that evolves high-order multivariate polynomials is developed. The polynomials are represented as tree-structured compositions of transfer polynomials allocated in the tree nodes and variables in the leaves. The current STROGANOFF uses a set of transfer polynomials in the nodes rather than the complete bivariate polynomial only. It conducts evolutionary search with a population of tree-like polynomials using the mechanisms of a genetic algorithm: fitness evaluation of the polynomial, fitness proportional selection of the most promising polynomials from the population, crossover and mutation of the trees.

The performance of two GP systems is examined on a stock market prediction task: the recent STROGANOFF (Nikolaev and Iba 2001), and a traditional Koza-style GP (Koza 1992). The implementation SGPC-1.1 (Tackett and Carmi 1994) of Koza-style GP which manipulates expressions of elementary functions is taken. The same preprocessing techniques and fitness function are used in both systems to facilitate the comparisons. The systems are trained and tested on a *Nikkei225* price average series from the Tokyo Stock Exchange (Iba and Sasaki 1999). The prediction risk from eventual use of the learned models is evaluated with statistical and economical measures.

The analysis of the experimental results indicates that: 1) the system STROGANOFF finds non-linear polynomials that describe better the directional changes in series movements, while SGPC discovers functional expressions that model better the continuous patterns in series movements; 2) using the novel fitness function helps to converge toward less overfitting solutions which generalize well, because not only STROGANOFF but also SGPC finds better forecasting models when equipped with this fitness function; 3) STROGANOFF produces models that outperform these from SGPC on differentially and rationally transformed series exhibiting magnified rates of change, and these polynomials yield highest profits from all experiments.

This paper is organized as follows. Section 1 introduces the polynomial models and their tree-like representation in STROGANOFF using a set of transfer polynomials. Section 2 describes the genetic mechanisms of the recent GP system STROGANOFF, and more precisely: the designed fitness function as well as the formula for estimating the polynomial coefficients, the crossover and the mutation operators. Section 3 gives the data transformation techniques and the embedding schemes for financial data preprocessing. After that, the experimental results on the correspondingly preprocessed series are provided. Finally, a brief discussion is made and a conclusion is derived.

1. Polynomial Models

Financial series modeling may be regarded as a multivariate regression problem. Given a series of raw observables (values): $..., v_t, v_{t+1}, v_{t+2}, ...$ sampled at discrete time intervals, the goal is to find out how unseen values depend on past values. The raw observables v_t are usually converted into the range [0,1] by linear scaling $x_t = Normalize(v_t)$, and so the normalized series becomes: $..., x_t, x_{t+1}, x_{t+2}, ...$. The Takens' theorem (Takens 1981) justifies the use of *delayed* (lagged) vectors $\mathbf{x}$ to reconstruct the dynamic system that has generated the series:

$$x_{t+1} = f(\mathbf{x}) = f(x_{t-(m-1)\tau}, x_{t-(m-2)\tau}, ..., x_t) \tag{5.1}$$

where m is *embedding dimension*, and τ is *delay time*.

The series is viewed as a data set $D = \{(\mathbf{x}_i, y_i)\}_{i=1}^{N}$ of size N, where $\mathbf{x}_i = (x_{i1}, x_{i2}, ..., x_{id}) \in \mathcal{R}^d$ are vectors of the independent variables having dimension d, and $y_i = x_{t+1} \in \mathcal{R}$ are the dependent variable values. We seek analytical functions $y = f(\mathbf{x})$ of the class high-order multivariate *Kolmogorov-Gabor polynomials*:

$$f(\mathbf{x}) = a_0 + \sum_i a_i x_i + \sum_i \sum_j a_{ij} x_i x_j + \sum_i \sum_j \sum_k a_{ijk} x_i x_j x_k + ... \tag{5.2}$$

where a_i are term coefficients.

The Kolmogorov-Gabor polynomial is preferred as a universal format for function modeling with which one may approximate any continuous function mapping to an arbitrary precision, in average squared residual sense, if there are sufficiently large number of terms.

1.1. Tree-Like Polynomials in STROGANOFF

STROGANOFF (Iba et al. 1994; Nikolaev and Iba 2001) manipulates populations of such polynomials represented as *tree structures*. It conducts evolutionary search using the mechanisms of a genetic algorithm: fitness evaluation, fitness proportionate selection, crossover and mutation operators. The fitness evaluation of a polynomial involves: 1) estimating the polynomial coefficients in the tree nodes by least squares fitting; and 2) calculating the error of fit at the tree root.

The trees serve to compose low-order transfer polynomials allocated in the tree nodes, and independent variables in the leaves following the GMDH method (Ivakhnenko 1971). The transfer polynomial outcomes are fed forward to their parent nodes, where partial models are composed. Complete bivariate transfer polynomials: $a_0 + a_1 x_1 + a_2 x_2 +$

$a_3x_1x_2 + a_4x_1^2 + a_5x_2^2$ are used. At the tree root, the network output is interpreted as a high-order high-dimensional multinomial.

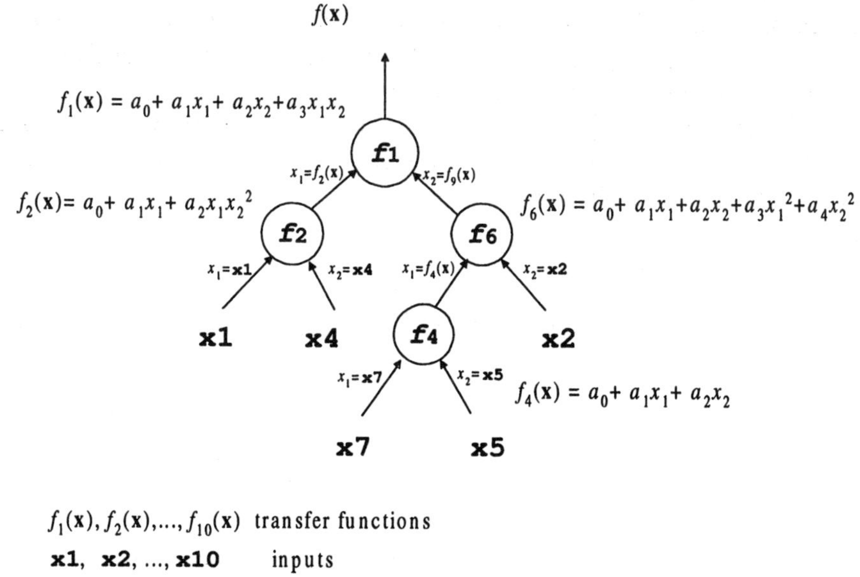

Figure 5.1. Tree-like polynomial in the enhanced STROGANOFF.

Table 5.1. The Set of Transfer Polynomials $\Phi = \{f_i\}_{i=1}^{10}$

Transfer Polynomials
$f_1(\mathbf{x}) = a_0 + a_1x_1 + a_2x_2 + a_3x_1x_2$
$f_2(\mathbf{x}) = a_0 + a_1x_1 + a_2x_1{x_2}^2$
$f_3(\mathbf{x}) = a_0 + a_1x_1 + a_2x_1x_2$
$f_4(\mathbf{x}) = a_0 + a_1x_1 + a_2x_2$
$f_5(\mathbf{x}) = a_0 + a_1x_1^2 + a_2x_2^2$
$f_6(\mathbf{x}) = a_0 + a_1x_1 + a_2x_2 + a_3x_1^2 + a_4x_2^2$
$f_7(\mathbf{x}) = a_0 + a_1x_1 + a_2x_1x_2 + a_3x_1^2$
$f_8(\mathbf{x}) = a_0 + a_1x_1 + a_2x_1x_2 + a_3x_2^2$
$f_9(\mathbf{x}) = a_0 + a_1x_2 + a_2x_1^2$
$f_{10}(\mathbf{x}) = a_0 + a_1x_1 + a_2x_2 + a_3x_1x_2 + a_4x_1^2 + a_5x_2^2$

1.2. Set of Transfer Polynomials

The current STROGANOFF uses a set of transfer polynomials rather than only the complete bivariate polynomial. Employing a set of a fixed number of transfer polynomials extends the flexibility of the models to fit the series, because in this way the non-linear interactions among the independent variables are captured more precisely.

An example tree-like structure that composes the polynomial model $f(\mathbf{x}) = f_1(f_2(x_1, x_4), f_6(f_4(x_7, x_5), x_2))$ using different transfer functions from the set in Table 5.1 is shown in Figure 5.1.

A small set $\Phi = \{f_i\}_{i=1}^{10}$ with *complete* and *incomplete* first-order and second-order polynomials in two variables is derived (Table 5.1) from the complete one (5.3) by elimination of terms:

$$f(\mathbf{h}(\mathbf{x})) = a_0 + \sum_{k=1}^{5} a_k h_k(\mathbf{x}) \tag{5.3}$$

where $\mathbf{x} = (x_i, x_j)$ is a binary vector, i,j are indices, and $h_k(\mathbf{x})$ are functions that produce the terms: $h_0(\mathbf{x}) = 1$, $h_1(\mathbf{x}) = x_i$, $h_2(\mathbf{x}) = x_j$, $h_3(\mathbf{x}) = x_i x_j$, $h_4(\mathbf{x}) = x_i^2$, $h_5(\mathbf{x}) = x_j^2$.

The advantage of these tree-structured polynomials is that they make possible the evaluation of complex, high-order models for acceptable time by composing simple transfer polynomials whose coefficients are computed rapidly by ordinary least-squares fitting.

2. GP Mechanisms

The GP system STROGANOFF organizes evolutionary search with a population of tree-like polynomials. The search is navigated by a special fitness function for proportionate selection of the most promising polynomials which are updated by mutation and crossover operators. The resulted offspring polynomials replace the worst models from the current population in similarity to the natural evolution.

2.1. The Fitness Function

The *fitness function* is a driving force of evolutionary search (Koza 1992). We use three criteria to design a fitness function for achieving *overfitting avoidance*: 1) an average error criterion that prefers more fit polynomials; 2) a regularization factor that tolerates smoother polynomials with higher generalization potential; and 3) a complexity penalty that favors short size polynomials. These criteria considered together help to balance the statistical bias with the statistical variance of the models, and contribute to the discovery of *accurate*, *predictive*, and *parsimonious* polynomials.

The series fitting is evaluated with the *rational average error* (RAE):

$$RAE = \frac{\sum_{i=1}^{N}(y_i - f(\mathbf{x}_i))^2}{\sum_{i=1}^{N}(y_i - y_{i-1})^2} + \lambda \sum_{j=1}^{A} a_j^2 \tag{5.4}$$

where λ is a regularization parameter, A is the number of all coefficients in the polynomial $f(\mathbf{x})$, a_j are their values, and N is the number of the data. The first term in this formula (5.4) shows error improvement over the random walk model. The second term is a regularizer that tolerates models with coefficients of small magnitude, that is smoother polynomials. The coefficients that minimize the error (5.4) are estimated with the following regularized least-squares formula:

$$\mathbf{a} = (\mathbf{H}^T\mathbf{H} + \lambda\mathbf{I})^{-1}\mathbf{H}^T\mathbf{y} \tag{5.5}$$

where $\mathbf{a}$ is a $(k+1) \times 1$ column vector with the coefficients of the polynomial $f(\mathbf{x}) = \mathbf{Ha}$, $\mathbf{H}$ is a $N \times (k+1)$,$1 \leq k \leq 5$, design matrix of row vectors $\mathbf{h}(\mathbf{x}_i) = (h_0(\mathbf{x}_i), ..., h_k(\mathbf{x}_i))$, $i = 1..N$, and $\mathbf{y}$ is a $N \times 1$ column vector with the desired outcomes.

Singular Value Decomposition is recommended to perform inversion of the the covariance matrix $(\mathbf{H}^T\mathbf{H}+\lambda\mathbf{I})$ since it avoids numerical problems arising from ill-conditioning (Press et al. 1992).

We design the fitness function according the *generalized cross-validation* (*GCV*) (Wahba 1990) principle to evaluate the level of fitting RAE as well as for the model complexity:

$$GCV = \frac{RAE}{(1 - A/N)^2} \tag{5.6}$$

where RAE is the rational average error (5.4), A is the number of polynomial coefficients, and N is the number of the data.

2.2. Crossover and Mutation

The genetic learning operators serve to sample the space of possible tree-structured polynomials. More precisely, such operators transform the underlying tree structures and, thus, they sample polynomials.

The *crossover* operator takes two parent trees, chooses randomly a cut point node in each tree, and swaps the subtrees rooted in the chosen nodes. The crossover is restricted by a predefined maximum tree size limit so that if an offspring tree of larger size results it is discarded. The maximum tree size is usually fixed by a design decision.

The *mutation* operator takes one tree, selects randomly a tree node and performs one of the following changes: 1) insertion of a random node before the selected one and with the subtree rooted at it as an immediate child, and a randomly generated terminal (leaf) as another child; 2) deletion of the selected node and replacing it by one of its children nodes; and, 3) substitution of the selected node by another randomly chosen one. These tree modifications are applied to binary trees of bivariate transfer polynomials in the nodes.

3. Preprocessing

Financial series *preprocessing* is performed in order to: 1) isolate significant information from the raw data; and, 2) determine how to assimilate this information through independent variables. Identification of variables that convey essential information is crucial for successful learning, since the raw observables are extremely noisy which limits their relevance for learning. Learning algorithms, like GP, can find dependencies among the variables but they can not decide what they have to describe.

Preprocessing the financial series for further learning of trend patterns in them includes two steps: *data transformation* and *embedding*.

3.1. Data Transformations

Several techniques for making a series learnable are studied in order to examine how they influence the GP of polynomials.

The *integral techniques* eliminate some noise frequencies from the series, while preserving the systematic information in it:

$$x_a = \frac{1}{l} \sum_{i=t-(l-1)/2}^{t+(l-1)/2} x_i \tag{5.7}$$

where l is the averaging period, and x_i are the series values. This is a kind of moving average filtering which makes the series smoother and the model search process easier (Makridakis and Wheelwright 1987).

The integral transformations, however, leave too much obscuring information like obvious tendencies. This can be alleviated by applying *differential transformations* (Deboeck and Cader 1994) as follows:

$$x_d = x_t - \frac{1}{l} \sum_{i=t-l-1}^{t} x_i \tag{5.8}$$

where x_d is the difference from the origin value x_t and the average from its nearest l neighbors.

The influence of the data volume can be overcome also by *rational transformations* that smash large magnitudes (Chen and Lu 1999):

$$x_r = \ln \frac{x_t}{x_{t-1}} \tag{5.9}$$

where x_t are x_{t-1} subsequent data from the series. The formula for rational manipulations (5.9) leads to series with sharper oscillations than these produced by differencing (5.8). These rational and differential data transformations are preferred in practice since they diminish the effects of nonstationarity and emphasize the rates of directional changes in time series (Deboeck and Cader 1994).

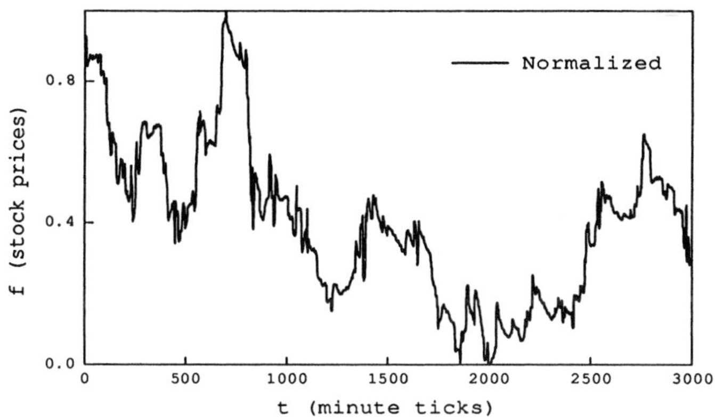

Figure 5.2. Original (normalized) financial series.

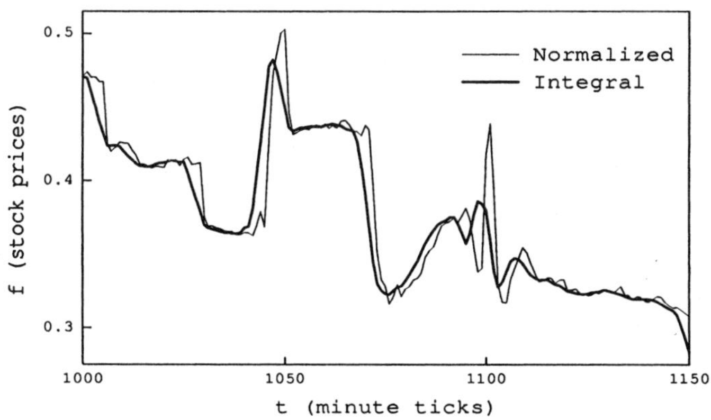

Figure 5.3. Corresponding sections from the integral and original series.

3.2. Embedding

The *embedding scheme* prescribes how to pass the series data to the model through variables. In other words, the embedding specifies on which historical data in the series the current data depends. Our concern is embedding by delay vectors with parameters $m = 10$, and $\tau = 1$.

A financial data series of price averages, the *Nikkei225* index from the Tokyo Stock Exchange, is taken. The data in this series are sampled at every minute from 9:00am to 12:00 pm, and from 1:00pm to 3:00 pm. The *training series* includes 3,000 points from the period April 1st, 1993 through April 17th, 1993. The *testing series* includes 30177 points from the period April 18th, 1993 through September 30th, 1993. The series is normalized, and four training series are derived from it:

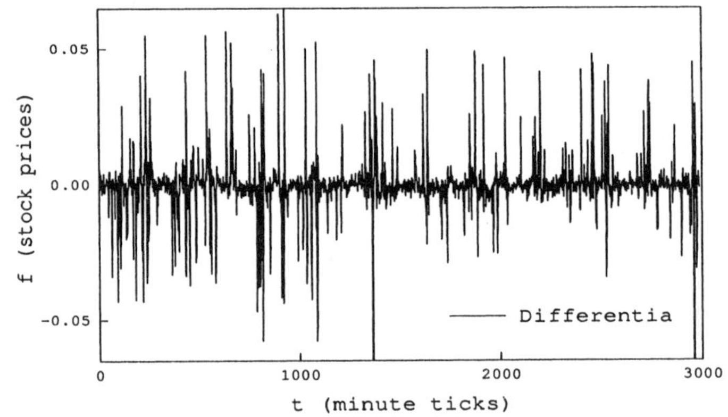

Figure 5.4. Differential financial series.

- *original series* (Figure 5.2) — the given series is taken directly using vectors $\mathbf{x} = (x_{t-m-1}, x_{t-m-2}, ..., x_t)$ in order to realize as to what degree it is difficult to learn;
- *integral series* (Figure 5.3) — each value is replaced by its moving average over a period $l = 5$ according to (5.7). Processing is done with vectors $\mathbf{x} = (x_{a-m-1}, x_{a-m-2}, ..., x_a)$. The integral series looks globally like the original one, but a detailed view reveals the smoothing;
- *differential series* (Figure 5.4) — each value is substituted by its variance from the mean of its neighbors within an interval $l = 3$ using formula (5.8), leading to delay vectors $\mathbf{x} = (x_{d-m-1}, x_{d-m-2}, ...,$ $x_d)$;
- *rational series* (Figure 5.5) — another series is obtained after rational transformations using formula (5.9), which series is consid-

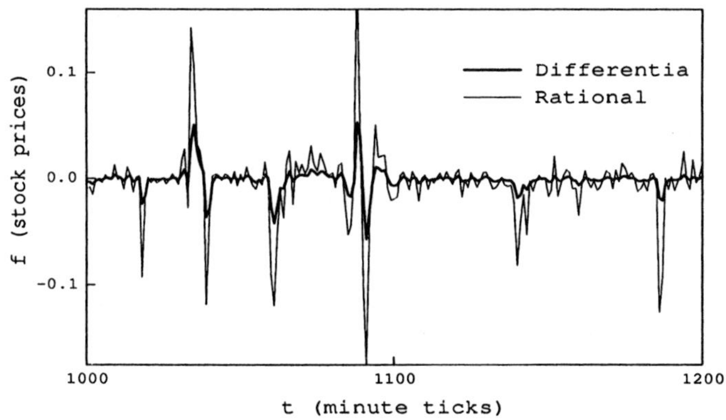

Figure 5.5. Corresponding sections from the rational and differential series.

ered with vectors $\mathbf{x} = (x_{r-m-1}, x_{r-m-2}, ..., x_r)$. The rational series looks similar to the differential one, however it features slightly magnified amplitudes than these of the differential series.

4. Experimental Results

With the employment of GP we attempt to identify nonlinear trends in noisy stock market price movements, assuming that such trends exist.

The measures taken to compare the performance of the derived models over the training and testing series are: mean squared error, hit percentage, and expected profit gain. The *mean squared error* (MSE) is the well known statistical measure defined as follows:

$$MSE = \frac{1}{N} \sum_{i=1}^{N} (y_i - f(\mathbf{x}_i))^2 \qquad (5.10)$$

where y_i is the outcome given with the i-th training data point, $f(\mathbf{x}_i)$ is the estimated outcome using the i-th data point, and N is the number of the data points in the series.

The *hit percentage* (*HIT*) shows how accurately the trend directions have been tracked by the model (Iba and Sasaki 1999):

$$HIT = \frac{N_{up\_up} + N_{down\_down}}{N} \qquad (5.11)$$

where N_{up_up} is the number of times when the estimated and the desired outcomes exhibit both upward raising tendency, and N_{down_down} is the number of times when the estimated and the given outcome exhibit are both falling. This is the percent of cases when the estimated movement direction coincides with the directional changes in the original series.

The expected $Profit$ is evaluated with a simple algorithm that generates buy/sell trade signals, with which the possible dividends from the model are calculated. We use the profit algorithm of Iba and Sasaki (1999), starting from an initial fixed amount of $1,000,000$ Japanese yen committed to each hypothetical trader during a particular test.

In this paper we report results from 25 runs conducted with the recent STROGANOFF and SGPC-1.1 (Tackett and Carmi 1994). The following parameters were used: $PopulationzeSize = 100$, $Generations = 300$, $MinTreeSize = 5$, and $MaxTreeSize = 40$. The regularization parameter is $\lambda = 0.005$. The same crossover, mutation, fitness function (5.6) and selection mechanisms are used in both systems to facilitate the comparisons. An essential difference between the studied GP is that the coefficients in STROGANOFF are estimated by least-squares (5.5), while the function parameters in SGPC are evolved.

The fitting accuracy of the evolved best models, in the sense of MSE, are estimated using the preprocessed *normalized training series* derived with formulae (5.7), (5.8), and (5.9). The testing results, in the sense of HITs and $Profits$ qualities, are estimated using the preprocessed *original testing series* without normalization.

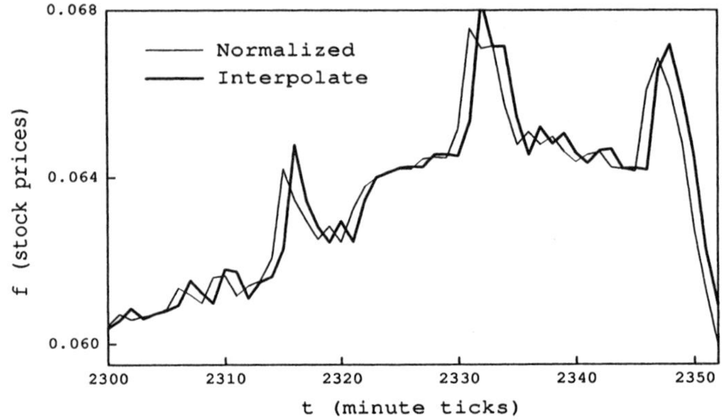

Figure 5.6. Interpolation by the best (original) polynomial from STROGANOFF.

Training with the Original Series. Both GP systems evolved well fitting models of the original series (Table 5.2). The interpolation and extrapolation of the series by the best polynomial from STROGANOFF in arbitrarily selected intervals are plotted in Figures 5.6 and 5.7. These plots show that the estimated polynomial outcome resembles closely not only the training series, but it also mimics the unseen data.

One can see in Table 5.2 that the achieved HITs with the best expression found by SGPC $HITs = 55.23\%$ are more than those of the best polynomial from STROGANOFF $HITs = 54.02\%$. The best polynomial

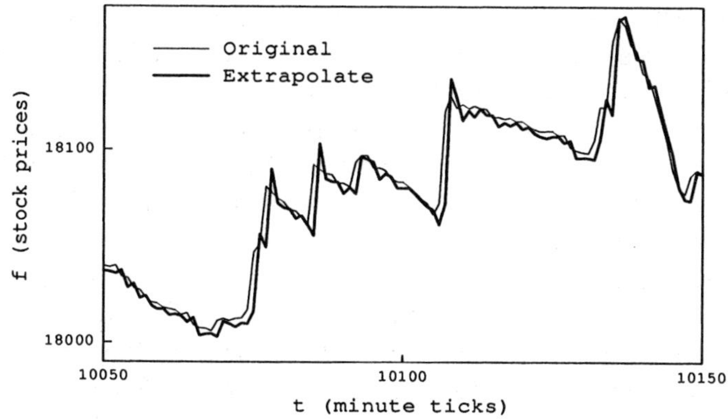

Figure 5.7. Extraplation by the best (original) polynomial from STROGANOFF.

from STROGANOFF, however, shows a higher $Profit = 14,714$. From an economic perspective the computed $Profits$ from the best solutions discovered by the two GP systems are not satisfactory. We are inclined to think that the two GP systems, and especially the traditional SGPC, discover overfitting models with very low MSE when trained with the original financial series. This strengthens the belief that preprocessing of the given financial data series may be envisioned a mean that could help to achieve more successful inductive learning of predictive models.

Table 5.2. Estimates of the Best Models Learned from the Original Series

	Accuracy	*Prediction*	
	(training)	(testing)	
	MSE	HIT	$Profit$
STROGANOFF	$1.39e-04$	54.02%	14,714
SGPC	$1.92e-06$	55.23%	11,230

Table 5.3. Estimates of the Best Models Learned from the Integral Series

	Accuracy	*Prediction*	
	(training)	(testing)	
	MSE	HIT	$Profit$
STROGANOFF	$4.09e-05$	53.77%	11,293
SGPC	$1.90e-06$	62.18%	28,987

Training with the Integral Series. The results derived after moving average filtering of the financial series were the best that were attained by SGPC on all series. Table 5.3 shows the accuracy of the best SGPC expression and relates it to the best polynomial found by STROGANOFF using the integral series.

The observation that SGPC evolves its best solution in mean squared error sense could be attributed to the fact that the integral series curvature is slightly smoother than that of the original series. The best functional expression found by SGPC has better MSE, HITs, and $Profit$ qualities than all of its best expressions over the other series. The best result from STROGANOFF on the integral series is its worst achievement. The interpolation capacity and the extrapolation potential of the best polynomial from STROGANOFF are given in Figures 5.8 and 5.9.

Unfortunately, the capacity to evolve a very accurate solution does not guarantee that this solution will be the best from an economic point of view. One observes that the best SGPC functional expression has highest $Profit$ than all other results achieved by SGPC, but not compared to the results from STROGANOFF (see also Tables 5.4 and 5.5). Concerning the undetected correlation between the MSE, the HITs and the $Profit$, we know from the theory of signal processing that moving average filtering can not protect well the inherent signal dynamics. With respect to the desired economic characteristics, such reasoning explains why the integral transformations did not help STROGANOFF to discover polynomials that yield optimal hits on unseen future data.

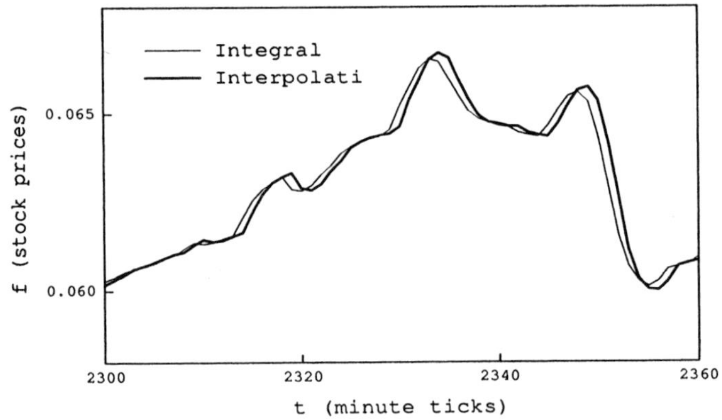

Figure 5.8. Interpolation by the best (integral) polynomial from STROGANOFF.

Training with the Differential Series. The highest number of HITs 74.66% achieved in all runs by the two GP systems exhibits the best polynomial from STROGANOFF on the differential series. Table 5.4 reveals that the accuracy in HITs and predictability in $Profit$s of the best STROGANOFF polynomial are clearly better than these of the best functional expression found by SGPC. While the $HITs$ are the best,

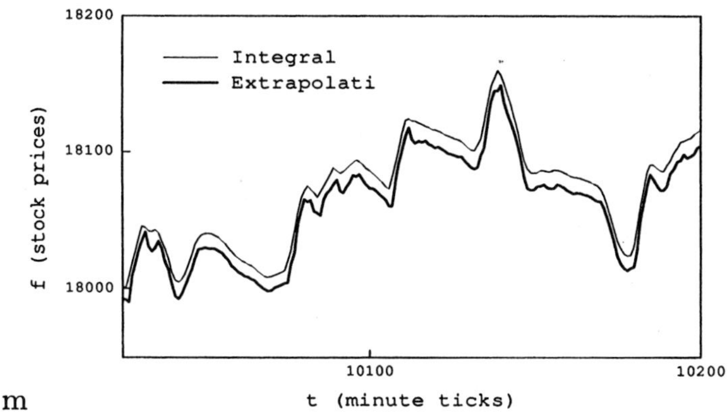

m

Figure 5.9. Extrapolation by the best (integral) polynomial from STROGANOFF.

the $Profit = 54,901$ is the second result after the one $Profit = 64,119$ attained on the rational series (Table 5.5).

Our differential transformation formula (5.8) with averaging over the three recent price data is suitable for identification of short and medium trends. If the goal is to perform long term forecasting the averaging should be adjusted to use more recent data, depending on the desired prediction interval (Deboeck and Cader 1994). The differential transformation allows to filter out noise and avoid close fitting of the series magnitudes which are not essential to forecasting. Figure 5.10 and 5.11 demonstrate the approximation abilities of the best polynomial from STROGANOFF found on the differential series.

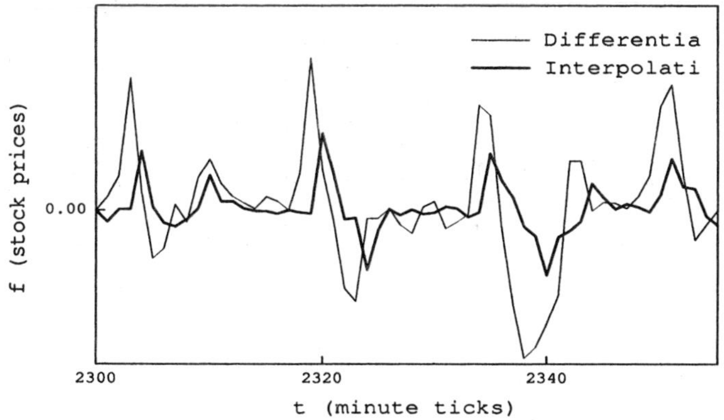

Figure 5.10. Interpolation by the best (differential) polynomial from STROGANOFF.

The results presented in Table 5.4 confirm that both GP systems are sensitive to such differential series transformations. One can see that

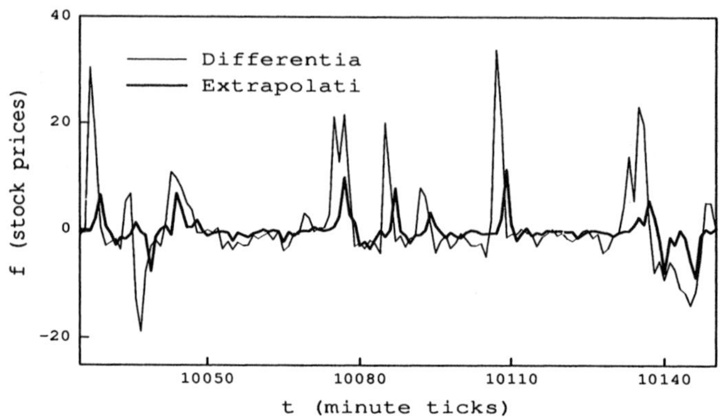

Figure 5.11. Extrapolation by the best (differential) polynomial from STROGANOFF.

SGPC evolved its worst expression on the differential series. Table 5.4 demonstrates that the polynomial accuracy on the training data MSE $=7.86e-05$ is not the best achievement of STROGANOFF since its best accuracy $MSE=4.09e-05$ is attained on the integral series. A highest degree of fitting the training data series, however, does not imply highest level of generalization. Overall, this brief analysis of the attained results by the studied GP systems allows us to judge that the evolved polynomials are better models of the differential series.

Table 5.4. Estimates of the Best Models Learned from the Differential Series

	Accuracy	*Prediction*	
	(training)	(testing)	
	MSE	HIT	$Profit$
STROGANOFF	$7.86e-05$	74.66%	54,901
SGPC	$1.94e-06$	49.94%	2,078

Table 5.5. Estimates of the Best Models Learned from the Rational Series

	Accuracy	*Prediction*	
	(training)	(testing)	
	MSE	HIT	$Profit$
STROGANOFF	$1.62e-03$	68.33%	64,119
SGPC	$2.37e-04$	52.76%	9,359

Training with the Rational Series. The conducted experiments with both GP systems revealed that the rational series is much more difficult to learn than the differential series. This can be attributed to the slightly higher frequency oscillations of the rational series curve compared to the those of the differential series. The rational series curve exhibits more sporadic characteristics, in the sense that it has more frequent occurrences of sharp spikes with higher amplitudes than those of the differential curve (Figure 5.5). Such sharp spikes do not lend themselves easily to modeling, they hinder the learning process.

The accuracy and prediction estimates of the best models evolved by the two GP systems are given in Table 5.5. The approximation quality of the best polynomial discovered by STROGANOFF in two arbitrary intervals is illustrated in Figures 5.12 and 5.13. This best polynomial produces the highest $Profit = 64,119$ compared to the best polynomials on the other transformed series.

We find that the differential and rational transformations really enable the system STROGANOFF to achieve most profitable results from an economic point of view, which is not the case for SGPC. Although the differential and rational transformations produce series with sporadic characteristics, that is they lead to series with much more frequent oscillations, such series seem to be more amenable to polynomial modeling and learning by GP. These two series transformations feature irregular characteristics that are difficult to capture by the functional expressions evolved by SGPC. Therefore, when the economical criteria are the objective, the system STROGANOFF should be preferred.

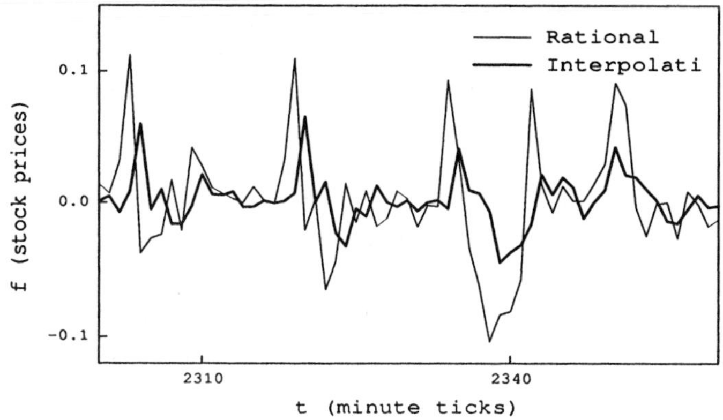

Figure 5.12. Interpolation by the best (rational) polynomial from STROGANOFF.

5. Discussion

The presented experimental results allow us to summarize that:

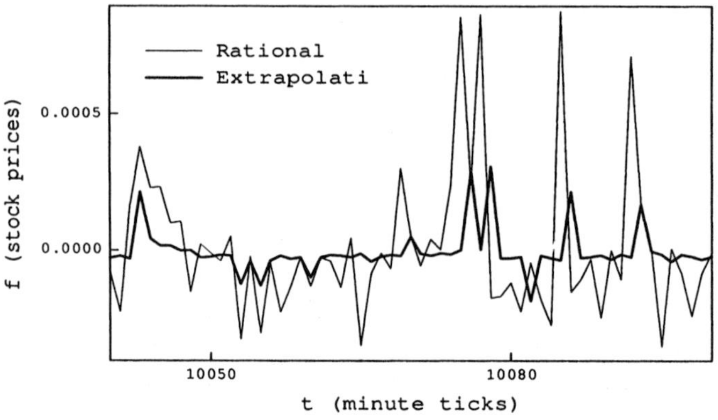

Figure 5.13. Extrapolation by the best (rational) polynomial from STROGANOFF.

1) the employment of the mean squared error MSE (5.10) only is not sufficient to decide which financial series model to select. Additional estimates of the model qualities that account not only for the relative fitting, but also for the number and for the magnitudes of the model coefficients are necessary. Such composite criteria are necessary in order to direct the search process toward highly predictive solutions;

2) the simple moving average data transformations may not help the GP of polynomials to evolve profitable polynomial models from given price movements, while they are helpful for GP systems with functional expression models like these evolved by SGPC. In this sense, one may reason that SGPC could evolve functional expressions that are more useful for modeling continuous patterns in time series;

3) the given financial series after preprocessing by differential and rational transformations are amenable to good approximation by genetically programmed polynomials. The discovered polynomials capture well the directional changes in the series, which confirms the advise of other researchers (Chen and Lu 1999) that such detrended transformations should be expected to enable learning of the economically important changes in the series;

4) we found that the GP of polynomials may be most beneficial for financial price modeling of differentially preprocessed series;

5) the defined fitness function enables to find better forecasting solutions than these in previous studies (Iba and Sasaki 1999) employing directly the mean squared error as a fitness function.

6. Conclusion

This paper presented empirical evidence, which indicate that GP of polynomials seems to be a promising paradigm to search for non-linear dependencies in financial data sequences. The relevance of the GP was illustrated on a series of stock market average prices. The results concern specifically the quality of polynomials evolved by GP, and should be expected to be valid for high-order multivariate polynomials only. We limited ourselves to a few basic transformations, but additional research will be performed on combinations of various data transformations through hybrid embedding techniques.

Further research is oriented toward tuning and extending the system STROGANOFF for other financial applications, like: option pricing, risk estimation, foreign currency exchange rates, etc., upon the assumption that the historical trends are correlated with the retrospective indicators from technical analysis point of view.

References

Allen, F. and R. Karjalainen (1999). "Using Genetic Algorithms to Find Technical Trading Rules," *Journal of Financial Economics*, 51(2), 245–271.

Bhattacharyya, S., O. Pictet, and G. Zumbach (1998). "Representational Semantics for Genetic Programming Based Learning in High-Frequency Financial Data," in J. R. Koza et al. (eds.), *Genetic Programming 1998: Proceedings of the Second Annual Conference of GP' 98*, 11–16. CA: Morgan Kaufmann.

Chen, S.-H. and C.-F. Lu (1999). "Would Evolutionary Computation Help in the Design of ANNs in Forecasting Foreign Exchange Rates?" in *Proceedings of the 1999 Congress on Evolutionary Computation, CEC99*, 267–273. Piscataway, NJ: IEEE Press.

Chen, S.-H. and C.-C. Ni (1998). "Evolutionary Artificial Neural Networks and Genetic Programming: A Comparative Study on Financial Data," in G. D. Smith, N. C. Steele, and R. F. Albrecht (eds.), *Artificial Neural Networks and Genetic Algorithms*, 397–400. Wien: Springer-Verlag.

Chen, S.-H. and C.-H. Yeh (1998). "Option Pricing with Genetic Programming," in J. R. Koza et al. (eds.), *Genetic Programming 1998: Proceedings of the Second Annual Conference of GP'98*, 32–37. CA: Morgan Kaufmann.

Colin, A. M. (1994). "Genetic Algorithms for Financial Modeling," in G. J. Deboeck (ed.), *Trading on the Edge. Neural, Genetic and Fuzzy Systems for Chaotic Financial Markets*, 148–173. New York: John Wiley & Sons.

Deboeck, G. J. and M. Cader (1994). "Pre and Postprocessing of Financial Data," in G. J. Deboeck (ed.), *Trading on the Edge. Neural,*

Genetic and Fuzzy Systems for Chaotic Financial Markets, 27–44. New York: John Wiley & Sons.

Iba, H. and T. Sasaki (1999). "Using Genetic Programming to Predict Financial Data," in *Proceedings of the 1999 Congress on Evolutionary Computation, CEC99*, 244–251. Piscataway, NJ: IEEE Press.

Iba, H., T. Sato, and H. de Garis (1994). "System Identification Approach to Genetic Programming," in *Proceedings of the First IEEE Conference on Evolutionary Computation*, 1, 401–406. Piscataway, NJ: IEEE Press.

Ivakhnenko, A. G. (1971). "Polynomial Theory of Complex Systems," *IEEE Transactions on Systems, Man, and Cybernetics*, 1(4), 364–378.

Koza, J. R. (1992). *Genetic Programming: On the Programming of Computers by Means of Natural Selection.* Cambridge, MA: MIT Press.

Koza, J. R. (1995). "Genetic Programming for Economic Modeling," in S. Goonatilaje and P. Treleaven (eds.), *Intelligent Systems for Finance and Business*, 251–269. London: John Wiley & Sons.

Lee, G. Y. (1999). "Genetic Recursive Regression for Modeling and Forecasting Real-World Chaotic Time Series," in L. Spector, W. B.Langdon, U.-M. O'Reilly and P. J. Angeline (eds.), *Advances in Genetic Programming*, III, 401–423. Cambridge, MA: MIT Press.

Makridakis, S. and S. C. Wheelwright (eds., 1987). *The Handbook of Forecasting.* New York: John Wiley & Sons.

Nikolaev, N. and H. Iba (2001). "Regularization Approach to Inductive Genetic Programming," *IEEE Transactions on Evolutionary Computation*, 5(4), 359–375.

Press, W. H., B. P. Flannery, S. A. Teukolski, and W. T. Vetterling (1992). *Numerical Recipes in C: The Art of Scientific Computing*, 2nd ed. Cambridge, England: Cambridge University Press.

Refenes, A.-P., Y. Abu-Mostafa, J. Moody, and A. Weigend (1996). *Neural Networks in Financial Engineering: Proceedings of the Third International Conference on Neural Networks in the Capital Markets.* London: World Scientific.

Santini, M. and A. Tettamanzi (2001). "Genetic Programming for Financial Time Series Prediction," in P. L. Lanzi et al. (eds.), *Proceedings of the 4th European Conference on Genetic Programming, EuroGP*, LNCS 2038, 361–370. Berlin: Springer.

Tackett, W. A. and A. Carmi (1994). "The Donut Problem: Scalability and Generalization in Genetic Programming," in K. E. Kinnear Jr. (ed.), *Advances in Genetic Programming*, 1, 143–176. Cambridge: MIT Press.

Takens, F. (1981). "Detecting Strange Attractors in Turbulence," in D. A. Rand and L.-S. Young (eds), *Dynamical Systems and Turbulence*, Lecture Notes in Mathematics, 898, 366–381. Berlin: Springer-Verlag.

Wahba, G. (1990). *Spline Models for Observational Data.* CBMS-NSF Regional Confererce Series 59. Philadelphia: SIAM Press.

Zapranis, A. and A.-P. Refenes (1999). *Principles of Neural Model Selection, Identification and Adequacy: With Applications to Financial Econometrics.* London: Springer-Verlag.

Zumbach, G., O. V. Pictet, and O. Masutti (2001). "Genetic Programming with Syntactic Restrictions Applied to Financial Volatility Forecasting," *Technical Report GOZ.2000-07-28.* Olsen Associates.

Chapter 6

NXCS: HYBRID APPROACH TO STOCK INDEXES FORECASTING

Giuliano Armano
DIEE, University of Cagliari
Piazza d'Armi, I-09123, Cagliari, Italy
armano@diee.unica.it

Michele Marchesi
DIEE, University of Cagliari
Piazza d'Armi, I-09123, Cagliari, Italy
michele@diee.unica.it

Andrea Murru
DIEE, University of Cagliari
Piazza d'Armi, I-09123, Cagliari, Italy
andreamurru@diee.unica.it

Abstract In this chapter, a hybrid approach for stock market forecasting is presented. It allows to develop a mixture of hybrid experts, each consisting of a genetic classifier and an associated artificial neural network. The resulting experts have been applied to stock market forecasting using technical trading rules as genetic inputs and other inputs—in particular past quotations—for the neural networks. In particular, the former are used to find quasi-stationary regimes within the financial data series, whereas the latter are assigned the task of making "context-dependent" predictions on the next day trend of the market. To this end, a novel kind of feedforward artificial neural network has been defined, allowing to implement suitable predictors without being compelled to exploit more complex neural architectures. Test runs have been performed on some well-known stock market indexes, also taking into account trading commissions. The tests pointed to the good forecasting capability of the proposed approach, which repeatedly outperformed the buy-and-hold strategy.

Keywords: Stock Market Forecasting, Time Series Prediction, Genetic Algorithms, eXtended Classifier Systems(XCSs), Artificial Neural Networks

Introduction

Stock market dynamics is characterized by a number of time series showing—usually over a broad time span (from years to decades)—the evolution of certain relevant information, e.g. stock indexes. These time series exhibit several features that make the prediction task a very difficult one. In fact, several studies claim that the serial correlation of financial time series is economically and statistically insignificant (e.g., Hawawini and Keim 1995). These studies seem to confirm the Efficient Market Hypothesis (EMH, Fama 1965), which stresses the ability of the market to rapidly assimilate any exogenous perturbation, so that every any imbalance is quickly discovered and counteracted by suitable changes on stock market prices. As a consequence, one may assume that stock market prices always reflect all information available to traders. On the other hand, several studies have doubt the soundness of the EMH assessing and documenting the predictability of financial time series (e.g., LeBaron 1991).

Considering the multiplicity of arguments in favor and against the EMH, one may conclude that the controversy actually lies in the delay required by the stock market to re-establish a new equilibrium on stock prices after a new public information has been made available to traders. In any case, most stock market investors and academics seem convinced that they can (at least statistically) predict stock price trends. Thus, many attempts have been made by both sides to model financial markets with a view to making effective predictions.

Unfortunately, it is particularly difficult to find suitable models that perform well over a long period of time. In fact, the number and type of input and state variables characterizing the model are unknown, or at least difficult to identify, the dependence on inputs can be highly non linear, and the model itself can be time-varying on different time spans. Furthermore, these aspects are strongly interrelated: economic variables often exhibit some type of correlation only under particular conditions (for example, stock quotations appear to be highly correlated in presence of large price variations); otherwise they are—or seem to be—independent (Salmon 2000). Accordingly, the corresponding time data series exhibit an apparent non-stationary behavior, in particular when stock markets move in and out of periods of market turbulence.

To deal with these intrinsic difficulties, human experts usually try to identify the current status / trend of the market by exploiting technical or fundamental analysis rules, and then take suitable actions according to some "immutable" or "ad-hoc" strategies, also taking into account a limited window of past quotations. Immutable strategies are characterized by rules with global scope and static extent, to be applied as soon as their "firing" conditions occur (e.g., buying when the current trend is bullish and selling when it is bearish). Conversely, ad-hoc strategies are characterized by rules with local scope and dynamic extent, i.e., whose validity holds only within a limited period of time and that are continuously checked and adapted to better fit the available data. Notably, human experts' decisions depend more on "ad-hoc" strategies than on "immutable" ones. This behavior seems to confirm the hypothesis that, assuming that a model exists able to capture the stock market dynamics, it is surely time-dependent.

In this chapter, a novel approach to stock market prediction, which synergistically exploits the potential of genetic and neural technologies to forecast the next day trend of the market, is presented. The overall system results in a mixture of hybrid experts, explicitly customized for financial time series prediction, each consisting of a neural predictor and an associated genetic classifier. Given an input, every neural predictor is allowed to take place in the decision process only if the current status / trend of the market enables its genetic classifier. It is worth pointing out that any hybrid expert contributes to the decision process in a way that mimics a human expert, due to the fact that its genetic classifier act as a "compound" technical analysis rule able to identify the stock market status / trend where the expert is considered more reliable. Furthermore, the reliability of each expert is updated according to its prediction capability, thus implementing a dynamic adaptation to the given environment.

Experiments have been carried out taking into account the daily quotations of three stock market general indexes, i.e., COMIT, S&P500, and NASDAQ. In particular, the module devised to make predictions has been trained and tuned (for each stock index) on about about 50% of the available data, being very careful to avoid any form of "data snooping" and considering also realistic transaction costs. The system was then tested, leaving its overall configuration unchanged. We compared the system's forecasting capabilities with the buy-and-hold (B&H) strategy; results are encouraging, demonstrating the validity of the approach.

The remainder of this chapter is organized as follows: in section 1 a short introduction to the state of the art in stock market forecasting with AI Techniques is given; in section 2 the proposed hybrid approach is described; in section 3 experimental results are discussed; in section 4 the conclusions are drawn.

1. Related Work

Advances in both analytical and computational methods have led to a number of interesting approaches to financial time series forecasting; e.g., linear auto-regressive models, principal component analysis, neural networks, genetic algorithms, and others (as starting points, see for examples, Hellström and Holmström 1997, Deboeck 1993, Hamilton 1994, Hardle et al. 1997, Kantz and Schreiber 1997).

Notwithstanding the multiplicity of existing proposals, in this section great emphasis is dedicated to connectionist and genetic approaches, as the proposed hybrid approach originates from them. Furthermore, the problem of how to identify quasi-stationary regimes is addressed; being—in our opinion—the most promising approach that allows coping with financial time series prediction without having to rely on an undifferentiated global model.

1.1. Neural Networks and Time-Series Prediction

Artificial Neural Networks (ANNs) have been widely used to deal with stock markets forecasting (Refenes et al. 1997; Refenes et al. 1997). In fact, they can identify highly non-linear models, have effective learning algorithms, handle noisy data, and use inputs of different kinds. Furthermore, systems based on complex non-linear models (e.g., exponential GARCH models,[1] Nelson 1991) show similar results, in terms of out-of-sample prediction performance, to those based on Multi-Layer Perceptron (MLP) architectures (Campbell et al. 1997).

A major weakness of MLP, from the point of view of time series forecasting, is the absence of an internal memory, making it difficult to define architectures able to capture the dynamics of time-varying data series on large observational windows. Even if standard MLPs can still be used (Castiglione 2001), a variety of more complex architecture with some form of internal memory has been proposed (see Campolucci et al. 1999 for a survey); in particular, recurrent ANNs (RANNs) have also proved to be effective in predicting financial data (Giles et al. 1997; Bengio and Gingras 1998).

Conceptually, using RANNs one may try to identify a global model able to capture most of the underlying process dynamics. However, from a practical point of view, several difficulties arise. In particular, as there is no guarantee that the network has been trained on all possible regimes, local overfitting is very likely to occur, thus giving rise to a model unable to make predictions on different regimes (LeBaron and Weigend 1997). In other words, RANNs trained on data belonging to a specific period perform reasonably well only if the test period is relatively similar. Note also that the similarity between two periods of time is typically guaranteed by sharing some kind of economic condition (e.g., bullish or bearish period), instead of being caused by temporal contiguity.

1.2. Genetic Algorithms and Time-Series Prediction

Genetic Algorithms (GAs) (Holland 1976; Goldberg 1989) are a family of computational models inspired by evolution. In a broader usage of the term, a GA architecture uses selection and recombination operators to handle a population of "chromosomes", each representing a potential solution—in form of a binary string—of a target problem. Usually, these chromosomes are randomly created, and undergo reproductive opportunities in such a way that better solutions are given more chances to reproduce than poorer ones.

Although GAs have been adopted in a multitude of different tasks, in this subsection we only recall some proposals that address the problem of financial time-series forecasting. Noever and Baskaran (1994) investigated the task of predicting trends and prices in financial time series, making experiments on the S&P500 stock market. Mahfoud and Mani (1996) addressed the general problem of predicting future performances of individual stocks. Their work is particularly relevant, as they make a comparison between GAs and ANNs applied to financial forecasting. According to their experiments—repeated on several stock markets—both approaches outperform the B&H strategy. A combined approach, obtained by averaging out GAs and ANNs outputs, is also experimented with positive results. GAs have also been used in a variety of hybrid approaches to financial time series prediction. For example, Muhammad and King (1997) devised evolutionary fuzzy networks to forecast the foreign exchange market, whereas Kai and Wenhua (1997) exploited GAs to train ANNs for predicting a stock price index.

1.3. The Problem of Regime Identification

Conventional time series are usually identified by adopting linear or non-linear global models, whose underlying process is considered stationary. Unfortunately, financial time series can hardly be thought of as being stationary, since predictions depend at least on the particular status/trend that holds while performing forecasting. Assuming that some form of non-stationarity has to be taken into account, conceptually a single global model could still be used to capture the whole process dynamics. Nevertheless, the behavior of financial time series can be better investigated by assuming that the underlying process exhibits the so-called piecewise stationarity, or multi-stationarity, in which one assumes that local stationarities (i.e., different regimes) hold, and that a rapid regime shifting typically occurs according to some exogenous events.

Within the time series community, several proposals have been made to overcome the problem of finding a segmentation of the input space that allows one to assume quasi-stationarity on a local basis. Below some relevant approaches devised to handle this problem are recalled. Tong and Lim (1980) introduced the Threshold AutoRegressive (TAR) model, a piecewise linear process whose central idea is to change the parameters of a linear AR model according to the value of an observable variable, called the threshold variable. If this variable is a lagged value of the time series, the model is called a Self-Exciting TAR (SETAR) model (Tsay 1989). Friedman (1991) devised the Multivariate Adaptive Regression Splines (MARS) model, which is able to represent process dynamics in the form of product spline basis functions. The number of basis functions and their parameters (product degree and knot locations) are automatically determined by the data. Lewis (1994) used MARS models to predict financial data, claiming that they are able to identify the underlying process better than classical AR models.

2. A Hybrid Approach for Stock Market Forecasting

In this section, we describe in detail the proposed hybrid approach for dealing with stock market forecasting, which lies in between a genetic and a neural approach. Firstly the "compliance" of the approach with reasonable hypotheses about the underlying process, as well as with the behavior of human experts, is discussed. Then, the corresponding system is described, together with all customizations devised to adapt it to financial time series prediction.

2.1. Regime Shifting and Human Experts' Behavior

As our objective is to perform experiments according to rather general hypotheses, we assume the stock market dynamics to be characterized by a multi-stationary process, with a rapid regime switching.[2] This assumption is also supported by the typical behavior observed in human experts, in particular when the amount of the corresponding investment is relevant. In fact, they first try to identify the current status/trend of the market, usually by means of suitable technical or fundamental analysis rules (Achelis 1995), and then apply locally-sound strategies to decide what action to take.

In our opinion, using a mixture of experts would be the most natural way of building an automatic procedure that takes into account the characteristics of the underlying model, as well as the will of reproducing the human experts' behavior. After briefly describing the mixtures of experts from an historical perspective, we will discuss the proposed approach as an implementation of these underlying concepts.

2.2. Using a Mixture of Experts to Identify Quasi-Stationary Regimes

Most of the early learning algorithms (e.g., CART (Breiman et al. 1984), CN2 (Clark and Niblett 1989), ID3 (Quinlan 1986) and its offspring C4.5 (Quinlan 1993)) apply the divide-and-conquer principle by recursively partitioning the input space until regions of roughly constant class membership are obtained. These algorithms yield a monolithic result by enforcing some reasonable heuristics for controlling the complexity of the search.

All these algorithms can also be reinterpreted in terms of multiple experts, despite the fact that this new perspective was not sufficiently well understood at the time of their development. In fact, any of the above algorithms can be seen as a tool for producing a set of "local experts", whose activation is mutually exclusive, due to the hard-partitioning mechanism enforced on the input space; i.e., for each region, only one expert is entrusted with making classification or prediction.

The k-Nearest Neighbor (k-NN) classification rule, a prototype-oriented technique for pattern classification (Cover and Hart 1967; Friedman et al. 1975), was another interesting proposal where the concept of multiple experts was implicitly adopted. Here, several prototypes usually hold for a given input space, each one being the "center" of a cluster of input vectors that share the same class label. Thus, in a sense, prototypes can be considered experts, each being able to measure how close an input vector is to it—according to a given distance metrics, which is usually domain-dependent. Performing a k-NN classification on an input vector consists of first selecting the k prototypes that get the best distance score,[3] and then enforcing a voting policy, usually a weighted majority rule.

The concept of multiple experts was explicitly introduced in the connectionist community by Jordan and Jacobs (Jacobs et al. 1991; Jordan and Jacobs 1992; Jordan and Jacobs 1994), whose mixtures of experts concur to produce the final output by means of suitable gating networks used to enforce a soft-partitioning mechanism able to guarantee weights normalization while blending experts outputs. Here, multiple experts are involved in the overall classification or prediction, due to the fact that regions may overlap, due to the adoption of soft- (instead of hard-) boundaries. Both experts and gating networks are implemented through linear models, the only non-linear mechanism being the one devised to perform outputs blending (i.e., softmax, McCullagh and Nelder 1989).

Starting from the work of Jordan and Jacobs, Weigend et al. devised non-linear gated experts (Weigend, et al. 1995; Weigend 1996), and applied them to several time series domains, including financial ones. The key features of these experts are the non-linearity of experts and gating networks, and the experts' ability to separately adapt to noisy data.

Our proposal falls into the category of mixtures of experts, although significant differences do exist compared to previous approaches. In particular, each expert of the proposed system is represented by a genetic classifier, together with a neural predictor. The genetic classifier has the task of identifying the regime where the neural predictor is assumed to work properly; hence it acts like a "guard" that—given an input—allows the neural predictor to be activated (or prevents it from being activated).

Before going into further detail, let us first briefly describe the framework (i.e., guarded experts) that encompasses our hybrid system.

2.3. Defining the Guarded Experts' Framework

Given an N-dimensional input space **I** and an output space **A**, a *guarded expert* $\tilde{\Gamma}$ is a partial function that maps a subset of **I** (say $I_g \subseteq I$) to **A**. Let us point out that **A** can be used to alternatively denote a set of class labels (for classification tasks) or a range in $\Re$ (typically, for prediction tasks). Being g the characteristic function associated with the set $I_g \subseteq I$ (i.e., $x \in I_g \Leftrightarrow g(x) = true$; *false* otherwise), a total function $\tilde{h} : I_g \to A$ can be defined that maps every $x \in I_g$ to a corresponding value in **A**, such that $x \in I_g \Rightarrow \tilde{h}(x) = \tilde{\Gamma}(x)$. Hence, the default underling semantics associated to a guarded expert $\tilde{\Gamma}$ applied to an input $x \in I$ is:

$$\tilde{\Gamma}(x) = if\ g(x)\ then\ \tilde{h}(x)\ else\ \bot \tag{6.1}$$

In other words, g acts as a *guard*, which, depending on the matching performed on the current input, controls the activation of the corresponding classifier or predictor, represented by the function $\tilde{h}$. For the sake of clarity, the notation:

$$\tilde{\Gamma} = \left\langle g, \tilde{h} \right\rangle \tag{6.2}$$

can be adopted to denote a guarded expert, where its corresponding guard and classifier / predictor are made explicit. Note that g is responsible for enforcing any given selection policy used to identify the context where $\tilde{h}$ can be properly used, whereas (if enabled) $\tilde{h}$ evaluates the correct response of the guarded expert according to the current input. In the above definition of guarded experts, no hypothesis is made about which part of the input x is actually used by g and $\tilde{h}$. Conceptually, g and $\tilde{h}$ can exploit any subset of the input space's features. It is worth pointing out, though, that a distinction between the inputs used for experts' selection (g) and for output evaluation ($\tilde{h}$) can be made without loss of generality. Hence, assuming the actual inputs of guards and classifiers / predictors to be x_1 and x_2, respectively, the following alternative semantics holds:

$$\tilde{\Gamma}(x) = if\ g(x_1)\ then\ \tilde{h}(x_2)\ else\ \bot \tag{6.3}$$

where $x = (x_1, x_2) \in I$.

When several guarded experts are put together, they form a population whose semantics is completely specified only when suitable implementation choices have been made about the underlying environment. To this end, specific (i) training strategies, (ii) region splitting and experts selection mechanisms, as well as (iii) voting policies or outputs blending mechanisms must be enforced.

As for training strategies, basically they can be either *batch* or *on-line*. The former are characterized by the fact that training is performed as an off-line activity, so that the overall population characteristics remain unchanged while the system is working on-line. The latter interleave learning and classification / prediction; i.e., in this case, a population of guarded classifiers / predictors is continuously updated according to the inputs it receives. For the sake of brevity, the issue of training strategies, which would actually require a separate chapter, is not addressed here. Let us simply recall that the typical training strategies for ANNs and GAs are error backpropagation and evolutionary selection schemes, respectively.

As for region splitting, let us point out that soft partitioning implies the presence of multiple experts, although the converse is not always true; in fact, there are systems where multiple experts are associated with hard region boundaries (e.g., GAs).

As for selection, several mechanisms have been devised in the literature; let us just recall, for the sake of brevity, those based on distance metrics (e.g., for k-NNs) and on binary conditions (e.g., for GAs). Note, anyway, that there are systems where all experts—at least in principle—take part in the voting activity, thus making experts' selection useless (e.g., Jacobs and Jordan mixtures of experts, and Weigend's non-linear gated experts).

As for voting, several policies have been proposed, able to evaluate the actual output according to the selected experts. Of course, a voting is required only when the *match set* has cardinality greater than 1 (i.e., multiple experts may be selected according to the given input). Basically, the most successful voting policies are: a) the weighted majority rule, which applies only to classifier systems, b) the weighted averaging rule, which applies to both classifiers and predictors.[4] In its simplest form, the weighted majority rule selects the class label that achieves the best score (relative majority), obtained by taking into account the supporting experts. A policy that requires the absolute majority can be adopted too. In this case, an input vector that achieves a best score $<$ 50% would be labeled as "unknown". The weighted averaging rule is typically adopted on classifier/prediction systems characterized by multiple experts that allow soft region boundaries; in this case, an outputs blending mechanism is actually required. In either case, each expert must have an associated weight (which usually gives information about its expected accuracy)—no matter whether a voting policy or outputs blending must be enforced.

It is now clear that a population of guarded experts is characterized by design choices aimed at ensuring a good trade-off between the generality of the system to be implemented and the need for satisfying the constraints imposed by the selected domain and goal. The next section illustrates all the significant choices made when trying to customize the guarded experts' framework to the task of financial time series prediction.

2.4. Implementing Guarded Experts

For modeling the stock market as a multi-stationary process, we adopted an on-line training strategy for implementing guarded experts, thus ensuring a highly reactive and adaptive capability of the resulting system. This strategy enforces an evolutionary mechanism, able to update the expected accuracy of each single expert according to its adaptation to the underlying environment.

Although the final output of the overall predictor could also be represented by a class label, such as taking a long or a short position, we decided to explicitly attempt to predict the next value of the time series, leaving to a strategic module the task of deciding what position to take. Our guarded experts have been specifically tailored for time series forecasting by implementing g and $\tilde{h}$ with a GA classifier and an ANN, respectively. In particular, ANNs are used to make predictions on a subset of the input sequence, whereas GA classifiers detect the occurrence of a regime similar to the one the ANN has been trained on. In this way, the potential of both GAs and ANNs are synergistically exploited to forecast the next day trend of the market. As Wilson's eXtended Classifier Systems (XCSs) (Wilson 1995) have actually been adopted to implement guards, we have called the corresponding framework NXCS, standing for Neural XCS. To the best of our knowledge, no previous work has been done on hybrid systems that use both XCSs and ANNs to perform time series prediction.

As for XCSs, they were devised based on Holland's classifiers systems (Holland 1986), which in turn are strictly related to the reinforcement-learning paradigm (Sutton and Barto 1988). XCSs retain most of the concepts illustrated in the classifier systems proposal, though with restricted expressive power. Since Wilson's initial work, several related studies have provided a deep theoretical understanding of the XCS approach (Kovacs 1996; Lanzi 1997; Wilson 1998; Kovacs 1999), and many extensions have been proposed (see, for example, Lanzi 1998, Lanzi and Perrucci 1999, and Wilson 1999).

An XCS architecture consists of a performance, reinforcement, and discovery component. In XCS terminology, inputs—represented by bit strings of fixed length—are named *messages*, and outputs *actions*. The system handles a population of classifiers; each represented by a condition-action pair, together with some parameters that characterize the internal state of each classifier. Conditions are represented by a ternary string of bits (i.e., "0" / *false*, "1" / *true*, "#" / *don't-care*), and have the same length as incoming messages. Only when a message is matched by a classifier's condition, is the corresponding action selected (see Appendix **A** for a brief introduction to XCSs).

It is worth pointing out that an NXCS expert can be described from both an evolutionary and a neural perspective. According to the former interpretation, it can be thought of as an extension of an XCS classifier, whose action part has been replaced with a suitable ANN. According to the latter interpretation, an NXCS can be considered an extension of a neural classifier/predictor, equipped with an additional guard that prevents the ANN from working outside the "context" identified by the guard itself. Thus, NXCSs are particularly suited to perform a synergistic integration between two sources of information (i.e., binary and numerical), by adopting—in our opinion—the most suitable, state-of-the-art, technology for each one. Notwithstanding its twofold nature (namely classifier or predictor depending on the application task), for the sake of simplicity, in the following we assume that NXCS experts be used only for prediction tasks.

2.5. Dealing with a Population of NXCS Experts

An NXCS is an evolutionary system whereby a population of experts, each characterized by a genetic guard and corresponding neural predictor, is raised in a typical XCS-like environment. In fact, the XCS architecture has been widely reused to train the NXCS population, and to split the input space for regime identification. In fact, NXCSs are characterized by a on-line training strategy, together with a hard region splitting (and selection) mechanism handled by genetic guards (thus, allowing a match set with cardinality greater than one).

On the other hand, the typical XCS voting policy (i.e., a weighted majority rule) has been turned—for NXCSs—into an outputs blending mechanism (i.e., a weighted averaging rule). In fact, the most natural way to produce the overall prediction of the next value of the stock market index is to average out the outputs of the selected experts. To give the reader a clearer picture of the underlying choices, Table 6.1 summarizes strategies, mechanisms, and policies for NXCSs, comparing them with other well-known systems.

The remainder of this subsection is devoted to illustrating how a population of NXCS experts is handled, highlighting the major differences that exist with XCSs.

2.5.1 Generating and Maintaining NXCS Experts.

Given an NXCS expert, its genetic guard supports the usual covering, crossover, and mutation operations, whereas its neural predictor is trained once,

Table 6.1. Comparison between the proposed system and other well-known systems, according to the most important implementation features (TS = Training Strategy, RS = Region Splitting, OV=OVerlapping, MSC = Match-Set Cardinality, SM = Selection Mechanism, VP/OB = Voting Policy / Outputs Blending).

System	TS	RS	OV	MSC	SM	VP/OB
CN2	batch	hard	no	=1	boolean rule	no
CART	batch	hard	no	=1	decision-tree path	no
C4.5	batch	hard	no	=1	decision-tree path	no
KNN	batch	soft	-	>1	distance metrics	weighted majority rule
mixtures of experts	batch on-line	soft	-	>1	no	weighted averaging with softmax
non-linear gated experts	batch	soft	-	>1	no	weighted averaging with softmax
GA	on-line	hard	yes	>1	binary condition	majority rule
XCS	on-line	hard	yes	>1	binary condition	weighted majority rule
NXCS	on-line	hard	yes	> 1	binary condition	weighted averaging[5]

immediately after being created, and left unaltered while the system is running.

Each NXCS expert has a set of associated parameters, to be used and updated while evaluating, step-by-step, the evolution of the population. In particular, for each NXCS expert $\tilde{\Gamma}$, let us recall the following parameters: (i) the *prediction* or *strength* p, i.e., the expected reward to the system for using $\tilde{\Gamma}$, (ii) the prediction error ϵ, i.e., the difference between expected and actual reward to the system for using $\tilde{\Gamma}$, and (iii) the *fitness* f, i.e., the degree of adaptation of $\tilde{\Gamma}$ to the environmental constraints. In XCSs, such parameters are all updated according to specific Widrow-Hoff rules (see Appendix **A**). In NXCSs, only the prediction error and fitness update follow this standard policy. As for the prediction parameter, we let it coincide with the *primary output* of the ANN designed to predict the next value of the time series.[6] The decision of enforcing a non-standard way of dealing with this parameter is mainly due to the fact that we are concerned with prediction tasks.

When a new NXCS expert has to be inserted into the population (due to a covering, crossover, or mutation operation), its neural predictor is generated with initial random weights and then trained using a set of examples obtained by selecting the inputs that are matched by its guard on a fixed-length window of past quotations. In this way, each neural predictor is trained on a different set of examples, according to the current time and to the corresponding guard.

2.5.2 NXCS Mechanisms for Experts Selection and Outputs Blending. In an XCS classifier system, experts' selection coincides with match-set formation. An XCS classifier belongs to the match set if the current input vector satisfies its condition. The action set is then formed by selecting those classifiers that support the winning action, evaluated applying the so-called fitness-weighted majority rule (i.e., each classifier contributes to strengthening the supported action according to its current fitness).

On the other hand, in an NXCS used for prediction tasks, the action set is no longer needed, due to the fact that the value to be forecasted actually results from blending the outputs of all predictors that belong to the match set. This is done by enforcing a fitness-weighted averaging rule, i.e., outputs blending is performed according to the fitness of each expert. Thus, given a match set M, the forecasted next value ν is evaluated according to the formula:

$$v = \frac{\sum\limits_{c \in \mathbf{M}} p_c \cdot f_c}{\sum\limits_{c \in \mathbf{M}} f_c} \tag{6.4}$$

where p_c and f_c are the prediction and the fitness of each NXCS expert c belonging to the match set. Note that ν is estimated by performing a weighted averaging on the prediction parameter, due to the fact that—as

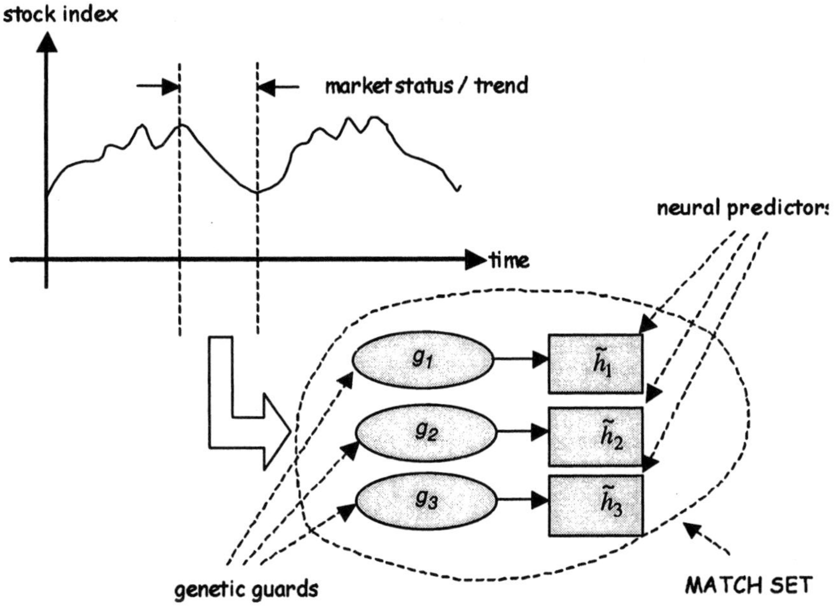

Figure 6.1. Identifying a regime and forming the match set.

already pointed out— p coincides with the primary output of the ANN devised to perform the prediction.

2.6. Customizing NXCS Experts for Stock Market Forecasting

According to the general choice made for NXCS, genetic guards and neural predictors have been supplied with different information, to better exploit their capabilities of dealing with binary and numerical inputs, respectively. As technical analysis rules are commonly used by financial traders to predict transitions between different stock market status/trends, we assumed that their firing conditions could be sensible inputs for the partitioning system, whereas the task of predicting future prices clearly requires information about past quotations. Hence, NXCS experts have been tailored to deal with stock market forecasting in the following way:

- Genetic guards have been entrusted with partitioning the input space while matching binary inputs, each bit representing the firing condition of an existing technical analysis rule. In this way, each guard is actually assigned the role of a "compound" technical analysis rule automatically created by the system;

- Neural predictors have been implemented by defining a novel kind of feedforward ANN, able to handle the short-term dynamics of the process on a weekly base. The ANN consists of an MLP (Rumelhart and McLelland 1986), followed by a module inspired to cascade correlation architectures (Fahlmann and Lebiere 1990).

2.6.1 Embodying Technical Trading Rules into NXCS Guards. Although—in general—the problem of segmenting the input space is still an open issue (see Sun and Peterson (1999) for an interesting discussion on the topic of input space partitioning from a multiple experts perspective), in the proposed system, guards are specifically entrusted with regime identification. Figure 6.1 illustrates the ability to identify the current regime through a suitable selection mechanism enforced by genetic guards.

For incorporating some domain knowledge in form of binary inputs to be processed by NXCS guards, an 8-bits vector has been adopted—each bit being built upon the value of some technical analysis *indicators*. The list of indicators involved in the definition of the binary input is reported in Table 6.2, under the assumption that $q(t)$ is the stock quotation at day t, and that MA (i.e., Mobile Average) is a generic "low-pass" filtering operator defined as:

$$MA_N(x\,(t)) = \frac{1}{N} \cdot \sum_{i=0}^{N-1} x(t-i) \tag{6.5}$$

Table 6.3 shows the binary features representing an input vector to be processed by the guards of NXCS experts. In the proposed system, each expert tests the input features according to the rule encoded within its guard and then (if enabled) participates to the task of predicting the next value of the stock index. In other words, the features tested by the guard of an NXCS expert are only used to identify a suitable "context", in which the corresponding neural predictor can be activated.

2.6.2 Customizing a Feedforward ANN for Stock Market Forecasting. Overfitting training data while performing poorly on test data is one of the major problems of learning algorithms, in particular when noisy data are involved. This problem is particularly relevant for economic time series, which are very noisy and require complex learning algorithms. Overfitting is also strongly related to non-stationarity: data may appear to be noisy when using inadequate models. On the

Table 6.2. Definitions of Technical Analysis Indicators

<table>
<tr><th>Indicator</th><th>Definition</th></tr>
<tr><td>Difference of Averages</td><td>$$DOA_{N1,N2}(t) = MA_{N1}\left[q\right](t) - MA_{N2}\left[q\right](t)$$</td></tr>
<tr><td>Rate of Change</td><td>$$ROC_N(t) = \frac{q(t) - q(t-N)}{q(t-N)}$$</td></tr>
<tr><td>Relative Strength Index</td><td>$$RSI_N(t) = \left(1 + \frac{MA^+_N\left[neg\right](t)}{MA^+_N\left[pos\right](t)}\right)^{-1}$$
where:
$$pos(t) = \max\left(0, \frac{q(t) - q(t-1)}{q(t)}\right)$$
$$neg(t) = \max\left(0, \frac{q(t-1) - q(t)}{q(t)}\right)$$
MA^+ differs from MA in the fact that it disregards non-positive values (thus, only positive values in the *pos* and *neg* sequences are taken into account).</td></tr>
<tr><td>Convexity</td><td>$$CV_N(t) = \sum_{i=1}^{N} q(t - N + i) - r(t - N + i)$$
where r is the line connecting $q(t-N)$ and $q(t)$</td></tr>
<tr><td>Up trend</td><td>$UP_N = 1$ if local minima in [t-N,t] form an increasing sequence; 0 otherwise.
Being $q_k = q(t-k)$, a local minimum occurs if $q_k = \min(q_{k-1}, q_k, q_{k+1})$ for $0 < k < N$. Note that $q(t)$ and $q(t-N)$ are considered minima.</td></tr>
<tr><td>Down trend</td><td>$DW_N = 1$ if local maxima in [t-N,t] form a decreasing sequence; 0 otherwise.
Being $q_k = q(t-k)$, a local maximum occurs if $q_k = \max(q_{k-1}, q_k, q_{k+1})$ for $0 < k < N$. Note that $q(t)$ and $q(t-N)$ are considered maxima.</td></tr>
</table>

Table 6.3. Binary Inputs to Be Matched by the Guard of an NXCS Expert

	Binary Input	Note
1	$DOA_{1,30}(t) > 0$ and $DOA_{1,30}(t-1) < 0$	We just crossed a valley.
2	$DOA_{1,30}(t) < 0$ and $DOA_{1,30}(t-1) > 0$	We just surmounted a peak.
3	$RSI_{15}(t) < 0{,}3$	Too many sales vs. purchases.
4	$RSI_{15}(t) > 0{,}7$	Too many purchases vs. sales.
5	$UP_5(t) = 1$	Bullish period.
6	$DW_5(t) = 1$	Bearish period.
7	$UP_5(t\text{-}1) = 1$ and $UP_5(t) \neq 1$	Current bullish period has just finished.
8	$DW_5(t\text{-}1) = 1$ and $DW_5(t) \neq 1$	Current bearish period has just finished.

other hand, partitioning the input sequence reduces the number of learning examples for each model, thus increasing the risk of overfitting each smaller training set. As our aim was to keep the neural architecture as simple as possible, for computational reasons, we first took into account single-output MLPs. Unfortunately, when noisy data are involved, the small amount of information contained in a single output is typically inadequate to distinguish non-linearities from noise. A possible solution could be to predict multiple outputs, to provide more information for characterizing the output dynamics. However, Weigend and Zimmermann (1998) pointed out that this approach only results in a slight performance increase over single-output MLPs. In fact, the BP algorithm on a standard MLP keeps the information relative to each point separate, so that the derivatives of the cost function relatively to each output are independent, and outputs can interact only through the hidden units. For this reason, they proposed a multi-layer architecture explicitly designed to exploit all available information and to minimize the risk of overfitting. Their feedforward ANN forecasts dynamic variables, such as derivatives and curvatures of prices on different time spans, instead of directly forecasting future prices. These variables are subsequently combined in an interaction layer (with constant weights), which outputs several estimates that—properly averaged—give the final prediction.

In this chapter, we adopt an approach similar to the one proposed by Weigend and Zimmermann, who devised a solution which increases the

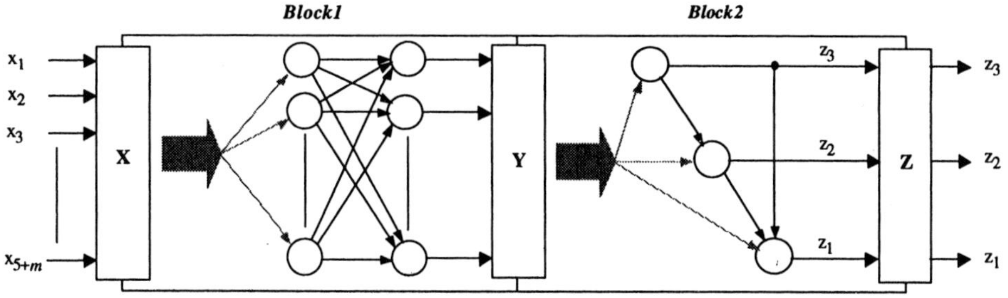

Figure 6.2. A feedforward ANN for stock market forecasting.

information flow by predicting multiple interacting outputs. The kind of feedforward ANN defined and used by NXCS to deal with the problem of stock market forecasting is shown in Figure 6.2. The first block of this architecture is a standard MLP, whereas the second block is structurally similar to a cascade correlation module, although the resemblance does not imply the use of a cascade correlation algorithm for training. In fact, as the modified ANN is still a feedforward network, backpropagation is actually used to train it. The adopted ANN is used for predicting three future quotations,[7] the key difference with a standard MLP being that each output is also an input for each neuron that predicts previous quotations. In particular, the output neuron that predicts the next stock index quotation (i.e., the primary output z_{t+1}) takes as input every other forecasted quotation (z_{t+2}, z_{t+3}). Note that, in this architecture, the derivatives of the standard Mean Square Error cost function are not independent with respect to each output, thus helping to reduce the influence of noise.

The adopted ANN is fed with a vector of 5+m numerical features. In particular, the input of a single ANN is represented by the value of 5 technical analysis indicators, together with the quotations of the last m days. To reduce the effect of outliers, the quotations are pre-processed referring them to a previous point in time, in a logarithmic scale, using the following equation:

$$dl_N(t) = sign\left[q(t) - q(t-N)\right] \cdot \ln\left(\frac{q(t)}{q(t-N)} + 1\right) \tag{6.6}$$

where *q(t)* and *q(t-N)* are the quotation at day t and t-N, respectively.

Table 6.4. Inputs to the ANN

	Numerical Input	Note
1	$\frac{DOA_{1,30}(t)}{MA_{30}(t)}$	Difference of averages (on 30 days and normalized with MA_{30}).
2	$MA_3[ROC_{15}(t)]$	3 weeks rate of change (averaged with MA_3).
3	$CV_{10}(t)$	2 weeks convexity.
4	$RSI_{15}(t)$	3 weeks relative strength index.
5	$MA_{10}[\sigma_{30}(t)]$	Monthly standard deviation of the relative price variations (averaged with MA_{10}).
6	$dl_5(t)$	Last quotations (k=0,1,...,4), pre-processed by equation (6.6) to reduce the effects of outliers.
7	$dl_5(t\text{-}1)$	
8	$dl_5(t\text{-}2)$	
9	$dl_5(t\text{-}3)$	
10	$dl_5(t\text{-}4)$	

Table 6.4 shows the inputs to the ANN embedded within an NXCS expert. Inputs 1-5 denote useful *indicators*, whereas inputs 6-10 denote past quotations, pre-processed using equation 6.6 (m=N=5). The quantity $\sigma_{30}(t)$ represents the standard deviation of the relative price variations, computed for the last 30 days.

3. Experimental Results

Assessing the performance of a financial forecasting system is not an easy task for various reasons. In fact, since data are non-stationary, the significance of the results obtained in a test period is not easy to quantify. Moreover, variables traditionally handled by learning algorithms, such as mean square error or percent of correct classifications, do not have direct economic significance. Lastly, stock markets have a number of constraints and costs that cannot be overlooked.

To test our approach, we implemented a forecasting system that guarantees economic significance of the results. The overall system is shown in Figure 6.3. It is composed of three modules: (i) a predictor, (ii) a strategic and accounting module, and (iii) a stock market simulator.

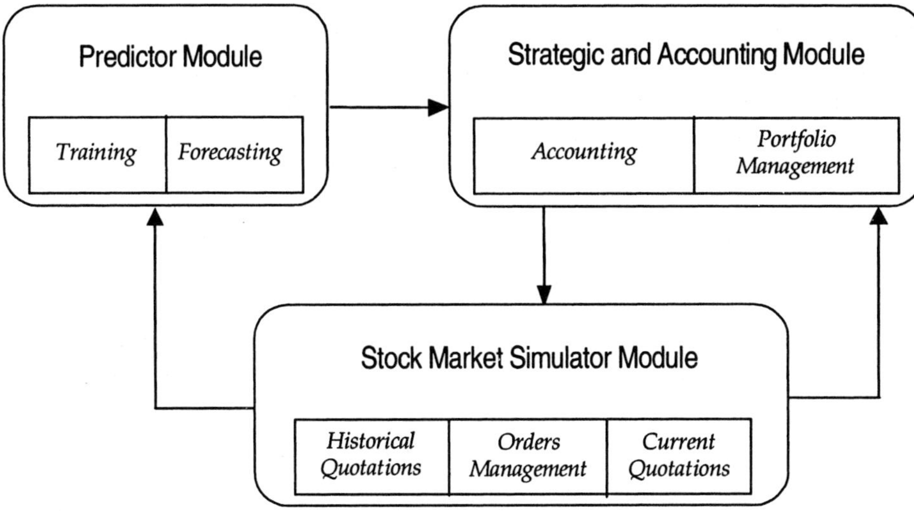

Figure 6.3. Overall system architecture.

The predictor represents the core of the entire system and is responsible for forecasting future quotations. It has been implemented by an NXCS, as described in the previous section.

The strategic and accounting module plays a marginal role, due to the presence of a single traded asset (the stock index) for each test. Presently, its main responsibility is to turn any prediction made by the NXCS module into an actual trading activity. To illustrate the simple mechanism that has been implemented, let us first recall that the overall prediction of the NXCS module ranges in [-1,1]. That is why, the decision of taking a long or short position is derived by simply thresholding the NXCS-module prediction with zero (i.e., a positive value is associated with a long position and vice versa). Furthermore, it is responsible for suspending any buy or sell activity as soon as the reliability of the NXCS predictor, evaluated on a fixed-length window of recent predictions, becomes too low. In this way, transaction costs are saved. Of course, the strategic module restarts its trading operations as soon as the predictor becomes reliable again.

The stock market simulator is a simple virtual market, maintaining the daily quotations of the forecasted index. We considered an index in place of a stock, because of the major impact that incoming news about economic results (or other relevant information about a company) can have on the quotation of a single stock. Hence, the impact of exogenous events is less traumatic on a stock market index, which performs a "filtering" versus any occurring shocks in specific stocks. Furthermore, a global index is an interesting benchmark for assessing the EMH; in fact, it fits exactly the B&H strategy of all quoted stocks, and this strategy cannot be outperformed for a long period of time assuming that the EMH holds.

The stock market simulator is also responsible for applying transaction costs that have a fixed part (10 euro) and a variable part (0.1%) on the value of the stock option bought or sold. The fixed part represents a trading fee, whereas the variable part represents the spread between the quotation of the stock index and its actual buy or sell price.[8] Note that switching from a short to a long position, and vice versa, is a two-step operation and therefore—as in a real market—both fixed and variable costs are doubled. Furthermore, only one daily operation is allowed in this market, and stock prices do not change as a consequence of this operation.

Historical quotations—required while training the predictor—are stored and made accessible by the market simulator. The system performance has been tested on three stock indexes:

- the COMIT index of the Italian stock market, from May 21, 1992 to April 20, 2000 (2000 items);
- the S&P500 index of the New York stock market, from June 2, 1993 to May 3, 2001 (2000 items);
- the NASDAQ index, from June 2, 1993 to May 3, 2001 (2000 items).

In the tests, a trader can buy or sell derivatives on the stock index concerned, and both short and long positions can be taken over the index. So that our virtual investor can gain from both a fall and rise of the stock index quotation, we defined two different kinds of derivative assets—to be used alternatively depending on the decision taken by the strategic module. The first should be used when a long position has to be taken, since it produces exactly the same returns as the stock index quotation. The second should be used when a short position has to be taken, since it produces the returns of the stock index quotation, with inverted sign. These derivative assets, which approximate the behavior of two different kinds of futures, will be denoted appending a "+" or "-" to the name of the stock index (e.g., S&P500+ and S&P500-).

For each stock index, the first 1,000 quotations were used as a training set (800 days) and validation set (subsequent 200 days), to find the best configuration for the NXCS predictor. This configuration includes: (i)

Table 6.5. Major Parameter Values of NXCS System Used in the Simulations

	Parameter	COMIT	S&P500	NASDAQ
XCS	*Maximal Population Size*	1000	1000	1000
	Learning Rate	0.0165	0.0083	0.0063
	Accuracy Fall-Off (α)	0.0007	0.08	0.055
	Accuracy Threshold (ε_0)	0.93	0.60	0.65
ANN	*Input Layer*	10	10	10
	Hidden Layer 1	8	8	8
	Hidden Layer 2	3	3	3
	Output Layer	3	3	3
	Learning Rate	0.001	0.001	0.001
	Momentum	0.1	0.1	0.1

the parameters derived from XCS, and (ii) the parameters of the neural architecture. The most significant system parameters are reported in Table 6.5 (see Appendix **A** for the definitions of accuray fall-off and accuracy threshold).

For each stock index, we performed several tuning sessions, assessing different system configurations. Each simulation started with an initial "cash" of 100,000 euro and we used a merit factor ω,[9] defined as follows:

$$\omega = \frac{CV_T \cdot \sigma_{B\&H}}{CV_{B\&H} \cdot \sigma_T} \tag{6.7}$$

where CV is the cumulative value of the portfolio, and σ is the standard deviation of the daily returns series, representing the risk. Subscripts T and *B&H* refer to the system being tested and to the B&H strategy, respectively. The configurations reported in Table 6.5 are the best we could achieve (i.e., the one that maximized ω) using the validation set. Once the best configuration was found, we tested the system on the subsequent 1,000 quotations.

The strategic and accounting module simulated two different trading strategies, called *defensive* and *aggressive*. In the defensive strategy, the system either has all its assets in cash or invested in the COMIT+, S&P500+, or NASDAQ+ derivative assets (depending on the stock index currently being experimented). If the predictor suggests taking a long position, either the system invests all its capital in the proper derivative assets, or, if it has already invested, it maintains its position. Conversely, if the predictor suggests taking a short position, the system

Table 6.6. Comparing Economic Results for COMIT, S&P500, or NASDAQ Stock Market Indexes (TS=Trading Strategy, CV=Cumulative Value, TR=Total Returns, σ=Standard Deviation, S= Sharpe Ratio, and ω=Merit Factor)

Stock Index	TS	CV	TR	σ	S	ω
COMIT	B&H	290,974	190.974%	1.404%	0.0832	1.0000
	Defensive	355,720	255.720%	1.137%	0.1173	1.5090
	Aggressive	381,999	281.999%	1.399%	0.1028	1.3174
NASDAQ	B&H	158,334	58.334%	2.279%	0.0315	1.000
	Defensive	215,431	115.431%	2.054%	0.0476	1.5096
	Aggressive	252,480	152.480%	2.272%	0.0521	1.5997
S&P500	B&H	148,077	48.077%	1.302%	0.0367	1.000
	Defensive	199,293	99.293%	1.215%	0.0628	1.4418
	Aggressive	255,022	155.022%	1.297%	0.0787	1.7284

sells all its stock, or, if it has already only cash assets, it maintains its position. This strategy cannot obtain returns during market falls, but for the days it stays out of the market, an annual interest rate of 4% is guaranteed to the virtual trader.

The aggressive strategy behaves like the defensive one while taking a long position, whereas to take a short position it buys the COMIT-, S&P500-, or NASDAQ- derivative assets. Such derivative assets guarantee a gain/loss equal to the variation of the corresponding indexes, but with inverted sign.[10] In this way, the system never has cash assets (except at the very beginning of the simulation), and can obtain higher returns by correctly forecasting falls in the stock quotation. It is worth pointing out that, when the system switches from one derivative asset to the other (e.g., from COMIT+ to COMIT-), the transaction costs are computed twice. Figure 6.4, Figure 6.5, and Figure 6.6 show the overall economic results (in euro) obtained adopting the NXCS defensive and aggressive strategies, against the B&H on the selected indexes.

Both defensive and aggressive trading strategies have an advantage over the B&H strategy, even though they pay considerable commissions. Furthermore, from Table 6.6, the aggressive strategy shows a slight advantage in terms of risk, evaluated by the standard deviation of relative price variations over the entire length of the simulation. The resulting Sharpe ratio (S) and merit factor (ω) show, in a risk-adjusted perspective, the superiority of both strategies over the B&H.

Table 6.7 shows the NXCS predictor performance in terms of percentage of correct classifications, compared with the B&H strategy. Note

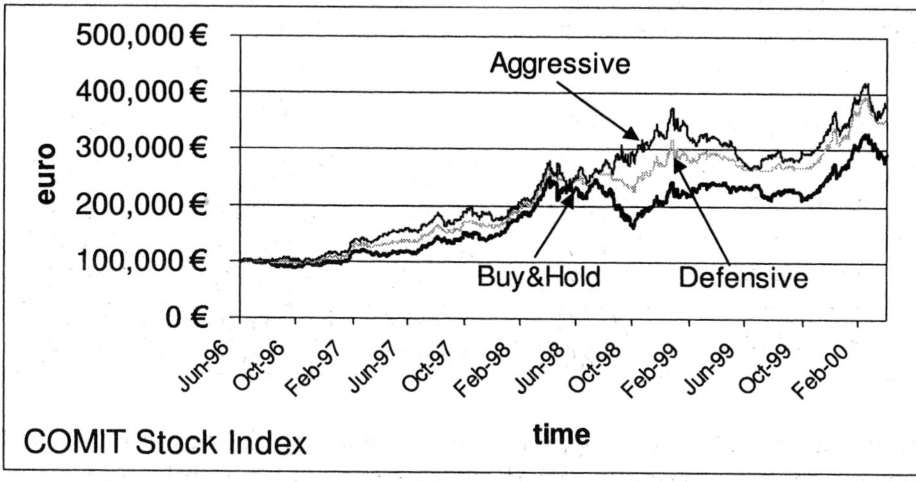

Figure 6.4. Economic Results on the COMIT index.

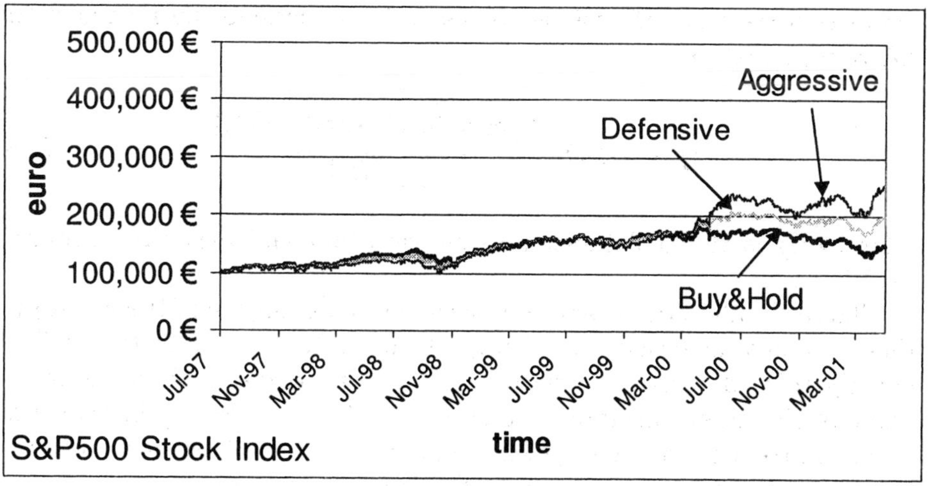

Figure 6.5. Economic Results on the S&P500 index.

Table 6.7. Comparing Predictive Capabilities (**TS**=Trading Strategy, **CC**=percentage of correct classifications, **ip%**= Ideal Profit)

Stock Index	TS	CC	CC > 1%	ip%
COMIT	B&H	55.500%	59.845%	11.204%
	NXCS	55.167%	61.140%	17.121%
NASDAQ	B&H	57.400%	53.065%	04.340%
	NXCS	56.900%	54.291%	10.690%
S& P500	B&H	54.700%	54.447%	04.904%
	NXCS	55.500%	59.030%	14.126%

that, for the B&H strategy, a correct classification is simply a day with a positive index variation. B&H was observed to be slightly superior, in terms of correct classifications, to the NXCS. To investigate this result, in apparent contrast with the above economic results (see Table 6.6), we assumed that the difference could derive from less significant variations and we computed new results using just the days with a percent index variation > 1%. As conjectured, disregarding these minimal variations, the NXCS predictor performs better than the B&H trading strategy. A further indicator (Ideal Profit, *ip%*) has also been defined, which evaluates the correct classifications versus an ideal classifier by means of the following equation:

$$ip\% = 100 \cdot \frac{\sum sign\left[d(t) \cdot \hat{d}(t)\right] \cdot |d(t)|}{\sum |d(t)|} \tag{6.8}$$

where *d(t)* is the real index percent variation and $\hat{d}(t)$ the estimated variation.

The *ip%* indicator clearly compares the performance of the given predictor versus an oracle, and ranges from -1 to +1. Note that ip% is strictly related with total returns, as it jeopardizes, with the absolute value of the index variation, the opportunities to make a gain that have been unattended by the system, and vice versa. This new indicator confers a significant advantage to the NXCS predictor.

4. Conclusions

In this chapter, a novel approach to stock market forecasting has been illustrated, from both the conceptual and technical points of view. Conceptually, a divide-and-conquer strategy has been devised, based

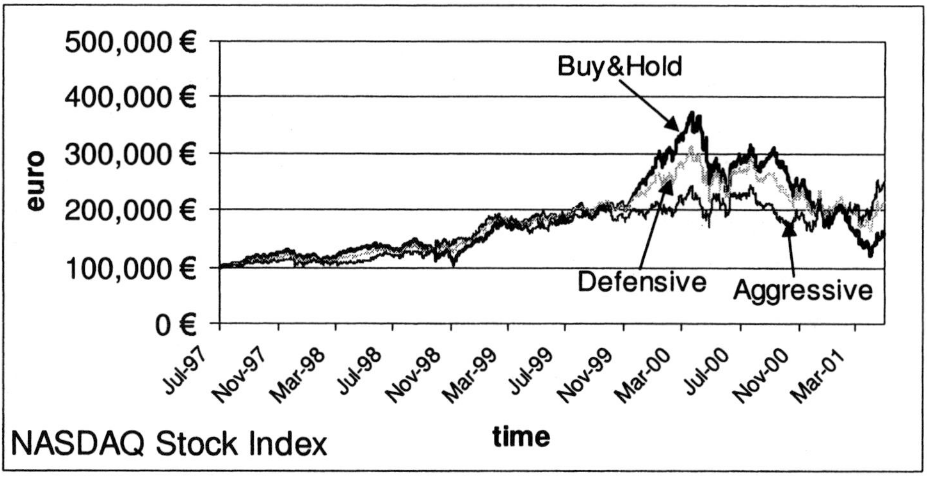

Figure 6.6. Economic Results on the NASDAQ index.

on the idea that segmenting the input space according to a significant amount of domain knowledge would facilitate the prediction task within each element of the partitioned space. To support this perspective, a new hybrid approach—called NXCS—has been defined, which synergistically integrates XCSs with ANNs. An NXCS module has been used as a predictor within a system that simulates trading operations on a stock market. The NXCS module has been customized to fit the needs of financial time series prediction by suitably taking into account technical analysis indicators, as well as previous quotations. A feedforward ANN, especially suited for financial time series forecasting, has been devised and adopted for the NXCS. The resulting system has proved to perform very well in the selected application task, obtaining outcomes superior to the B&H strategy on several stock market indexes.

Appendix: A Short Introduction to eXtended Classifier Systems

An eXtended Classifier System (XCS) consists of a performance, reinforcement, and discovery component. In XCS terminology, inputs—represented by bit strings of fixed length—are named *messages*, and outputs *actions*. The system handles a population of *classifiers*, each represented by a condition-action pair, together with some parameters that characterize the internal state of each classifier. Conditions are represented by a ternary string of bits (i.e., "0" / *false*, "1" / *true*, "#" / *don't-care*), and have the same length as incoming messages. Only when a message is matched by a classifier's condition, is the corresponding action selected—thus generating a

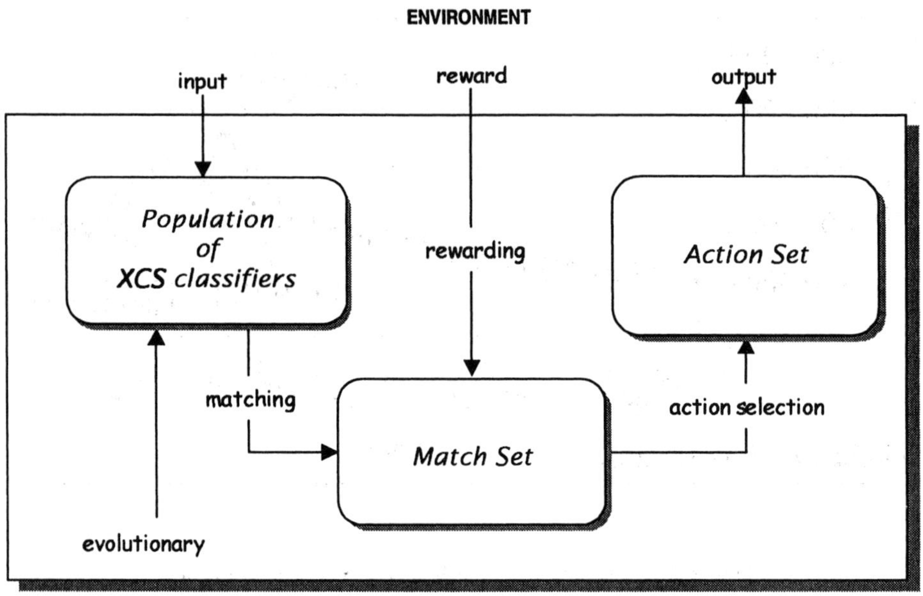

Figure 6.A.1. The XCS basic scheme.

context-dependent behavior. The overall functional architecture of an XCS is shown in Figure 6.A.1.

Given a classifier c, its most important parameters are:
p_c = *prediction* (or *strength*), expected reward to the system for using c;
ϵ_c = *prediction error*, difference between expected and actual reward to the system for using c;
f_c = *fitness*, degree of adaptation of c to the environmental constraints.
In absence of ambiguity, we will write p, ϵ, and f to denote p_c, ϵ_c, and f_c, respectively.

As for the performance component, each time the system receives a message from the environment, a *match set* ($\boldsymbol{M}$) is created containing all classifiers whose conditions match the incoming message. Then the system has to select what action to take, according to its voting policy. The most common voting policy is the fitness-weighted majority rule, evaluated using a *prediction array* ($\boldsymbol{P}$). Let M_i be the subset of $\boldsymbol{M}$ containing the classifiers whose action is i; the i-th element of the prediction array, say P_i, is calculated by means of the formula:

$$P_i = \frac{\sum_{c \in \mathbf{M}_i} p_c \cdot f_c}{\sum_{c \in \mathbf{M}_i} f_c} \tag{6.A.1}$$

Note that P_i is an estimate of the reward the system can obtain for choosing the corresponding action.

Table 6.A.1. Widrow-Hoff Rules Used for Updating the Most Important Parameters of an XCS Classifier (The Parameter β Controls the "Inertial" Behavior of the Update)

Operation	**Formula**	**Where ...**
Prediction Update	$p_{t+1} = p_t + \beta \cdot (r - p_t)$	r is the overall actual reward obtained by the system from the environment.
Prediction Error Update	$\varepsilon_{t+1} = \varepsilon_t + \beta \cdot (\lvert p_{t+1} - r \rvert - \varepsilon_t)$	
Fitness Update	$f_{t+1} = f_t + \beta \cdot (k' - f_t)$	k' is the relative accuracy of a classifier, evaluated by normalizing its accuracy (k) over the action set corresponding to the selected action i^*.

As for the reinforcement component, once the best action i^* is selected, and a reward r has been obtained by the system from the environment, all the classifiers in M_{i^*} (the *action set*) are updated according to the Widrow-Hoff rules reported in Table 6.A.1.

The accuracy and the relative accuracy (k and k', respectively) of a classifier in M_{i^*} can be evaluated as follows:

$$k = \exp\left[\ln(\alpha) \cdot \frac{\varepsilon - \varepsilon_0}{\varepsilon_0}\right] \tag{6.A.2}$$

$$k' = \frac{k}{\sum_{c \in \mathbf{M}_{i*}} k_c} \tag{6.A.3}$$

The parameters α and ϵ_0 (i.e., *accuracy fall-off* and *accuracy threshold*, respectively) are set once, before running an XCS experiment. Note that $k = \alpha$ when the prediction error $\epsilon = 2\epsilon_0$.

As for the discovery component, new classifiers can be generated in two different ways:

- When no classifier exists that covers the incoming message (or the classifiers covering it have low fitness), a new classifier able to cover the message is created, computing its parameters by averaging over the whole population of classifiers;

- Given a match set $\boldsymbol{M}$, when the average time elapsed from the last mating (evaluated on the classifiers in $\boldsymbol{M}$) exceeds a predefined threshold, the system selects two classifiers in $\boldsymbol{M}$, with a probability proportional to each classifier's fitness. Then, crossover and mutation operators are applied to the selected classifiers, thus yielding two offspring, whose parameters are computed as the average of their parents.

When the maximum permissible size for the given population has been attained, a classifier must be selected for deletion. The probability of being deleted increases for classifiers with low fitness, and is proportional to the average cardinality of the match sets they have been involved with in the past. In this way, classifiers that frequently occur in large match sets have a higher probability of being deleted. It is worth pointing out that the reward resulting from the outside environment is shared among the classifiers belonging to the action set. In this way, separate niches tend to emerge, each of them hosting some (at least one) "specialized" classifiers. It is easy to see that high fitness can only be achieved for classifiers that make few mistakes within their niches. Furthermore, the fitness-sharing mechanism that distributes rewards among classifiers in each niche, together with the deletion criterion, tend to limit the number of classifiers that can survive in a niche.

Notes

1. GARCH = Generalized AutoRegressive Conditional Heteroskedasticity.
2. An example of rapid regime shifting is the well-known herding behavior.
3. The distance score is typically used to assess the degree of accuracy of a prototype against a given input vector.
4. Without going into unnecessary details, let us only note that the weighted averaging rule can be used for evaluating a-posterior probabilities (one for each possible action / class label) starting from the outputs of a set of classifiers.
5. For prediction tasks.
6. The concept of primary output stems from the consideration that ANNs may have multiple outputs, in which case an input is arbitrarily selected as being primary. For time series forecasting, this primary output coincides with the next predicted value of the time series.
7. Without loss of generality, the percent variation of the current price is actually predicted, with a range in [-1,1]. The same range applies also to the overall prediction, due to the adopted outputs blending mechanism.
8. In real markets, the quotation of a stock is the price of the last contract. At any time, there is a buy price (the best buy offer) and a sell price (the best sell offer).
9. The absolute value of the investment is relevant, due to the presence of a fixed part in transaction costs. Furthermore, this relatively small value is compatible with the hypothesis that prices do not change as consequence of buy/sell operations.
10. In a real market we can buy or sell futures on the index to obtain a similar result. In this chapter we do not consider the peculiarities of advanced financial instruments.

References

Achelis, S. B. (1995). *Technical Analysis from A to Z*, 2nd ed. Chicago: Irwin Professional Publishing,

Bengio Y. and F. Gingras (1998). "Handling Asynchronous or Missing Data with Recurrent Networks," *International Journal of Computational Intelligence and Organizations*, 1(3), 154–163.

Breiman, L., J. Friedman, R. Olshen, and C. Stone (1984). *Classification and Regression Trees.* Belmont, CA: Wadsworth International Group.

Campbell, J. Y., A. W. Lo, and A. C. MacKinlay (1997). *The Econometrics of Financial Markets.* Princeton, New Jersey: Princeton University Press.

Campolucci, P., A. Uncini, F. Piazza, and B. D. Rao (1999). "On-Line Learning Algorithms for Locally Recurrent Neural Networks," *IEEE Transactions on Neural Networks*, 10(2), 253–271.

Castiglione, F. (2001). "Forecasting Price Increments Using an Artificial Neural Network," *Advances in Complex Systems*, 4(1), 45–56.

Clark, P. and T. Niblett (1989). "The CN2 Induction Algorithm," *Machine Learning Journal*, 3(4), 261–283.

Cover T. M. and P. E. Hart (1967). "Nearest Neighbor Pattern Classification," *IEEE Transactions on Information Theory*, 13(1), 21–27.

Deboeck G. (1993). "Neural, Genetic, and Fuzzy Approaches to Design of Trading Systems," *Proceedings of the 2nd Annual International Conference on AI Applications on Wall Street: Tactical and Strategic Computing Technologies*, 184–193.

Fahlmann S. E. and C. Lebiere (1990). *The Cascade-Correlation Learning Architecture*, Technical Report CMU-CS-90-100, Carnegie Mellon University.

Fama E. F. (1965). "The Behavior of Stock Market Prices," *The Journal of Business*, 38, 34–105.

Friedman, J. H. (1991). "Multivariate Adaptive Regression Splines," *Annals of Statistics*, 19(1), 1–141.

Friedman J. H., F. Baskett, and L. J. Shustek (1975). "An Algorithm for Finding Nearest Neighbors," *IEEE Transactions On Computers*, 24(10), 1000–1006.

Giles, C. L., S. Lawrence, and A. Chung Tsoi (1997). "Rule Inference for Financial Prediction using Recurrent Neural Networks," *Proceedings of IEEE/IAFE Conference on Computational Intelligence for Financial Engineering (CIFE)*, 253–259.

Goldberg, D. E. (1989). *Genetic Algorithms in Search, Optimization and Machine Learning.* Addison-Wesley.

Hamilton, J. D. (1994). *Time series analysis.* Princeton University Press.

Hardle, W., H. Lutkepohl, and R. Chen (1997). "A Review of Nonparametric Time Series Analysis," *International Statistical Review*, 65, 49–72.

Hawawini, G. and D. B. Keim (1995). "On the Predictability of Common Stock Returns: World-wide Evidence," in R.A. Jarrow, V. Maksimovic and W.T. Ziemba (eds.), *Handbook in Operations Research and Management Science*, 9, 497–544. Amsterdam: North-Holland.

Hellström, T. and K. Holmström (1997). *Predicting the Stock Market*, Technical Report IMa-TOM-1997-07. Department of Mathematics and Physics, Malardalen University, Sweden.

Holland, J. H. (1976). "Adaptation," in R. Rosen and F. M. Snell (eds.), *Progress in Theoretical Biology*, 4, 263–293. New York: Academic Press.

Holland, J. H. (1986). "Escaping Brittleness: The possibilities of General-Purpose Learning Algorithms Applied to Parallel Rule-Based Systems," in R.S. Michalski, J. Carbonell, and M. Mitchell (eds.), *Machine Learning II*, 593–623. Morgan Kaufmann.

Jacobs, R. A., M. I. Jordan, S. J. Nowlan, and G. E. Hinton (1991). "Adaptive Mixtures of Local Experts," *Neural Computation*, 3, 79–87.

Jordan, M. I. and R. A. Jacobs (1992). "Hierarchies of Adaptive Experts," in J. Moody, S. Hanson, and R. Lippman (eds.), *Advances in Neural Information Processing Systems*, 4, 985–993. Morgan Kaufmann.

Jordan, M. I. and R. A. Jacobs (1994). "Hierarchical Mixtures of Experts and the EM Algorithm," *Neural Computation*, 6, 181–214.

Fu, K. and W. Xu (1997). "Training Neural Network with Genetic Algorithms for Forecasting the Stock Price Index," *Proceedings of the 1997 IEEE International Conference on Intelligent Processing Systems*, 1, 401–403.

Kantz, H. and T. Schreiber (1997). *Nonlinear Time Series Analysis*, Cambridge Nonlinear Science Series 7. Cambridge University Press.

Kovacs, T. (1996). *Evolving Optimal Populations with XCS classifier Systems*, MSc. Dissertation. University of Birmingham, UK.

Kovacs, T. (1999). "Strength or Accuracy? A Comparison of Two Approaches to Fitness," A. S. Wu (ed.), *Second International Workshop on Learning Classifier Systems during GECCO99*, 258–265.

Lanzi, P. L. (1997). "A study of the Generalization Capabilities of XCS," *Proceedings of the Seventh International Conference on Genetic Algorithms (ICGA97)*, 418–425. San Francisco, CA: Morgan Kaufmann.

Lanzi, P. L. (1998). "Adding Memory to XCS," *Proceedings of the IEEE Conference on Evolutionary Computation (ICEC98)*, 609–614.

Lanzi, P. L. and A. Perrucci (1999). "Extending the Representation of Classifier Conditions Part II: From Messy Coding to S-Expressions,"

Proceedings of the Genetic and Evolutionary Computation Conference (GECCO '99), 1, 345–353. Morgan Kaufmann.

LeBaron, B. (1991). *Technical Trading Rules and Regime Shifts in Foreign Exchange*, Technical Report. Department of Economics, University of Wisconsin, Madison, WI.

LeBaron, B. and A. S. Weigend (1997). "A Boostrap Evaluation of the Effect of Data Splitting on Financial Time Series," *IEEE Transactions on Neural Networks*, 9, 213–220.

Lewis, P. A. W., B. K. Ray, and J. G. Stevens (1994). "Modelling Time Series Using Multivariate Adaptive Regression Splines," in A. S. Weigend and N. A. Gershenfeld (eds.), *Time Series Prediction: Forecasting the Future and Understanding the Past*, 296–318. Addison Wesley.

Mahfoud, S. and G. Mani (1996). "Financial Forecasting Using Genetic Algorithms," *Applied Artificial Intelligence*, 10(6), 543–565.

McCullagh, P. and J. A. Nelder (1989). *Generalized Linear Models*. London: Chapman and Hall.

Muhammad, A. and G. A. King (1997). "Foreign Exchange Market Forecasting Using Evolutionary Fuzzy Networks," *Proceedings of the IEEE /IAFE 1997 Computational Intelligence for Financial Engineering*, 213–219.

Nelson, D. B. (1991). "Conditional Heteroskedasticity in Asset Returns: A New Approach," *Econometrica*, 59, 347–370.

Noever, D. and S. Baskaran (1994). "Genetic algorithms trading on the S&P500," *The Magazine of Articial Intelligence in Finance*, 1(3), 41–50.

Quinlan, J. R. (1986). "Induction of Decision Trees," *Machine Learning*, 1, 81–106.

Quinlan, J. R. (1993). *Programs for Machine Learning*. Los Altos, CA: Morgan Kaufmann.

Refenes, A. P. N., A. N. Burgess, and Y. Bentz (1997). "Neural Networks in Financial Engineering: A Study in Methodology", *IEEE Transactions on Neural Networks*, 8(6), 1222–1267.

Refenes, A. P. N., A. Zapranis, and G. Francis (1997). "Stock performance modeling using neural networks: A Comparative Study with Regression Models," *Neural Networks*, 7(2), 375–388. Elsevier.

Rumelhart, D. E. and J. L. McLelland (1986). *Parallel Distributed Processing, Explorations in the Microstructure of Cognition*, 1, Foundations. MIT Press.

Salmon, M. (2000). "Measuring Dependency in Non-Gaussian Financial Assets," *Fifth International Meeting on Quantitative Methods in Applied Sciences.*

Sun, R. and T. Peterson (1999). "Multi-Agent Reinforcement Learning: Weighting and Partitioning," *Neural Networks*, 12(4-5), 727–753.

Sutton, R. S. and A. G. Barto (1998). *Reinforcement Learning: An Introduction.* Cambridge, MA: MIT Press.

Tong, H. and K. S. Lim (1980). "Threshold Autoregression, Limit Cycles and Cyclical Data," *Journal of Royal Statistics Society*, 42, 245–292.

Tsay, R. (1989). "Testing and Modeling Threshold Autoregressive Processes," *Journal of the American Statistical Association*, 84, 431–452.

Weigend, A. S. (1996). "Time Series Analysis and Prediction Using Gated Experts with Application to Energy Demand Forecast," *Applied Artificial Intelligence*, 10, 583–624.

Weigend, A. S. and H. G. Zimmermann (1998). "Exploiting Local Relations as Soft Constraints to Improve Forecasting," *Journal of Computational Intelligence in Finance*, 6, 14–23.

Weigend, A. S., M. Mangeas, and A. N. Srivastava (1995). "Nonlinear Gated Experts for Time Series: Discovering Regimes and Avoiding Overfitting," *International Journal of Neural Systems*, 6, 373–399.

Wilson, S. W. (1995). "Classifier Fitness Based on Accuracy," *Evolutionary Computation*, 3(2), 149–175.

Wilson, S. W. (1998). "Generalization in the XCS classifier System," *Proceedings of the Third annual Genetic Programming Conference*, 665–674. San Francisco, CA: Morgan Kaufmann.

Wilson, S. W. (1999). "Get Real! XCS with Continuous-Valued Inputs," in L. Booker, S. Forrest, M. Mitchell, and R. Riolo (eds.), *Festschrift in Honor of John H. Holand*, 111–121. Center of Study of Complex Systems, The University of Michigan, ANN Arbor, MI.

III

TRADING

Chapter 7

EDDIE FOR FINANCIAL FORECASTING

Edward P. K. Tsang
University of Essex
edward@essex.ac.uk

Jim Li
University of Essex
jli@essex.ac.uk

Abstract EDDIE is a genetic-programming based system for channelling expert knowledge into forecasting. FGP-2 is an implementation of EDDIE for financial forecasting. The novelty of FGP-2 is that, as a forecasting tool, it provides the user with a handle for tuning the precision against the rate of missing opportunities. This allows the user to pick investment opportunities with greater confidence.

Keywords: Genetic Programming, Financial Forecasting, Precision, Genetic Decision Trees

Introduction

The Forecasting Problem and Its Difficulties

In machine forecasting, one is often given a series of observations over a set of monitored variables$\{x_1, x_2, \cdots, x_n\}$, and asked to find the regularity in the data in order to predict the value of a dependent variable y. For example, given three years' of records of the daily closing prices, trade volumes, changes in interest rate, market indices, etc., one may attempt to predict whether a share price will rise or fall in the following week. This prediction task is difficult for many reasons. In our research, we focus on the following two:

1. Variables identification: Can the observed variables explain the dependent variables? In other words, is y determined by a function of the x_i's? Taking more variables into consideration may incur higher observation cost. Finding the right x_i's in a domain is a problem for the *human expert*. Can the expert be helped to do his/her job more efficiently?

2. Variables interactions: Even if we are sure that y is a function of the x_i's, how could this function be found? Many *machine learning techniques* are designed to address this problem. Real life forecasting problems are difficult because the x_i's are rarely independent of each other. For example, a company's share price may be affected by, among many other things, the interest rate and the company's sales volume; the sales volume could be affected by money supply, which is affected by the interest rate. Combinatorial explosion prevents one from examining combinations of all the factors and all possible interactions between them.

EDDIE — A Forecasting Tool

EDDIE (which stands for Evolutionary Dynamic Data Investment Evaluator) is an interactive tool, designed at University of Essex, to help analysts to search the space of interactions and make financial decisions (Tsang et al. 1998). Given a set of variables, EDDIE attempts to find interactions among variables and discover non-linear functions (addressing point 2 above). By using genetic programming, EDDIE generates decision trees, which can be understood by human experts. Human expertise is channelled into EDDIE through human feedback to the system. In this process, EDDIE helps the human expert to experiment with different variables (x_i's) more easily (addressing point 1 above).

FGP is an implementation of EDDIE for financial forecasting. FGP has been applied to a variety of financial forecasting problems with demonstrated accuracy (Li and Tsang 1999; Tsang et al. 2000). It generates Genetic Decision Trees (GDTs), which enables the program to explain how a forecast was arrived at. This allows the users to judge whether the reasons for the prediction are sound or not.

An example GDT generated for daily valuations of S&P 500 is shown below:

(IF(NOT($PMV\_50 > -38.97492$))
  (IF($Vol\_50 > 26.43218$)
    (IF(AND(NOT(OR($TRB\_50 > -124.21202$)($Vol\_50 < 42.15001$)))
        (AND($PMV\_12 < 31.53422$)(NOT($Filter\_63 > 104.84150$))))
      Buy
      Don't Buy)
    Buy)
  Don't Buy)

In this GDT, PMV_50, Vol_50 , PMV_12 and $Filter\_63$ are technical indicators (e.g. Vol_50 is "50 days volatility", measured by standard deviation of the last 50 day's closing prices). Details of these indicators will not be elaborated here as The main aim of this paper is to argue that EDDIE is a useful forecasting tool. Whether the rules generated are good or not depends on the quality of the input indicators, and this GDT was generated using low quality text-book indicators. [1]

In this paper, we focus on a novel feature of FGP-2, in that it provides the user with a handle for tuning the precision against the rate of missing opportunities, as it will be explained below.

1. Forecasting Performance Criteria

Prediction accuracy is naturally important to forecasting. In this section, we define precision and explain why it is important in some forecasting applications. As a forecasting tool, FGP-2 is designed to give the user control over the precision of a forecast.

Without lost of generality, we shall describe a specific forecasting task. We shall use it to describe the criteria for measuring forecasting success. Suppose one is asked to forecast whether an index will rise by $r\%$ within the next n periods. Each day can be classified into a *positive position*, where the target return will be achieved, or a *negative position*, where the target return will not be achieved. Given a prediction and the reality (in hindsight), one may construct a contingency table as shown in Table 7.1. The following analysis applies to all two-class classification prediction problems in general.

Here we define some measures of success in forecasting. The *rate of correctness* (RC) in a prediction is the number of all correct predictions over the total number of predictions. The *rate of failure* (RF) is the proportion of positions that were wrongly predicted positive (FP) over the number of positive predictions (N+). The precision is 1 - RF, i.e. the proportion of positive positions that were correctly predicted. The *rate*

Table 7.1. A Contingency Table for a Two-class Classification Prediction Problem

# of True Negative (TN)	# of False Positive (FP)	Actual Negative O_-=TN+FP
# of False Negative (FN)	# of True Positive (TP)	Actual Positive O_+=FN+TP
Predicted Negative N_-=TN+FN	Predicted Positive N_+=FP+TP	Total # of Predictions $T=N_++N_-=O_++O_-$

of missing chances (RMC) is the number of wrongly predicted negative (FN) over the number of actual positives (O_+):

$$\mathbf{RC} = \frac{TP+TN}{T}; \quad \mathbf{RF}(1-precsion) = \frac{FP}{N_+}; \quad \mathbf{RMC} = \frac{FN}{O_+}$$

Ideally one would like RC to be 1. How close RC could approach 1 is limited by the nature of the data (e.g. are all the relevant variables used? is there noise?) as well as the quality of the forecasting algorithm.

In financial forecasting, a positive prediction may lead to investment. If this prediction is wrong, the investor will not be able to achieve the return desired. Such mistakes could be costly. Therefore, we assume that the user would want control over RF, if possible. (In other applications, it may be important not to miss any opportunities, hence reduce FN.)

It should be noted that RF may trivially be reduced to 0 if the system makes no recommendation to buy. However, a system that never recommends any "buy" will not be useful to any investor. Therefore, in reality, one would like to reduce RF (the principle goal) without significantly increasing RMC (which is seen as a constraint to the target forecasting system). FGP-2 was built with the following mission: *FPG-2 Mission: to enable the user to reduce RF by increasing RMC, or vice versa, without significantly affecting RC.* In other words, RC will not be used as the objective function here. The objective is to reduce RF. However, RC is considered by EDDIE as a constraint.

2. FGP-2 —Trading Precision with the Rate of Missing Opportunities

2.1. FGP-1: Brittle Results with a Linear Fitness Function

In a forecasting problem, RC is what one would like to improve. In previous papers, we have presented the effectiveness of FGP-1, an early implementation of EDDIE, in achieving reasonably high RC (Li and

Tsang 1999; Tsang and Li 2000). In the preceding section, we explained that RC may not be the only criterion for measuring forecasting performance. In some applications, one may want to reduce RF (or 1 - precision).

FGP-1 used $f_{(1)}$ (Equation (7.1)) as its fitness function. It allows the users to reflect their preference by means of adjusting the weights w_rc, w_rmc and w_rf.

$$f_{(1)} = w\_rc * RC - w\_rmc * RMC - w\_rf * RF \qquad (7.1)$$

One possibility of achieving low RF is to assign a high value to w_rf in FGP-1, which we studied thoroughly at an early stage. We fixed w_rc to 1 (to prevent FGP-1 from reducing RF to 0 by generating rules that produce no "buy" signals) and w_rmc to 0. (We have tried an alternative, which is to set w_rmc to 1; no better results were found). We attempted to vary w_rf to a value between 0 and 1. Our experiments show that FGP-1's performance can be very sensitive to the three weights; in other words, performance of FGP-1 was brittle. This is elaborated below.

We found in our experiments that if the value that we choose for w_rf is too close to 1, FGP-1 achieved lower RF by making no positive recommendations at all; we shall refer to this as the "no-positive-prediction problem". If the value that we choose for w_rf is too low, the performance of FGP-1 is no different from FGP-1 that sets w_rf to 0; we shall refer to this as the "no-effect problem."

There is often some constant a (0.62 in our experiments) such that if the value of w_rf deviates slightly from a on either side, one of the above two problems occur. When w_rf was set to the fine-tuned critical value (0.62), FGP-1 did not generate effective decision trees reliably. In our experiments, only two out of ten runs generated a few correct positive positions on the test period; the remaining 8 runs either showed the no-positive-prediction or the no-effect problem.

According to our experience, the weighted fitness function is satisfactory. Firstly, it is likely that the critical value for w_rf varies from one data set to another. We found this value (0.62) for the test data set, but there is no guarantee that it will work for unseen data. Secondly, as explained above, even when w_rf is set to the fine tuned critical value, some of the GDTs generated suffered from the no-positive-prediction and some suffered from the no-effect problem. Therefore, if one picks one of these GDTs and apply it to unseen data, its performance is difficult to predict (because the sensitive w_rf that is good for the test data may or may not be good for the unseen data). Ideally, one would like a system that generates GDTs which performance is not too sensitive to parameter setting. Besides, one would hope that given one set of parameters, the system generates GDTs with similar (reliable) performance.

2.2. FGP-2: Putting Constraints into EDDIE

Our mission is to reduce RF, possibly at the price of RMC. To do so, we introduced a new parameter to FGP, $\Re = [P_{min}, P_{max}]$, which defines the minimum and maximum percentage of recommendations that we instruct FGP to make in the training data (like most machine learning methods, the assumption is that the test data exhibits similar characteristics). We call the new fitness function $f_{(2)}$.

If one aims to make accurate forecasts for a given series, then choosing appropriate values for $\Re$ and the weights for $f_{(2)}$ remains a non-trivial task. In this paper, our focus is to first examine whether RF can be reduced by any choice of $\Re$. Then we shall see if RF can be adjusted without affecting the overall forecasting correctness (RC).

Efficacy of the constraint in fitness function is first demonstrated by the following experiment, where we took $\Re = [35\%, 50\%]$, $w\_rmc = 0$ and $w\_rf = 1$. We ran FGP 10 times. Results are showed in Table 7.2. With $f_{(2)}$, FGP-2 does not exhibit the brittleness in FGP-1, as demonstrated by the relatively small standard deviation (STD) in RC. For reference, we have included the AARR (Average Annualised Rate of Return) and RPR (Ratio of Positive Returns) in Table 7.2. RPR measures the proportion of times when FGP-2's recommendation gives a positive return, even when the target $r\%$ has not been achieved. Both AARR and RPR are for reference only, as they are not used to train FGP-2.

To see the effect of the constrained fitness function, we compare the above results with those generated by FGP using RC only as the fitness function (i.e. $f_{(1)}$ with $w\_rmc = w\_rf = 0$). Results are listed in Table 7.3. From Table 7.3, we can see that by using $f_{(2)}$, the mean RF is reduced from 43.51% to 40.06%. For reference, the mean AARR rises from 55.07% (Table 7.3) to 63.38% (Table 7.2) and the mean RPR rises from 65.86% to 69.95%. The price to pay for a lower RF is that more opportunities were missed: the mean RMC rises from 46.77% to 65.74%. The mean RC only slightly decreases from 54.19% to 53.72%. To determine whether result differences are statistically significant, the two-tailed paired t-test was applied on the null hypothesis that the mean performances of two groups were not statistically different under each of the five criteria. Shown in Table 7.4 are t-values and their corresponding p-values under each criterion. The results indicate that by using the constrained fitness function with $\Re = [35\%, 50\%]$, FGP-2 generates decision

Table 7.2. FGP-2 Results on Test Data Using the Constrained Fitness Function with $\Re = [35\%, 50\%]$

RULES	RF	RMC	RC	AARR	RPR	Number of Recommendations
GDT 1	40.34%	64.02%	53.92%	60.68%	70.59%	357
GDT 2	41.22%	62.67%	53.66%	63.83%	67.55%	376
GDT 3	40.12%	66.22%	53.66%	61.98%	70.96%	334
GDT 4	40.06%	66.39%	53.66%	62.60%	70.78%	332
GDT 5	40.25%	67.40%	53.39%	64.02%	69.66%	323
GDT 6	41.03%	59.46%	54.27%	58.26%	69.29%	407
GDT 7	41.47%	66.39%	52.95%	62.99%	67.35%	340
GDT 8	39.94%	68.75%	53.30%	63.98%	69.16%	308
GDT 9	39.82%	66.55%	53.74%	63.41%	71.73%	329
GDT 10	36.40%	69.59%	54.63%	72.02%	72.44%	283
MEAN	40.06%	65.74%	53.72%	63.38%	69.95%	338.9
STD	1.41%	2.99%	0.48%	3.53%	1.67%	34.7

Table 7.3. FGP Results on Test Data Using the General Fitness Function ($w\_rc = 1, w\_rmc = w\_rf = 0$)

RULES	RF	RMC	RC	AARR	RPR	Number of Recommendations
GDT 1	41.11%	44.59%	56.56%	57.82%	66.61%	557
GDT 2	43.89%	44.93%	54.10%	52.40%	66.09%	581
GDT 3	42.35%	53.55%	54.27%	55.04%	68.97%	477
GDT 4	45.02%	43.07%	53.22%	54.96%	64.60%	613
GDT 5	44.09%	44.09%	54.01%	52.33%	63.68%	592
GDT 6	44.58%	52.53%	52.69%	54.02%	65.88%	507
GDT 7	43.33%	50.51%	53.92%	54.90%	65.57%	517
GDT 8	43.61%	38.85%	55.07%	60.34%	66.36%	642
GDT 9	43.36%	45.27%	54.54%	53.82%	65.21%	572
GDT 10	43.79%	50.34%	53.57%	55.09%	65.58%	523
MEAN	43.51%	46.77%	54.19%	55.07%	65.86%	558.1
STD	1.11%	4.71%	1.07%	2.42%	1.39%	51.7

trees with statistically better RF, AARR and RPR at a significant level of $\alpha = 0.001$, though they have statistically worse RMC. These decision trees do not show any statistically different for RC (p-value is 0.2612). That is, RC has not been compromised as RF is reduced.

Table 7.4. t-statistics for Comparing Mean Performances of two Groups (Using RC Only versus Using the Constrained Fitness Function with $\Re = [35\%, 50\%]$)

Criteria	For RF	For RMC	For RC	For AARR	For RPR
t values	-4.64	6.33	-1.16	4.69	4.71
p values($\alpha = 0.001$)	0.0002.5	0.000005	0.261247	0.000182	0.000175

3. Empirical Evaluation of the Constrained Function

3.1. Objective of the Experiments

To test FGP-2's usefulness as a tool for tuning precision against missing opportunities, we tested it on historic data. We should re-iterate that FGP-2's prediction accuracy is limited by whether or not the predicted value is a function of the observed variables (as we pointed out in Section 1.1, point 1). When no such function exists, neither FGP-2 nor any other comparable algorithms would be able to make accurate predictions. The primary objective of the test results presented below is not to demonstrate that FGP-2 can predict DJIA accurately. (In fact, the results presented below do not represent the most accurate predictions that FGP-2 has ever made.) Instead, the primary objective is to observe whether FGP-2 can be used to trade precision with the rate of missing opportunities. We would only conclude that FGP-2 achieves what it is designed to achieve if one could instruct it to improve precision at the cost of increasing the rate of missing opportunities, or vice versa.

3.2. Experimental Data

In this section, we present a typical set of test results by FGP-2, based on daily closing prices of the Dow Jones Industrial Average (DJIA) Index. Other indices and share prices have been used with similar results. Experiments presented in this paper were carried out on DJIA daily closing index from 07/04/1969 to 09/04/1981, a total of 3,035 trading days, as illustrated in Figure 7.1. We took the data from 07/04/1969 to 11/10/1976 (1,900 trading days) as training data, and the period from 12/10/1976 to 09/04/1981 (1135 trading days) as testing data. For the purpose of analysis, we chose $r = 2.2$ and $n = 21$ days, which give roughly 50% of positive positions in both the training and test periods. Details of the experiemental setup are shown in Table 7.5

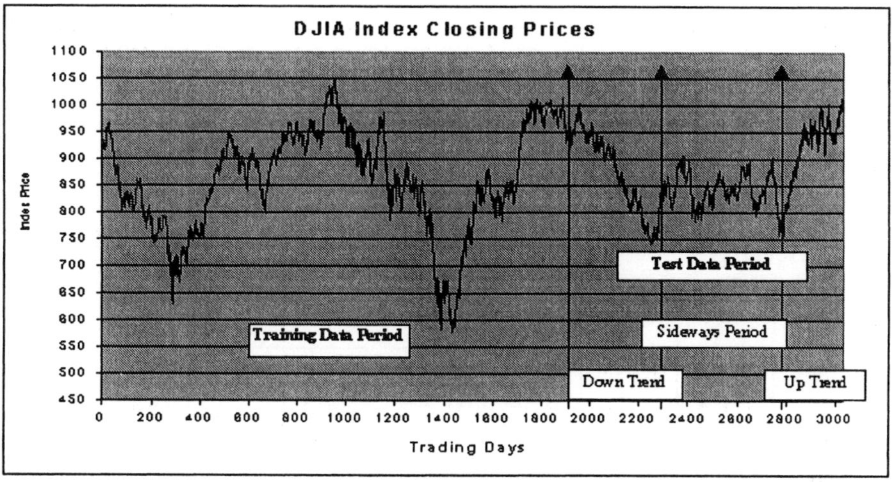

Figure 7.1. Dow Jones Industrial Average (DJIA) index daily closing prices from 07/04/1969 to 09/04/1980 (3035 trading days)

Table 7.5. Parameters Used in FGP-2 for Experiments

Target	To find GDTs with low RF, with return of 2.2% with 22 days
Input terminals	Six technical indicators plus Real as thresholds
Prediction terminals	{0, 1}: 1 representing "Positive"; 0 representing "Negative".
Non-terminals	If-then-else , And, Or, Not, $>$, $\geq$, $<$, $\leq$, $=$.
Crossover rate	0.9
Mutation rate	0.01
Population size	1,200
Maximum no. of generations	30
Termination criterion	Generation limit or Time limit, whichever reached first
Selection strategy	Tournament selection, Size = 4
Max depth of individual programs	17
Max depth of initial individual programs	4
Maximum run time (hours)	2
Hardware and operating system	Pentium PC 200MHz running Windows 95 with 64M RAM.
Software	Borland C++ (version 4.5)

3.3. Experimental Results

To further explore the impact of the constraint $\Re$ on reducing RF, we took five additional non-overlapping $\Re$s in the fitness function respectively. The five mutually exclusive $\Re$s are [5%, 10%], [10%, 15%], [15%, 20%], [20%, 35%] and [50%, 60%]. For each $\Re$, we run FGP-2 ten times using the parameters $w\_rc = w\_rf = 1$ and $w\_rmc = 0$. We calculated the mean performances on test data with respect to RF, RMC, RC, RPR, AARR and the mean number of positive recommendations. All experimental results are showed in Table 7.6. The results are visualised in Figure 7.2.

Table 7.6. The Effect of the Constraint $\Re$ on the Mean Performances of FGP-2

$\Re$ [%,%]		RF	RMC	RC	AARR	RPR	Number of Recommendation
[5, 10]	Mean	13.48%	99.14%	48.19%	224.03%	92.22%	6.2
	SD	14.85%	0.63%	0.26%	229.24%	10.86%	4.8
[10,15]	Mean	28.60%	94.05%	49.70%	136.81%	82.95%	49.3
	SD	6.22%	1.65%	0.76%	30.52%	4.40%	13.1
[15,20]	Mean	31.02%	85.69%	51.74%	99.58%	79.02%	125.1
	SD	5.21%	6.41%	1.67%	25.50%	5.47%	62.7
[20,35]	Mean	36.00%	75.25%	53.41%	75.68%	73.61%	229.8
	SD	2.59%	5.50%	1.19%	9.55%	3.41%	55.1
[35,50]	Mean	40.06%	65.74%	53.72%	63.38%	69.95%	338.9
	SD	1.41%	2.99%	0.48%	3.53%	1.67%	34.7
[50.65]	Mean	46.73%	45.47%	51.31%	52.26%	62.57%	606.2
	SD	1.37%	10.40%	1.64%	1.63%	1.67%	115.4

Figure 7.2 shows that RF decreases gradually as $\Re$ is reduced. The lowest RF (13.48%) is obtained by using the smallest $\Re$ [5%, 10%] whereas the highest RF (46.73%) is obtained by using the biggest $\Re$ [50%, 65%]. The six mean RFs in the graph suggest that tightening the constraint ($\Re$) in the fitness function may lead to a lower RF. Reduction in RF obviously benefited RPR and AARR in the test data. RPR rises from 57.16% to 92.85%. AARR increases dramatically from 40.32% to 300.33%. The results obtained by using the tightest constraints [5%, 10%] provide the most reliable recommendations, with a failure rate of 13.48%. The price to pay for using a constraint of this tightness is that it makes fewer positive recommendations, which leads to higher rate of missing chances (RMC). If we reduce $\Re$ beyond a certain point, no positive recommendations will be made by FGP-2. Results in this experiment suggest that $\Re$ is a useful handle for tuning RF against RMC in FGP.

We also tested FGP-2 on different market conditions, namely, (a) down-trend period from 12/10/1976 to 12/04/78 (378 trading days); (b) side-way-trend period from 13/04/1978 to 27/03/1980 (496 trading days); and (c) up-trend period from 28/03/1980 to 09/04/81 (261 trad-

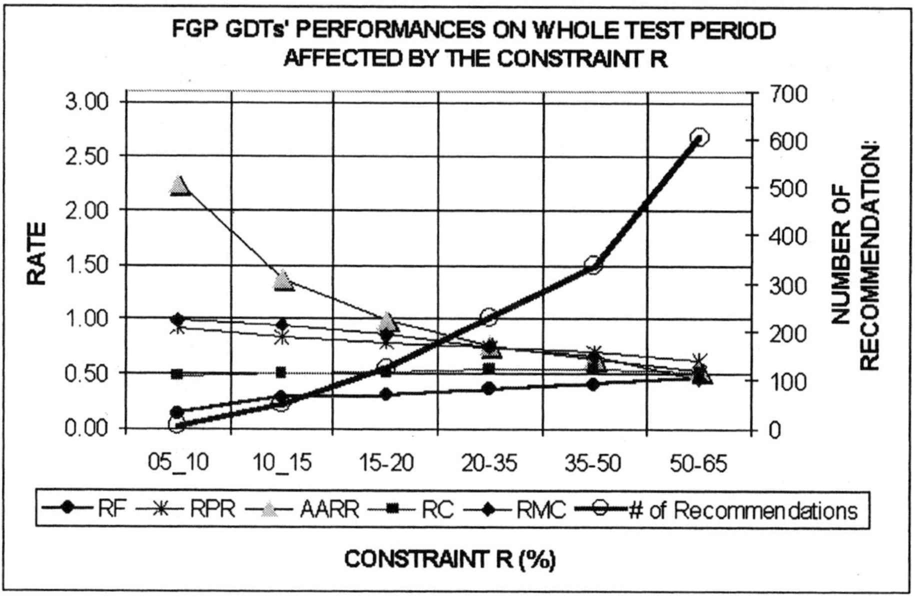

Figure 7.2. Visualization of the effect of the constraint $\Re$ on the mean performances of FGP-2

ing days), as illustrated in Figure 7.2. Results obtained were consistent with those shown above. FGP-2 was tested extensively in other data sets. For simplicity, details of values are not presented here.

4. Comparison Study

Up to this point, we only tested FGP-2 with the constrained fitness function on a financial index. Should it still be effective and applicable to individual stock data? How does FGP-2 compare with other methods? To partially answer these questions, we referred to Saad et al (1998) in which three specially developed Neural Networks, (i.e. Time Delay (TDNN) , Recurrent (RNN) and Probabilistic (PNN)), and a linear classifier were employed to address a similar prediction problem. They also have the goal of achieving low false alarm.

We compared performances based on predictions with $r = 2\%$ and $n = 22$; i.e. daily predictions on whether a return of 2% or more can be achievable within the next 22 trading days. We tested the above algorithms on ten stocks:

- Apple (AAPL), IBM(IBM), Motorola(MOT), Microsoft (MSFT): representing the technology group which generally has high volatility

- American Express(AXP), Well Fargo(WFC): representing the banks
- Walt Disney Co. (DIS), McDonald (MCD): representing consumer stocks
- Public Svc New Mexico (PNM), Energras (V.EEG): representing cyclical stocks

These data series vary in their starting dates, but all ended by 06/03/1997. Following (Saad et al. 1998), the last 100 days were chosen as the test data for each stock.

In the experiments, we ran FGP-2 10 times for each data set. For each run, we took 500 trading days just before the final 100 days as training data, and took a constraint $\Re$ = [20%, 30%] for most data sets except for AAPL, PNM and V.EEG, for which we took a constraint $\Re$ = [10%, 20%]. The $\Re$ values were chosen to reflect the percentage of positive positions in the data. The termination condition was set to 50 generations. Since FGP-2 is a probabilistic technique, it was run ten times for each share. For each share, we picked the best decision tree generated in FGP-2's ten runs for the purpose of comparison, as the same was done for the three different neural networks reported in (Saad et al. 1998).

Table 7.7 lists the performance of the three different NNs, a linear classifier and FGP-2 on 10 stocks. The "Total" column summarises the total number of predicted positive positions on all 10 stocks. The last column, "Ave.", reports the average rate of failure over 10 stocks. On average, the NNs out-performed the linear classifier in RF (7.56%, 3.05% and 3.61% as opposed to 18.62%). The average RF for all the GDTs generated was 5.08% (much better than the RF achieved by the linear classifier, 18.62%). The average RF by the best GDT was 1.29%, which was lower than any of the NNs.[2] The best-found GDT found 385 positive signals totally, which is slightly more than 372 found by the linear classifier. This shows that RMC has not been compromised by FGP-2 in its attempt to reduce RF. On individual shares, the RF by the best GDT found by FGP-2 was at least as good as the RF found by the NNs in 8 out of the 10 shares.

5. Conclusion and Future Research

How accurate a forecasting program can be is limited not only by the algorithm that it uses, but also by the quality of the data. In previous papers, we have reported the capability of FGP-2 in finding patterns in historical data when patterns exist. In this paper, we argue that, depending on the application, a user may want a forecasting program to

Table 7.7. Performance Comparisons among NNs., a Linear Classifier and FGP-2 in Terms of RF and N_+ (The Total Number of Predicted Positive Positions)

Stocks			AAPL	IBM	MOT	MSFT	AXP	WFC	DIS	MCD	PNM	V.EEG	Total	Ave.
Profit Opp. (r=2%;n=22)			**62**	**72**	**81**	**87**	**92**	**85**	**74**	**73**	**50**	**70**	**746**	**74.6**
PPN	Total N+		51	25	48	49	20	45	19	4	63	14	338	
	RF (%)		7.84	4.00	18.75	4.08	0.00	4.44	0.00	0.00	36.50	0.00		7.56
TDNN	Total N+		10	9	27	61	17	19	7	6	22	8	186	
	RF (%)		0.00	0.00	0.00	0.00	0.00	0.00	0.00	0.00	18.00	12.50		3.05
RNN	Total N+		16	22	33	46	49	29	48	53	35	37	368	
	RF (%)		0.00	0.00	3.03	2.17	0.00	0.00	0.00	5.66	17.14	8.11		3.61
The Linear Classifier	Total N+		82	24	87	17	10	22	2	32	20	76	372	
	RF (%)		31.71	20.83	18.39	0.00	0.00	13.64	0.00	21.88	60.00	19.74		18.62
Mean and STD of 10 GDTs	Total N+	Mean	18.5	68.7	20.7	26.8	38.3	66.6	20.1	40.2	23.4	49.4	373	
		STD	9.9	3.9	5.1	6.2	9.9	11.1	3.1	1.8	5.9	9.6		
	RF (%)	Mean	9.16	10.15	1.33	3.10	3.72	8.20	0.40	0.00	13.07	4.83		5.08
		STD	5.66	1.13	2.82	2.47	3.10	2.33	1.30	0.00	12.30	3.90		
The Best GDT	Total N+		4	70	28	33	39	69	22	43	28	49	385	
	RF (%)		0.00	8.57	0.00	0.00	0.00	4.35	0.00	0.00	0.00	0.00		1.29

sacrifice investment opportunities for high precision, or vice versa. Our mission is to develop a forecasting tool to help users achieve their preferred performance. FGP-2 allows us to favour precision or investment opportunities through adjusting the tightness of a constraint in the objective function. The effectiveness of FGP-2 in influencing the precision is supported by our experiments.

Many other issues are relevant to the practicality of FGP-2. First of all, it must be established that in the series to be predicted, past patterns will repeat themselves in the future. Secondly, predictions only improve one's odds statistically. One needs to know how to use the predictions to invest one's money, so as to reduce risk. Moreover, if EDDIE were to be asked to recommend buying cautiously, one must have a viable policy for investing one's capital when it is idle. These issues will be left to future research.

Acknowledgments

Full credit should be given to James Butler who invented the acronym EDDIE. He implemented EDDIE-1 and tested it on horse racing. Many colleagues and students at University of Essex have contributed to the EDDIE project through discussion. In particular, we would like to thank Paul Scott, Jeff Reynolds, Tung-Leng Lau, Robert Pugh, Tony Lawson (MSc student), Sheri Markose, Abdel Salhi, Hakan Er, Xin Yao and Qingfu Zhang. This project is partly supported by two Research Promotion Funds by the University of Essex. Jin Li was supported by the

Overseas Research Scholarship and the University of Essex Scholarship. The DJIA index data was kindly supplied by Professor Blake LeBaron of University of Wisconsin-Madison. Ten stock data were kindly supplied by Emad W. Saad of the Texas Tech University.

Notes

1. For the record, this rule recommended 33 "buy" over 500 trading days between 22 May 1998 and 21 April 2000. All 33 occasions resulted in a gain of 4% within 22 days, which was what the GDT was the target return during the training over the preceding 1000 days.

2. The favourable results by FGP may be partly due to the rather bullish market over test period in which over 50% of the position are positive for all the shares; e.g. 87% of the positions were positive for MSFT and 92% for AXP.

References

Li, J. and E. P. K. Tsang (1999a). "Improving Technical Analysis Predictions: An Application of Genetic Programming," *Proceedings of The 12th International Florida AI Research Society Conference*, 108–112.

Li, J. and E. P. K. Tsang (1999b). "Investment Decision Making Using FGP: A Case Study," *Proceedings of the 1999 Congress on Evolutionary Computation*, 1253–1259. IEEE Press.

Tsang, E. P. K., J. Li, and J. M. Butler (1998). "EDDIE Beats the Bookies," *International Journal of Software, Practice and Experience*, 28(10), 1033–1043. Wiley.

Saad, E. W., D. V. Prokhorov, and D. C. Wunsch II (1998). "Comparative Study of Stock Trend Prediction Using Time Delay, Recurrent and Probabilistic Neural Networks," *IEEE Transactions on Neural Networks*, 9(6), 1456–1470.

Tsang, E. P. K. and J. Li (1983). "Combining Ordinal Financial Predictions with Genetic Programming", in K. S. Leung, L.-W. Chan, and H. Meng (eds.), *Intelligent Data Engineering and Automated Learning—IDEAL 2000: Data Mining, Financial Engineering, and Intelligent Agents, Lecture Notes in Computer Sciences*, 532–537. Springer.

Chapter 8

FORECASTING MARKET INDICES USING EVOLUTIONARY AUTOMATIC PROGRAMMING

A Case Study

Michael O'Neill
University of Limerick
Ireland
michael.oneill@ul.ie

Anthony Brabazon
University College Dublin
Ireland
anthony.brabazon@ucd.ie

Conor Ryan
University of Limerick
Ireland
conor.ryan@ul.ie

Abstract This study examines the potential of an evolutionary automatic programming methodology, Grammatical Evolution, to uncover a series of useful technical trading rules for market indices. A number of markets are analysed; these are the UK's FTSE, Japan's Nikkei, and the German DAX. The preliminary findings indicate that the methodology has much potential.

Keywords: Evolutionary Automatic Programming, Grammatical Evolution, Market Indices, Technical Trading Rules, FTSE, DAX, Nikkei

Introduction

The objective of this study is to determine whether an evolutionary automatic programming methodology, Grammatical Evolution (GE), is capable of uncovering useful technical trading rules for a number of market indices.

We will begin by providing the background and motivation to this line of research, followed by descriptions of the problem domain and of the evolutionary automatic programming approach adopted. A demonstration of how GE can be applied to index trading will provided by conducting experiments on data from three different markets. The chapter will end with some discussion and conclusions.

Technical Analysis

A market index is comprised of a weighted average measure of the price of individual shares which make up that market. The value of the index represents an aggregation of the balance of supply and demand for these shares. Some market traders, technical analysts, believe that prices move in trends and that price patterns repeat themselves (Murphy 1999). If we accept the premise that there are rules, although not necessarily static rules, underlying price behaviour, it follows that trading decisions could be enhanced through use of an appropriate rule induction methodology such as Grammatical Evolution (GE).

Although controversy exists amongst financial theorists regarding the veracity of the claim of technical analysts, recent evidence has suggested that it may indeed be possible to uncover patterns of predictability in price behaviour. Brock, Lakonishok, and LeBaron (1992) found that simple technical trading rules had predictive power and suggested that the conclusions of earlier studies that technical trading rules did not have such power were premature. Other studies which indicated that there may be predictable patterns in share price movements include those which suggest that markets do not always impound new information instantaneously (Chan, Jegadeesh , and Lakonishok 1996), and that stock markets can overreact as a result of excessive investor optimism or pessimism (Dissanaike 1997). The continued existence of large technical analysis departments in international finance houses is consistent with the hypothesis that technical analysis has proven empirically useful.

Potential for Application of Evolutionary Automatic Programming

As noted by Iba and Nikolaev (2000), there are a number of reasons to suppose that the use of an evolutionary automatic programming (EAP) approach can prove fruitful in the financial prediction domain. EAP can conduct an efficient exploration of the search space and can uncover dependencies between input variables, leading to the selection of a good subset for inclusion in the final model. Additionally, use of EAP facilitates the utilisation of complex fitness functions including discontinuous, non-differentiable functions. This is of particular importance in the financial domain as the fitness criterion may be complex, usually requiring a balancing of return and risk. EAP, unlike, for example, basic neural net approaches to financial prediction, does not require the ex-ante determination of optimal model inputs and their related transformations. Another useful feature of EAP is that it produces human-readable rules that have the potential to enhance understanding of the problem domain.

Motivation for Study

This study was motivated by a number of factors. Much of the existing literature concerning the application of genetic algorithms (GAs) or GP to the generation of technical trading rules (allen; bauer; neely; Deboeck 1994) concentrates on the US and to a lesser extent the Japanese stock markets. Published research on this area is both incomplete and scarce. To date, only a limited number of GA / GP methodologies and a limited range of technical indicators have been considered. This study addresses these limitations by examining index data drawn from a number of markets, that is, the UK, German, and Japanese stock markets, and by adopting a novel evolutionary automatic programming approach.

The chapter is organised as follows. Section 1 discusses the background to the technical indicators utilised in this study. Section 2 describes the evolutionary algorithm adopted, Grammatical Evolution (O'Neill and Ryan 2001; O'Neill 2001; Ryan et al. 1998). Section 3 outlines the data and function sets used. The following sections provide the results of the study followed by a discussion of these results and finally a number of conclusions are derived.

1. Background

As with any modelling methodology, issues of data pre-processing need to be considered. Rather than attempting to uncover useful technical trading rules for an index using raw current and historical price information, this information is initially pre-processed. The objective of these pre-processing techniques is to uncover possible useful trends and other information in the time-series of the raw index data whilst simultaneously reducing the noise inherent in the series.

Technical Indicators

The development of trading rules based on current and historic market price information has a long history (Brown, Goetzmann , and Kumar 1998). The process entails the selection of one or more technical indicators and the development of a trading system based on these indicators. These indicators are formed from various combinations of current and historic price information. Although there are potentially an infinite number of such indicators, the financial literature suggests that certain indicators are widely used by investors (brock; Murphy 1999; Pring 1991).

Four groupings of indicators are given prominence in prior literature:

i. Moving average indicators

ii. Momentum indicators

iii. Trading range indicators

iv. Oscillators

Given the large search space, an evolutionary automatic programming methodology has promise to determine both a good quality combination of, and relevant parameters for, trading rules drawn from individual technical indicators.

We intend to use of each of these groupings as our model is developed. Our initial study on the FTSE dataset (O'Neill et al. 2001) included only a moving average indicator. This study also includes momentum, and trading range volatility indicators.

Moving Average Indicators.

The simplest moving average systems compare the current share price or index value with a moving average of the share price or index value over a lagged period, to determine how far the current price has moved from an underlying price trend. As they smooth out daily price fluctuations, moving averages can heighten the visibility of an underlying trend. A variation on simple moving average systems is to use a moving average convergence divergence (MACD) oscillator. This is calculated by taking the difference of a short run and a long run moving average.

In a recursive fashion, more complex combinations of moving averages of values calculated from a MACD oscillator can themselves be used to generate trading rules. For example, a nine day moving average of a MACD oscillator could be plotted against the raw value of that indicator. A trading signal may be generated when the two plotted moving averages cross. Moving average indicators are trend following devices and work best in trending markets. They can have a slow response to changes in trends in markets, missing the beginning and end of each move. They tend to be unstable in sideways moving markets, generating repeated buy and sell signals (whipsaw) leading to unprofitable trading. Trading systems using moving averages trade-off volatility (risk of loss due to whipsaw) against sensitivity. The objective is to select the lag period which is sensitive enough to generate a useful early trading signal but which is insensitive to random noise.

Momentum.

The momentum of a security is the ratio of a time-lagged price to the current price ($\frac{Price_t}{Price_{t-x}}$). The belief underlying this indicator is that strongly trending shares tend to continue to move in that direction for a period of time as more investors buy or sell the trending share. There is recent evidence that momentum trading strategies can work, particularly when investing in smaller firms. Technical analysts consider that price momentum can foretell a price turning point as momentum will tend to peak before the price peaks.

Trading Range Breakout systems.

In these systems, a signal is usually generated if a price breaks out of a defined range. A simple example of a trading rule would be to buy a share when it exceeds its previous high in the last four weeks and conversely to sell a share if it falls below its previous four week low. A more complex approach is to plot an envelope at ± 'x' standard deviations above and below a moving average. Penetration of the bands by the current day's price indicates a possible price trend reversal.

A description of the evolutionary automatic programming system used to evolve trading rules now follows.

2. Grammatical Evolution

Grammatical Evolution (GE) is an evolutionary algorithm that can evolve computer programs in any language (O'Neill and Ryan 2001; O'Neill 2001; Ryan et al. 1998). Rather than representing the programs as syntax trees, as in GP (koza), a linear genome representation is used. Each individual, a variable length binary string, contains in its codons (groups of 8 bits) the information to select production rules from a Backus Naur Form (BNF) grammar. As such, GE adopts a biologically-inspired, genotype-phenotype mapping process.

At present, the search element of the system is carried out by an evolutionary algorithm, although other search strategies with the ability to operate over variable-length binary strings have been used (O'Sullivan and Ryan 2002). In particular, future advances in the field of evolutionary algorithms can be easily incorporated into this system due to the program representation.

2.1. The Biological Approach

The GE system is inspired by the biological process of generating a protein from the genetic material of an organism. Proteins are fundamental in the proper development and operation of living organisms and are responsible for traits such as eye colour and height (Lewin 2000).

The genetic material (usually DNA) contains the information required to produce specific proteins at different points along the molecule. For simplicity, consider DNA to be a string of building blocks called nucleotides, of which there are four, named A, T, G, and C, for adenine, tyrosine, guanine, and cytosine respectively. Groups of three nucleotides, called codons, are used to specify the building blocks of proteins. These protein building blocks are known as amino acids, and the sequence of these amino acids in a protein is determined by the sequence of codons on the DNA strand. The sequence of amino acids is very important as it plays a large part in determining the final three-dimensional structure of the protein, which in turn has a role to play in determining its functional properties.

In order to generate a protein from the sequence of nucleotides in the DNA, the nucleotide sequence is first transcribed into a slightly different format, that being a sequence of elements on a molecule known as RNA. Codons within the RNA molecule are then translated to determine the sequence of amino acids that are contained within the protein molecule. The application of production rules to the non-terminals of the incomplete code being mapped in GE is analogous to the role amino acids

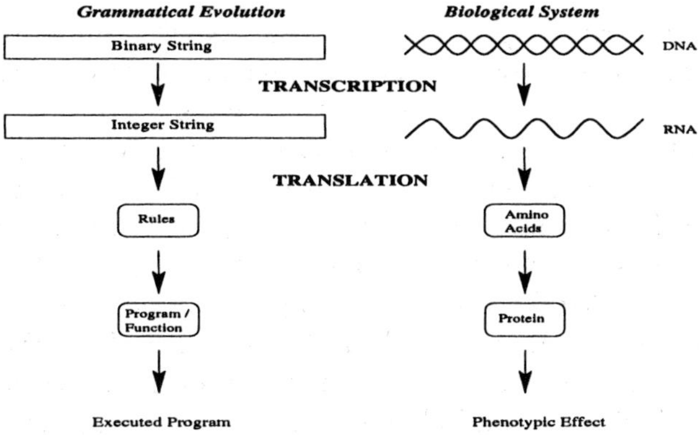

Figure 8.1. A comparison between the grammatical evolution system and a biological genetic system. The binary string of GE is analogous to the double helix of DNA, each guiding the formation of the phenotype. In the case of GE, this occurs via the application of production rules to generate the terminals of the compilable program. In the biological case by directing the formation of the phenotypic protein by determining the order and type of protein subcomponents (amino acids) that are joined together.

play when being combined together to transform the growing protein molecule into its final functional three-dimensional form.

The result of the expression of the genetic material as proteins in conjunction with environmental factors is the phenotype. In GE, the phenotype is a computer program that is generated from the genetic material (the genotype) by a process termed a genotype-phenotype mapping. This is unlike the standard method of generating a solution (a program in the case of GE) directly from an individual in an evolutionary algorithm by explicitly encoding the solution within the genetic material. Instead, a many-to-one mapping process is employed within which the robustness of the GE system lies.

Figure 8.1 compares the mapping process employed in both GE and biological organisms.

2.2. The Mapping Process

When tackling a problem with GE, a suitable BNF (Backus Naur Form) grammar definition must first be decided upon. The BNF can be either the specification of an entire language or, perhaps more usefully, a subset of a language geared towards the problem at hand.

In GE, a BNF definition is used to describe the output language to be produced by the system. BNF is a notation for expressing the grammar of a language in the form of production rules. BNF grammars consist of `terminals`, which are items that can appear in the language, e.g. binary operators **+**, **-**, unary operators **Sin**, constants **1.0** etc. and `non-terminals`, which can be expanded into one or more terminals and non-terminals. For example from the grammar detailed below, **<expr>** can be transformed into one of four rules, i.e it becomes **<expr><op><expr>**, **(<expr><op><expr>)** (which is the same as the first, but surrounded by brackets), **<pre-op>(<expr>)**, or **<var>**. A grammar can be represented by the tuple $\{N, T, P, S\}$, where N is the set of non-terminals, T the set of terminals, P a set of production rules that maps the elements of N to T, and S is a start symbol which is a member of N. When there are a number of productions that can be applied to one element of N the choice is delimited with the '|' symbol. For example,

```
N = { <expr>, <op>, <pre_op> }
T = { Sin, +, -, /, *, X, 1.0, (, ) }
S = <expr>
```

And P can be represented as:

```
(A) <expr> ::= <expr> <op> <expr>       (0)
             | ( <expr> <op> <expr> )   (1)
             | <pre-op> ( <expr> )      (2)
             | <var>                    (3)
(B) <op> ::= +           (0)
           | -           (1)
           | /           (2)
           | *           (3)
(C) <pre-op> ::= Sin
(D) <var> ::= X          (0)
            | 1.0        (1)
```

The compilable code produced will consist of elements of the terminal set T. The grammar is used in a developmental approach whereby the evolutionary process evolves the production rules to be applied at each stage of a mapping process, starting from the start symbol, until a complete program is formed. A complete program is one that is comprised solely from elements of T.

As the BNF definition is a plug-in component of the system, it means that GE can produce code in any language thereby giving the system a unique flexibility.

For the above BNF, Table 8.1 summarizes the production rules and the number of choices associated with each.

The genotype is used to map the start symbol onto terminals by reading codons of 8 bits to generate a corresponding integer value, from which an appropriate production rule is selected by using the following map-

Table 8.1. The Number of Choices Available from Each Production Rule

Rule no.	Choices
A	4
B	4
C	1
D	2

ping function:

$$Rule = Codon\ Value\ MOD\ No.\ Rule\ Choices$$

Consider the following rule from the given grammar i.e., given the non-terminal *op*, which describes the set of binary operators that can be used, there are four production rules to select from.

```
(B) <op> :: =    +        (0)
               | -        (1)
               | /        (2)
               | *        (3)
```

If we assume the codon being read produces the integer 6, then

$$6\ MOD\ 4\ =\ 2$$

would select rule (2). Each time a production rule has to be selected to transform a non-terminal, another codon is read. In this way the system traverses the genome.

During the genotype-to-phenotype mapping process, it is possible for individuals to run out of codons, and in this case we wrap the individual and reuse the codons. This is quite an unusual approach in EAs, as it is entirely possible for certain codons to be used two or more times. This technique of wrapping the individual draws inspiration from the gene-overlapping phenomenon that has been observed in many organisms (Lewin 2000).

In GE, each time the same codon is expressed it will always generate the same integer value, but, depending on the current non-terminal to which it is being applied, it may result in the selection of a different production rule. We refer to this feature as intrinsic polymorphism. What is crucial, however, is that each time a particular individual is mapped from its genotype to its phenotype, the same output is generated. This is the case because the same choices are made each time. However, it is possible that an incomplete mapping could occur, even after several wrapping events, and in this case the individual in question is given the lowest possible fitness value. The selection and replacement mechanisms then operate accordingly to increase the likelihood that this individual is removed from the population.

An incomplete mapping could arise if the integer values expressed by the genotype were applying the same production rules repeatedly. For example, consider an individual with three codons, all of which specify rule 0 from below,

```
(A) <expr> :: =  <expr><op><expr>          (0)
                 |(<expr><op><expr>)        (1)
                 |<pre-op>(<expr>)          (2)
                 |<var>                     (3)
```

even after wrapping the mapping process would be incomplete and would carry on indefinitely unless stopped. This occurs because the nonterminal `<expr>` is being mapped recursively by production rule 0, i.e., it becomes `<expr><op><expr>`. Therefore, the leftmost `<expr>` after each application of a production would itself be mapped to a `<expr><op><expr>`, resulting in an expression continually growing as follows: `<expr><op><expr><op><expr><op><expr>` etc.

Such an individual is dubbed invalid as it will never undergo a complete mapping to a set of terminals. For this reason we impose an upper limit on the number of wrapping events that can occur. It is clearly essential that stop sequences are found during the evolutionary search in order to complete the mapping process to a functional program. The stop sequence being a set of codons that result in the non-terminals being transformed into elements of the grammars terminal set.

Beginning from the left hand side of the genome then, codon integer values are generated and used to select rules from the BNF grammar, until one of the following situations arise:

i. A complete program is generated. This occurs when all the non-terminals in the expression being mapped are transformed into elements from the terminal set of the BNF grammar.

ii. The end of the genome is reached, in which case the *wrapping* operator is invoked. This results in the return of the genome reading frame to the left hand side of the genome once again. The reading of codons will then continue, unless an upper threshold representing the maximum number of wrapping events has occurred during this individual's mapping process.

iii. In the event that a threshold on the number of wrapping events has occurred and the individual is still incompletely mapped, the mapping process is halted, and the individual is assigned the lowest possible fitness value.

To reduce the number of invalid individuals being passed from generation to generation, a steady state replacement mechanism is employed. One consequence of the use of a steady state method is its tendency to maintain fit individuals at the expense of less fit, and in particular, invalid individuals.

3. Problem Domain & Experimental Approach

We describe an approach to evolving trading rules using GE. This study uses daily data for the UK FTSE 100, the German DAX, and the Japanese NIKKEI stock indices.

The FTSE data is drawn from the period 26/04/1984 to 4/12/1997, the training data set was comprised of 365 days from the first day plus an additional 75 days at the beginning of this time to allow for the time lag introduced with technical indicators such as the moving average. The remaining data is divided into five hold out samples totaling 2125 trading days, see Figure 8.2.

The DAX and NIKKEI data are drawn from the period 01/01/1991 to 03/12/1997, with the initial 440 days becoming the training data set as before. The remaining data is divided into two hold out samples in each case. These periods can be seen in Figures 8.3 and 8.4

The division of the hold out period into a number of segments is undertaken to allow comparison of the out of sample results across different market conditions, in order to assess the stability and degradation characteristics of the developed model's predictions.

The rules evolved by GE are used to generate one of three signals for each day of the training or test periods. The possible signals are *Buy*, *Sell*, or *Do Nothing*. Permitting the model to output a Do Nothing signal reduces the hard threshold problem associated with production of a binary output. This issue has not been considered in a number of prior studies. A variant on the trading methodology developed in Brock et al. (1992) is then applied.

If a buy signal is indicated, a fixed investment of \$1,000 (arbitrary) is made in the market index. This position is closed at the end of a ten day (arbitrary) period.

On the production of a sell signal, an investment of \$1,000 is sold short and again this position is closed out after a ten day period. This gives rise to a maximum potential investment of \$10,000 at any point in time (the potential loss on individual short sales is in theory infinite but in practice is unlikely to exceed \$1,000). The profit (or loss) on each transaction is calculated taking into account a one-way trading cost of 0.2% and allowing a further 0.3% for slippage. The total return generated by the developed trading system is a combination of its trading return and its risk free rate of return generated on uncommitted funds.

The rate adopted in this calculation is simplified to be the average interest rate over the entire data set. For the FTSE the rate is 8.5%, the DAX 6.0% and for the NIKKEI it is 1.5%.

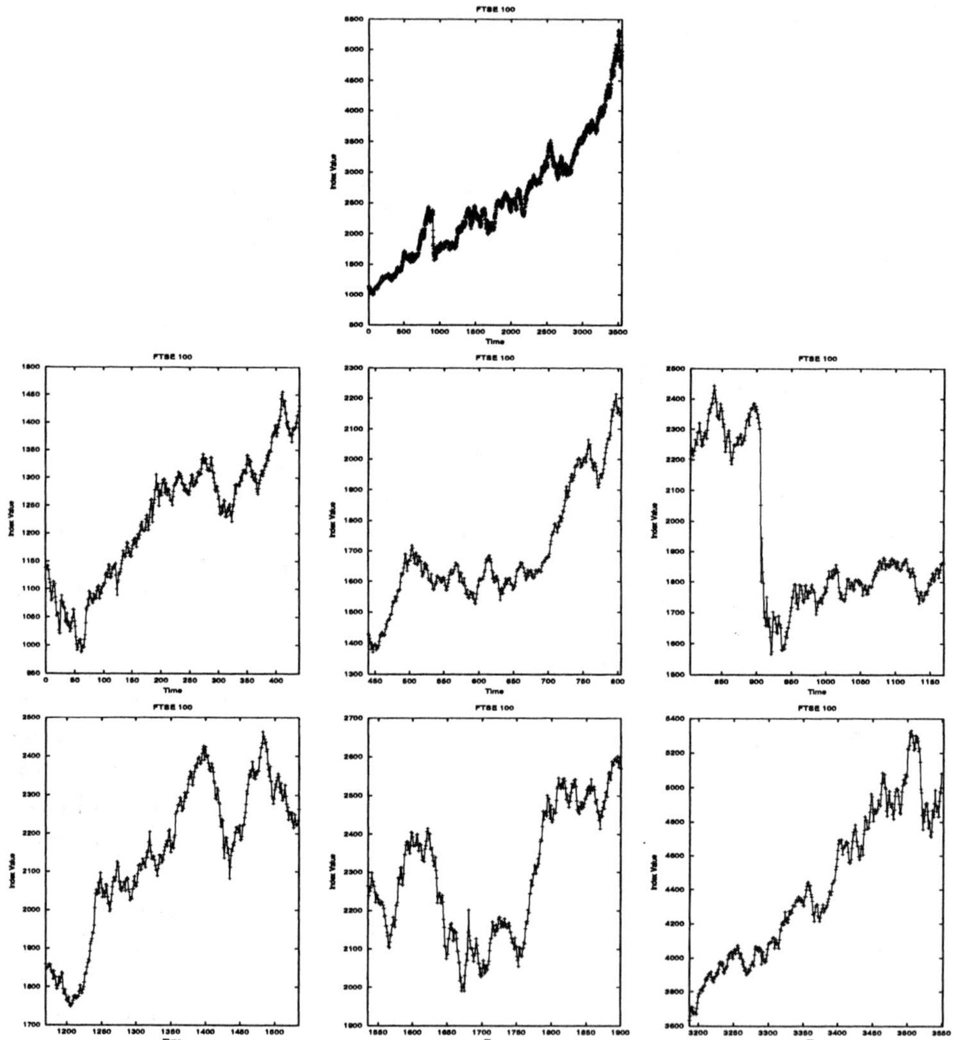

Figure 8.2. A plot of the FTSE 100 over the entire data set (top), over the training period (middle-left), over the first two test periods. Days 365 to 730 (middle-center), and days 730 to 1095 (middle-right), and the third, fourth & fifth test periods (bottom row, from left to right).

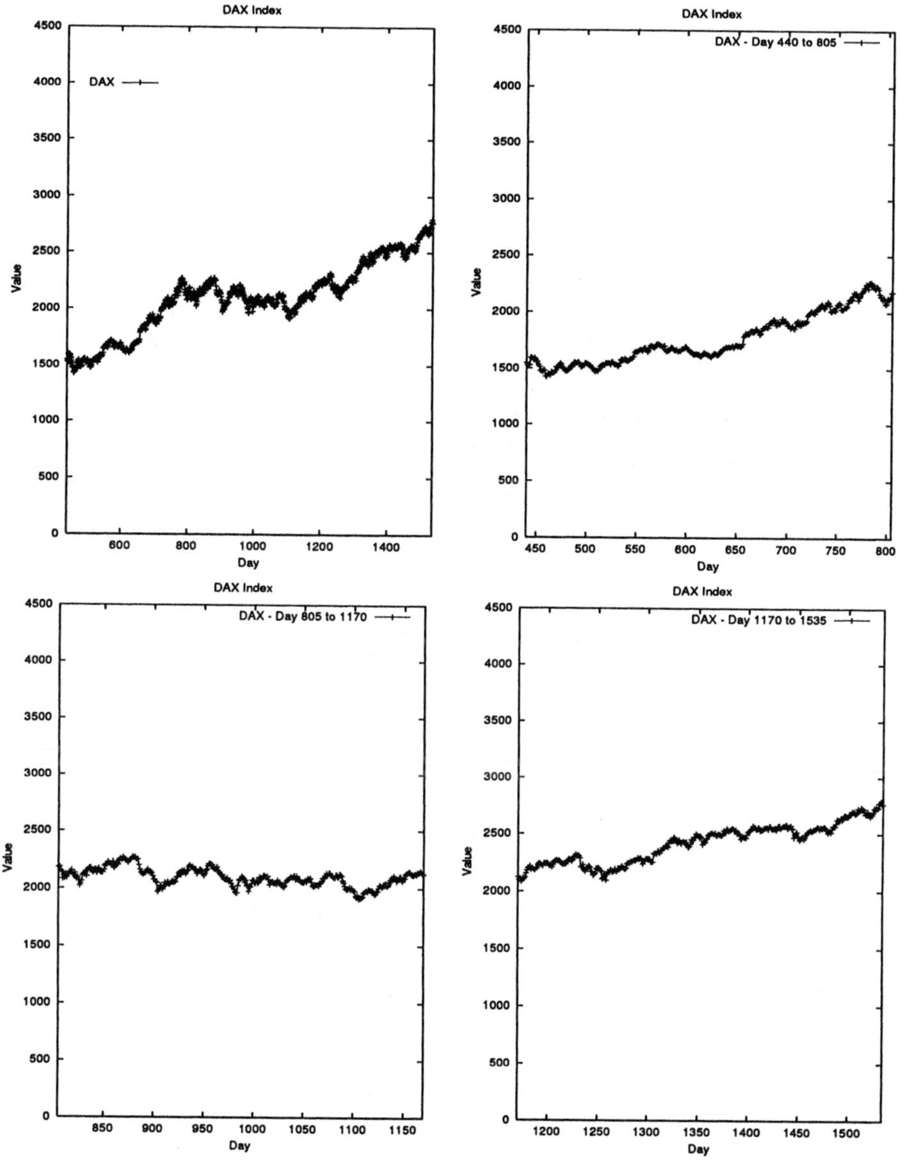

Figure 8.3. A plot of the DAX over the entire data set (top left). Taken from this data set period illustrated we can see the training period (top right), and the two test periods (bottom left and right respectively).

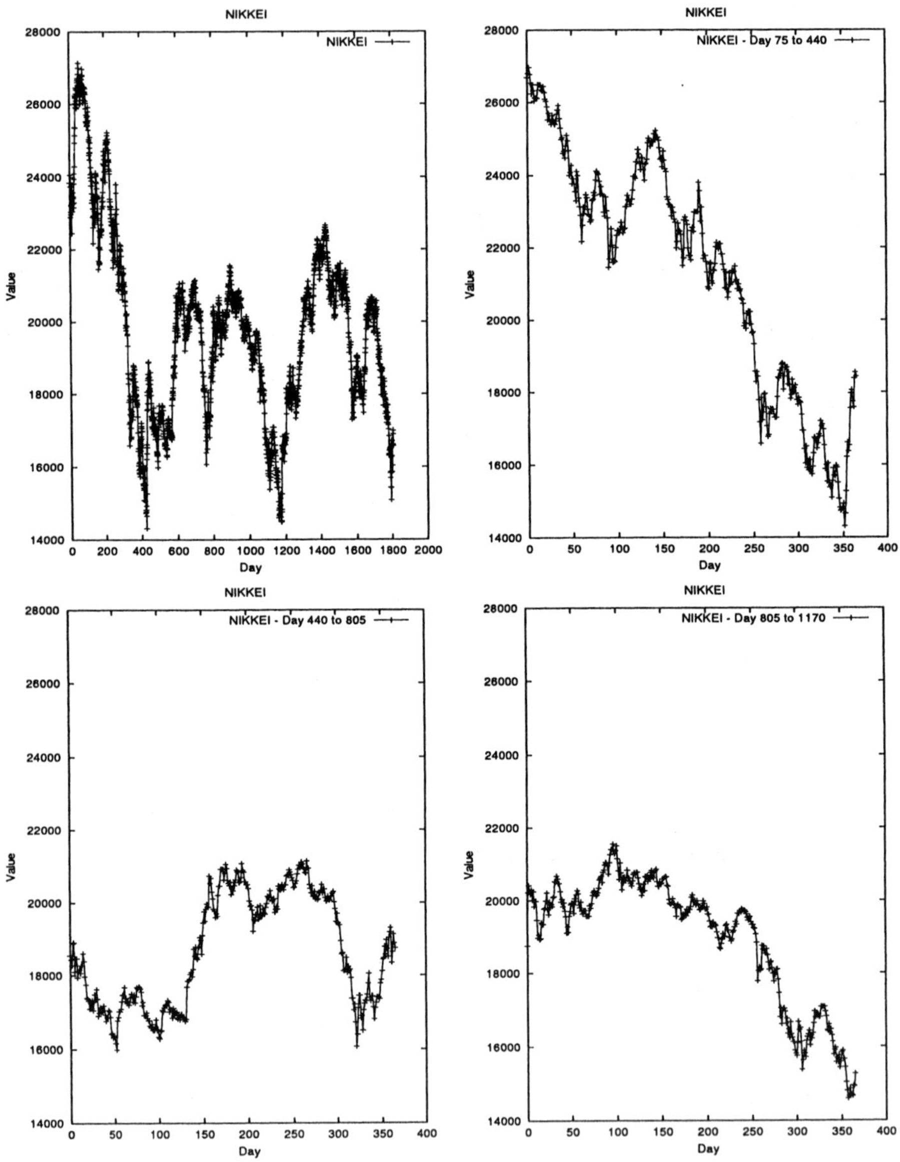

Figure 8.4. A plot of the Nikkei over the entire data set (top left), over the training period (top right), over the two test periods (bottom left and right respectively).

As well as the moving average, the momentum and trading range volatility technical indicators are adopted in these preliminary experiments, as can be seen in the grammar, given below.

```
N={<code>,<expr>,<fopbi>,<fopun>,<matbi>,<relbi>,<var>,<int>}
T={p,=,(,),f_and,f_or,f_not,+,-,*,>,<,>=,<=,scale,ma,day,1,2,3,4,5,10}
S=<code>
P={ <code> ::=  p = <expr> ;

<expr> ::= <fopbi> (<expr>, <expr>) | <fopun> (<expr>) | <expr><matbi><expr>
         | <expr><relbi><expr> | <var>

<fopbi> ::= f_and | f_or

<fopun> ::= f_not

<matbi> ::= + | - | *

<relbi> ::= > | < | >= | <=

<var> ::= <int> | day | ma(<int>,day) | momentum(<int>,day)  | trb(<int>,day)

<int> ::= 1 | 2 | 3 | 4 | 5 | 10 }
```

In addition to the technical indicators the grammar also allows the use of the binary operators `f_and`, `f_or`, the standard arithmetic operators, and the unary operator `f_not`, and the current days index value `day`. The operations `f_and`, `f_or`, and `f_not` return the minimum, maximum, of the arguments, and 1 - the argument, respectively.

The signals generated for each day, Buy, Sell, or Do Nothing, are post-processed using fuzzy logic. The trading rule, a fuzzy trading rule, returns values in the range 0 to 1. We use pre-determined membership functions, in this case, to determine what the meaning of this value is. The membership functions adopted were as follows:

$$Buy = Value < .33$$
$$DoNothing = .33 >= Value < .66$$
$$Sell = .66 >= Value$$

3.1. Data Preprocessing

The values of the indices changed substantially over the training and testing period. Before the trading rules were constructed, these values were normalised using a two phase preprocessing. Initially the daily values were transformed by dividing them by a 75 day lagged moving average. These transformed values are then normalised using linear scaling into the range 0 to 1. This procedure is a variant on that adopted by Allen and Karjalainen (1999) and Iba and Nikolaev (2000).

3.2. Selection of Fitness Function

A key decision in applying an EAP methodology to construct a technical trading system is to determine what fitness measure should be adopted. A simple fitness measure such as the profitability of the system both in and out of sample is inadequate as it fails to consider the risk associated with the developed trading system. The risk of the system can be estimated in a variety of ways. One possibility is to consider market risk, defined here as the risk of loss of funds due to a market movement. A measure of this risk is provided by the maximum drawdown (maximum cumulative loss) of the system during a training or test period. This measure of risk can be incorporated into the fitness function in a variety of formats including: (return/maximum drawdown) or return - 'x'(maximum drawdown), where 'x' is a pre-determined constant dependent on an investor's psychological risk profile. For a given rate of return, the system generating the lowest maximum drawdown is preferred.

This study incorporates drawdown in the fitness function by subtracting the maximum cumulative loss during the training period from the profit generated during that period. This is a conservative approach which will encourage the evolution of trading systems with good return to risk characteristics. This will provide a more stringent test of trading rule performance as high risk/high reward trading rules will be discriminated against.

4. Results

Thirty runs were performed on each of the three datasets using a population size of 500 individuals over 100 generations. Bit mutation at a probability of 0.01, and a variable-length one-point crossover probability of 0.9 was adopted. A comparison of the best individuals evolved for each dataset to the benchmark buy and hold strategy can be seen in Table 8.2, Table 8.3, and Table 8.4, for the FTSE, DAX and NIKKEI datasets respectively.

In the case of the FTSE and NIKKEI datasets the evolved rules produce a superior performance to the benchmark strategy, while performance over the DAX dataset is not as strong. The poorer performance for the DAX market can be attributed to over-fitting of the evolved rules to the training data. For each of the evolved trading rules the associated risk is less than that of the benchmark strategy as can be seen in the average daily investment figures reported.

Table 8.2. A comparison of the buy and hold benchmark to the best evolved individual for the FTSE dataset.

Trading Period (Days)	**Buy & Hold** Profit (US$)	**Best-of-run** Profit(US$)	**Best-of-run** Avg. Daily Investment
Train (75 to 440)	3071	3156	3219
Test 1 (440 to 805)	5244	1607	1822
Test 2 (805 to 1170)	-1376	4710	3151
Test 3 (1170 to 1535)	1979	2387	6041
Test 4 (1535 to 1900)	1568	-173	3274
Test 5 (3196 to 3552)	3200	2221	3767
Total	13686	13908	

Table 8.3. A comparison of the benchmark buy and hold strategy to the best evolved individual on the DAX dataset.

Trading Period (Days)	**Buy & Hold** Profit (US$)	**Best-of-run** Profit(US$)	**Best-of-run** Avg. Daily Investment
Train (440 to 805)	3835	3648	7548
Test 1 (805 to 1170)	-41	-1057	8178
Test 2 (1170 to 1535)	3016	469	8562
Total	6831	3060	

Table 8.4. A comparison of the benchmark buy and hold strategy to the best evolved individual on the NIKKEI dataset.

Trading Period (Days)	**Buy & Hold** Profit (US$)	**Best-of-run** Profit(US$)	**Best-of-run** Avg. Daily Investment
Train (75 to 440)	-6285	3227	9247
Test 1 (440 to 805)	59	-1115	7164
Test 2 (805 to 1170)	-3824	633	9192
Total	-10050	2745	

5. Discussion

In evaluating the performance of any market predictive system, a number of caveats must be borne in mind. Any trading model constructed and tested using historic data will tend to perform less well in a live environment than in a test period for a number of reasons. Live markets have attendant problems of delay in executing trades, illiquidity, interrupted/corrupted data and interrupted markets. The impact of these issues is to raise trading costs and consequently to reduce the profitability of trades generated by any system. An allowance for these costs ("slippage") has been included in this study but it is impossible to determine the scale of these costs ex-ante with complete accuracy. In addition to these costs, it must be remembered that the market is competitive. As new computational technologies spread, opportunities to utilise these technologies to earn excess risk-adjusted profits are eroded. As a result of this technological "arms-race", estimates of trading performance based on historical data may not be replicated in live trading as other market participants will apply similar technology. This study ignores impact of dividends. Although a buy-and-hold strategy will generate higher levels of dividend income than an active trading strategy, the precise impact of this factor is not determinable ex-ante. It is notable that the dividend yield on most stock exchanges has fallen sharply in recent years and that the potential impact of this factor has lessened.

6. Conclusions & Future Work

Grammatical Evolution has been shown to successfully generate trading rules with a performance superior to the benchmark buy and hold strategy on two of the three datasets analysed. In addition the risk involved with the adoption of the evolved trading rules is less than the benchmark.

The risk of the benchmark buy-and-hold portfolio exceeded that of the portfolio generated by the technical trading rules because, the benchmark buy and hold portfolio maintains a fully invested position at all times in the market, whereas the portfolio generated by the evolved technical trading system averaged a capital investment of \$3,546, \$8,096, and \$8,534 over the trading periods on the FTSE, DAX, and NIKKEI datasets respectively.

The results clearly show that there is much potential in this model and that there is notable scope for further research utilising GE in this problem domain. Our preliminary methodology has included a number of simplifications, for example, we only considered a small set of primitive technical indicators. The incorporation of additional technical indicators may further improve the performance of our approach. One factor that can have a large impact on the performance of GE is the choice of grammar. In this case study only one grammar was investigated, and thus trading rules conforming to one syntactical structure were evolved. Future work will involve the investigation of more complex syntactical structures.

A number of additional extensions of this work are left for future development. One extension would be to incorporate a more complex model of learning (forgetting). Glassman (1973) suggested that the "fallibility of memory" (p. 88) may represent a useful adaptive device when faced with a dynamic environment. At present in our model, all historic data observations are given equal weighting which implicitly assumes model stationarity. By a suitable modification of the fitness function, whereby more recent data observations are assigned a higher weighting in the model construction process, model development could be biased towards more recent data (Refenes et al. 1997). The weighting parameter could also be evolved as a component of the developed model.

Another avenue for exploration is the utility of stacked or multi-stage models (Zhang et al. 1997). The construction of a single GE trading model, implicitly assumes that there is a dominant global error minimum and implicitly hopes that the constructed model approaches this minimum. Given the limitations of a problem domain in which input/output relationships are dynamic and where input data is incomplete and noisy, the error surface may not have this property. In such a topology there may be potential to improve predictive quality by building multiple models. To implement this approach, a series of GE models could be developed to produce trading signals, using non-homogeneous inputs. The predictions of the individual models could then be utilised as inputs to a second stage model which produces the final trading signal. This second-stage model could be developed using GE or alternatively could employ a different methodology such as neural networks.

References

Allen, F. and R. Karjalainen (1999). "Using genetic algorithms to find technical trading rules," *Journal of Financial Economics*, 51, 245–271.

Bauer R. (1994). *Genetic Algorithms and Investment Strategies.* New York: John Wiley & Sons.

Brock, W., J. Lakonishok, and B. LeBaron (1992). "Simple Technical Trading Rules and the Stochastic Properties of Stock Returns," *Journal of Finance*, 47(5), 1731–1764.

Brown, S., W. Goetzmann, and A. Kumar (1998). "The Dow Theory: William Peter Hamilton's Track Record Reconsidered," *Journal of Finance*, 53(4), 1311–1333.

Chan, L. K. C., N. Jegadeesh, and J. Lakonishok (1996). "Momentum strategies," *Journal of Finance*, 51(5), 1681–1714.

Deboeck G. (1994). "Using GAs to Optimise a Trading System," in G. Deboeck (ed.), *Trading on the Edge: Neural, Genetic and Fuzzy Systems for Chaotic and Financial Markets.* New York: John Wiley & Sons.

Dissanaike, G. (1997). "Do Stock Market Investors Overreact?," *Journal of Business Finance and Accounting* (UK), 24(1), 27–50.

Glassman, R. (1973). "Persistence and Loose Coupling in Living Systems," *Behavioral Science*, 18, 83–98.

Iba H. and N. Nikolaev (2000). "Genetic Programming Polynomial Models of Financial Data Series," in *Proceedings of CEC 2000*, 1459–1466. IEEE Press.

Koza, J. (1992). *Genetic Programming.* MIT Press.

Lewin, B. (2000). *Genes VII.* Oxford University Press.

Murphy, J. J. (1999). *Technical Analysis of the Financial Markets.* New York: New York Institute of Finance.

Neely, C., P. Weller, and R. Dittmar (1997). "Is Technical Analysis in the Foreign Exchange Market Profitable? A Genetic Programming Approach," *Journal of Financial and Quantitative Analysis*, 32(4), 405–428.

O'Neill M. (2001). "Automatic Programming in an Arbitrary Language: Evolving Programs with Grammatical Evolution;" Ph.D. thesis. University of Limerick.

O'Neill M. and C. Ryan (2001). "Grammatical Evolution," *IEEE Transactions on Evolutionary Computation.*

O'Neill M., A. Brabazon, C. Ryan, and J. J. Collins (2001). "Evolving Market Index Trading Rules Using Grammatical Evolution," *LNCS 2037, Applications of Evolutionary Computation: EvoWorkshops 2001*, 343–352. Springer.

O'Sullivan J. and C. Ryan (2002). "An Investigation into the Use of Different Search Strategies with Grammatical Evolution," in *Proceedings of EuroGP'2002*, In print.

Pring, M. (1991). *Technical Analysis Explained: The Successful Investor's Guide to Spotting Investment Trends and Turning Points.* McGraw-Hill.

Refenes, A. N., Y. Bentz, D. W. Bunn, A. N. Burgess, and A. D. Zapranis (1997). "Financial Time Series Modelling with Discounted Least Squares Backpropagation," *Neurocomputing*, 14, 123–138.

Ryan C., J. J. Collins, and M. O'Neill (1998). "Grammatical Evolution: Evolving Programs for an Arbitrary Language," *LNCS 1391, Proceedings of EuroGP'98*, 83–95. Springer-Verlag.

Zhang, J., E. B. Martin, A. J. Morris, and C. Kiparissides (1997). "Inferential Estimation of Polymer Quality Using Stacked Neural Networks," *Computers and Chemical Engineering*, 21(Supplement), 1025–1030.

Chapter 9

GENETIC FUZZY EXPERT TRADING SYSTEM FOR NASDAQ STOCK MARKET TIMING

Sze Sing Lam
School of Business & Administration
The Open University of Hong Kong
Homantin, Kowloon
Hong Kong
sslam@ouhk.edu.hk

Kai Pui Lam and Hoi Shing Ng
Department of Systems Engineering & Engineering Management
The Chinese University of Hong Kong
Shatin, N.T.
Hong Kong
kplam@se.cuhk.edu.hk, hsng@se.cuhk.edu.hk

Abstract Technical indicators are developed for monitoring the movement of stock prices from different perspective. They are widely used to define trading rules to assist investors to make the buy-sell-hold decision. Most of these trading rules are vague and fuzzy in nature. Since the stock prices are affected by the artificial news factors from time to time, investors cannot be the winner all the time on using the same set of trading rules. The weight (i.e., the significance) of a trading rule to investors is varying with time. The problem of determining the weight of the trading rules can be modelled as an optimization problem. In this paper, we propose a Genetic Fuzzy Expert Trading System (GFETS) for market timing. We apply the fuzzy expert system to simulate vague and fuzzy trading rules and give the buy-sell signal. The set of trading rules adopted in the system will vary with time and is optimized using Genetic Algorithm (GA). Two training approaches—incremental and dynamic—are designed and studied. The system was evaluated with the stocks in

NASDAQ market. Experimental results showed that the system can give reliable buy-sell signals and using the system to perform buy-sell can produce significant profit.

Keywords: Genetic Algorithm, Fuzzy Expert System, Technical Indicator, Trading Rules

Introduction

"Stock market news has gone from hard to find (in the 1970s and early 1980s), then easy to find (in the late 1980s), then hard to get away from. The financial weather is followed as closely as the real weather: highs, lows, troughs, turbulence, and endless speculation about what's next and how to handle it" (Lynch and Rothchild 2000). Therefore, "Buy low and then sell high" could only be a dream of stock market practitioners. Investors always ask: "when should we buy or sell?" To answer the question, it requires the prediction of reversal in price trends of stocks, which is difficult though not impossible.

Stock prices are affected by a number of known as well as unknown economic factors. This is further complicated by the participation of human investors whose decision follows their own set of rules or principles and the interpretation of economic factors can vary. An economic factor can be both negative and positive. Based on these views, famous scholar in financial area such as William Sharpe disbelieved that ones could time the market and gave a very pessimistic view of market timing approaches (Allen and Karjalainen 1999). Although market-timing techniques were criticized in the literature, many investors use them to provide supporting evidence for marking the buy-sell decision. Recent research suggested that it is possible to construct good trading rules based on some relatively simple relationships (Allen and Karjalainen 1999; Edmonds, Burkhardt, and Adjei 1996).

Technical indicators are designed for identifying the price trend using the historical data. They are frequently used by investors for deriving trading rules. Although these trading rules are mostly vague and fuzzy, they can be modelled as fuzzy rules of the fuzzy expert system. The inference mechanism of the fuzzy expert system can effectively stimulate these trading rules to give buy-sell signals. Unfortunately, these indicators may contain related or conflicting information and the trading rules may produce confusing buy-sell-hold suggestions. A selection process is required to extract the representative rules from a vast amount of trading rules. Genetic Algorithm (GA) provides a good mean for the selection process. Based on the historical data, the buy-and-sell signal generates by the fuzzy expert system. In this paper, GA will be integrated with the fuzzy expert system to form a stock marketing timing system. The system was evaluated using the data of the NASDAQ market. NASDAQ is the major hi-tech stock market in the world. The largest hi-tech companies always choose the NASDAQ market as their first posting market (http://www.nasdaq.com) because raising the fund in NASDAQ can improve the image of the companies. The NASDAQ's companies have similar characteristics. They are major player in their own industry. They have high growth rate. Their profit can increase hundred times each year and so is their stock price. However, their stock price can also plummet if they fail to keep the high growth rate. In general, the stock price of NASDAQ fluctuates greatly making it a challenging task to buy-sell in this market. In our experiment, we have evaluated the system using the NASDAQ's data for Microsoft, Oracle, CISCO, IBM and Starbucks.

The paper is organized as follows. Section 1 presents the basic design of GFETS and the evaluation results of this simple system. Sections 2 and 3 discuss about enhancing the system with incremental and dynamic training approaches, respectively. The performance evaluation of these training approaches are reported in their section. Lastly, the further research direction is given in Section 4.

1. Genetic Fuzzy Expert System for Market Timing

In this section, we will describe the methodology of using GFETS in market timing. We will first give a brief overview about technical analysis and then discuss about the modelling of trading rules in GFETS. Finally, we will illustrate how Genetic Algorithm (GA) can be applied to select the trading rules in GFETS.

1.1. Technical Analysis

Technical analysis is the study of the price and volume data of stocks for the purpose of identifying the changes in price trends in the market. The market is assumed to be a reflection of human investors' attitude towards the changing market factors. Although human relationships are extremely complex and never repeat in identical combinations, the recurrence of similar characteristics in the market is sufficient for forecasting future price trends. Indicators are developed to determine the trend and therefore play an importance role for the stock analysis. Generally, technical analysis techniques can be divided into three major areas including: trend analysis, momentum analysis and volatility analysis (

Richard, Bauer , and Julie 1999).

TREND ANALYSIS - used to indicate whether a new trend has appeared in the data or an existing trend has finished. For examples, Moving Average (MA) is an indicator for the trend analysis. A simple MA is the sum of stock prices divided by the total number of stock prices.

MOMENTUM ANALYSIS - measures the rate of change of data as opposed to the actual levels. For example, the Relative Strength Index (RSI) is considered to be most useful as an indicator of trend reversals when there is divergence between the RSI and the price lines.

VOLATILITY ANALYSIS - measures the rate of random change in market prices.

Trading or decision rules can be defined on the indicator to provide a signal to indicate to the technical analyst a course of action (e.g. buy, sell, hold). For example, a simple rule on MA say that the investor should buy if the closing price of a day crosses above its 5 days MA because this suggests that the price trend will up in short term. One or more of these trading rules can be combined using the knowledge-based approach to form a trading system. Depending on the indicator being used and the rationale behind its construction, trading rules can be well defined as well as vagueness. For example, a popular trading rule of RSI suggests investors should buy if the RSI is above 70. Since the interpretation of "above" is non-exact, the rule is vague unlike that of MA. This type of trading rules can be better modelled as fuzzy trading rules. Fuzzy trading rules can be combined into a fuzzy expert system to form a stock market timing system to give effective buy-sell signals.

1.2. Fuzzy Trading Rules

GFETS is a fuzzy expert system. Fuzzy expert system is a general system to perform approximate reasoning. It applies a set of if-then conditional statements or rules in the canonical form to map the input space to output space (Driankov, Hellendoorn , and Reinfrank 1993). Both spaces are specified as fuzzy concepts or linguistic variables, which are defined by membership functions. A membership function is a curve that defines how each point in the input space is mapped to a degree of membership between 0 and 1. Interpreting an if-then involves evaluating the antecedent (which involves fuzzifying the input and applying any necessary fuzzy operators) and applying that result to the consequent (known as implication). The output of each rule is fuzzy sets and normally aggregated into a single output fuzzy set, which is then defuzzified to give a single number. The interpretation of a fuzzy system is non-unique and problem specific. A lot of fuzzy operators, implication method, and defuzzification approaches exist in the literature (e.g. see Klir and Yuan 1995). Their selection and application are domain specific.

In GFETS, we apply the conventionally {min, max} operators for {AND, OR} operations. Since the output of the system is buy, sell, and hold signal, which is crisp in nature, we use Sugeno method of fuzzy inference. It is similar to the typical fuzzy expert system except that the output is crisp rather than fuzzy (Surgeno and Murakami 1985). For examples, a simple trading rule based on the KD-line can be written as follows:

- If %K is small, decision is buy.
- If %K is large, decision is sell.

where the %K is the KD-line input, and "mall" and "large" are fuzzy sets to qualify the input. Similarly, a trading rules for the weighted moving average can be expressed in the following form:

- If the daily closing price crosses above the 5 days weighted moving average, decision is buy.
- If the daily closing price crosses below the 5 days weighted moving average, decision is sell.

It differs from the KD-line's rules only by the input. Since the output membership functions are singleton and the implication method is multiplication and the aggregation operator just includes all the singletons, the output can be defuzzified using weighted average approach (see Driankov, Hellendoorn , and Reinfrank 1993). Based on this model, any fuzzy trading rule of popular indicators can be formulated and included in the knowledge base of GFETS for market timing.

1.3. Selection of Fuzzy Trading Rules Using Genetic Algorithm

As mentioned before, the fuzzy trading rules may be inconsistent and conflicting with each other because they utilize the different information obtained from the price and volume data. The reliability of the indicator may vary with time. Even good indicators can lose their validity when combined with the others in a trading system in a particular point of time. The fuzzy trading rules to be adopted for market timing must be adjusted to meet the dynamic changes in the market. In general, it would be difficult to select the most relevant fuzzy trading rules given so many fuzzy trading rules to be optimized for such large amount of price and volume data. In this paper, we propose a GA to solve the fuzzy trading rules selection problem.

GA is a very effective and robust searching algorithm invented by Holland (1975) based on Darwinian evolution hypothesis for biological system. A population of individual evolutes to adapt to the environment changes through randomized processes of selection, reproduction, sexual recombination, and mutation. Natural population improves over generations by evolution based on the principles of natural selection and survival of the fittest. A simple GA consists of the following steps:

```
Begin Initialize a population of trading rules chromosomes
    Evaluate each chromosome in the population
    WHILE (not termination-condition)
        Select chromosomes with bias by their fitness
        Create new chromosomes using the genetic operators
      Evaluate the new chromosomes and replace the old population
    End
End
```

For each fuzzy trading rule in the knowledge base of GFETS, we use a binary digit {0,1} to represent whether or not it will be activated (or used) when GFETS performs the market timing. A value of 1 indicates that the rule will be "on" (i.e., it is activated); a value of 0 indicates that it will be "off" (i.e., it is not activated). For a knowledge base with n fuzzy trading rules, we encode the "on/off" status of the rules using a chromosome of n binary digits. The value of the i-th digit gives the status of the i-th fuzzy trading rules in the knowledge base. For example, for a knowledge base with 7 fuzzy trading rules, a chromosome of "1100101" indicates that among the seven rules only rules 1, 2, 5 and 7 will be applied in the fuzzy inference in market timing.

Standard GA operators are used to select fuzzy trading rules from the fuzzy trading rules knowledge base as shown in Figure 9.1. Different combinations of fuzzy trading rule will be generated and evaluated in each generation using GA. Historical stock data such as the daily high, low and closing data will be the input of fuzzy inference system to make the buy-sell decision. The buy-sell signals will input to the trading system to simulate the market trading. The overall performance of the buy-sell using the fuzzy trading rules will be the feedback for the GA

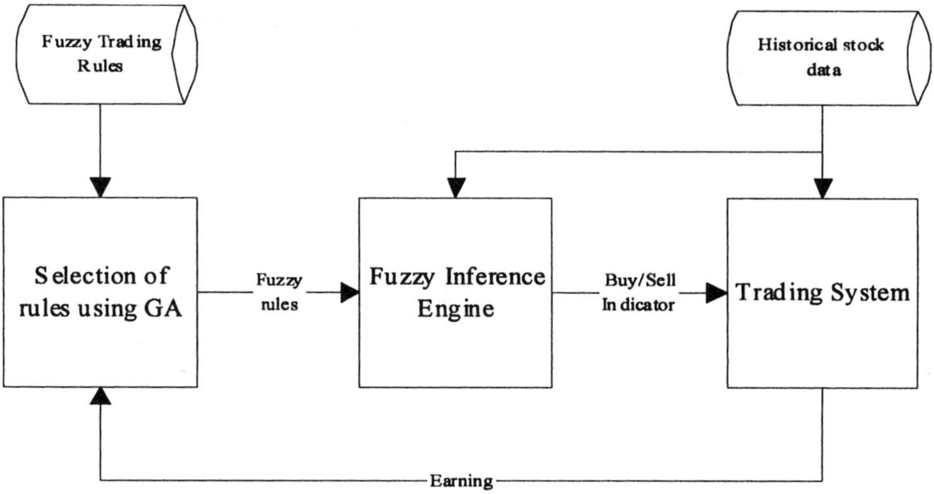

Figure 9.1. Selection of Fuzzy trading rules using GA.

to evaluate the fitness of the chromosome and generate next population. The fitness of the chromosome (or the performance of the fuzzy trading rule combination) is defined as the accumulated profit divided by the maximum cost of the buy transaction. After a number of generations, the subset of good fuzzy trading rules will be identified for timing the market in future.

1.4. Implementation

GFETS was developed using the MATLAB of Mathworks, Inc. The fuzzy expert system was implemented using the Fuzzy Logic Toolbox. The fuzzy trading rules, fuzzy variables and fuzzy system were defined using the interactive interface as shown in Figure 9.2. The GA was implemented based on the public domain GA toolbox of Houck *et al.* (1995). Since there does not exist any toolbox for technical analysis, we developed a toolbox for calculating the market indicators based on the given daily high, low, and closing price of stocks and predicting the market timing for buy-sell.

In the experiment, we have adopted the trading rules of the following twelve market indicators in the knowledge base:

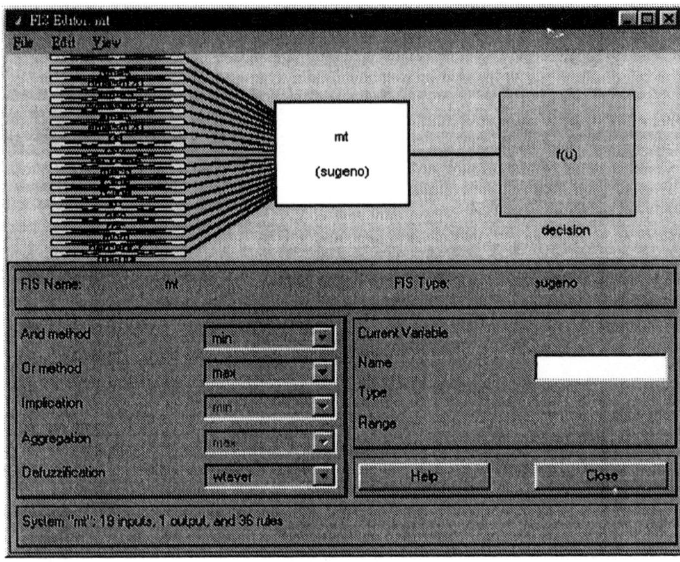

Figure 9.2. Defining Fuzzy System with Fuzzy Logic toolbox.

- Trend Analysis: Daily Moving Average (DMA), Weighted Moving Average (WMA), Exponential Moving Average (EMA), Directional Movement Index (DMI)
- Momentum Analysis: Relative Strength Index (RSI), Moving Average Convergence-Divergence (MACD), Fast and Slow Stochastic (KD-line), Oscillator (OSC), Rate of Change (ROC), Percent R (P-R)
- Volatility Analysis: Volatility Index (VI)

A total of 36 standard buy-sell rules related to these indicators are translated to fuzzy rules (see Appendix). We also coded the trading system using MATLAB to simulate the trading in the market using the selected fuzzy trading rules.

1.4.1 Performance Evaluation of GFETS. GFETS was evaluated with five NASDAQ stocks including Microsoft, Oracle, CISCO, IBM and Starbucks. The stock data from 03/11/1999 to 02/11/2001 was captured from NASDAQ' website (http://www.nasdaq.com). Figure 9.3 shows the closing price of the five stocks in the sampling period.

In the first experiment, the closing price, high price, low price and open price of the whole period were used as the input to GFETS for

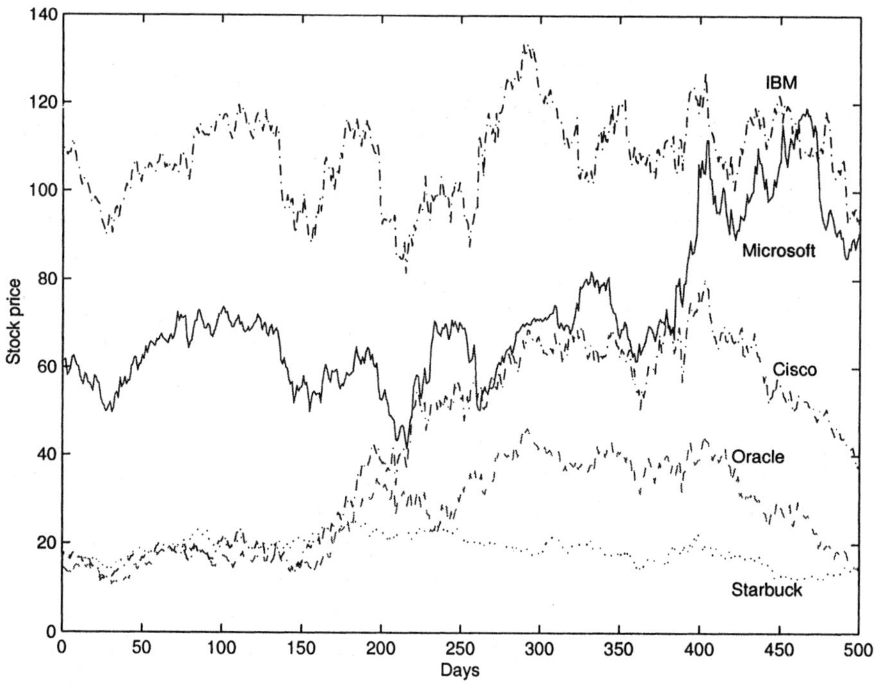

Figure 9.3. Closing price of Microsoft, Oracle, CISCO, IBM, and Starbucks (from 03/11/99 to 02/11/2001).

selecting the rules and also for performing the trading using the selected rules. This is to stimulate the situation of trading with perfect information of the market price using the fuzzy trading rules. The results were tabulated in Table 9.1, where *Gain* and *Capital* are the total gain and maximum capital of the investment, respectively; $Profit\%$ is the profit in percentage; $Total$ is the total number of buy-sell; $+ve$ and $+ve\%$ are respectively the number and the percentage of buy-sells that are profitable. For the investment in these five stocks at the past two years, the average return was 153.5%. The highest return was 217% for CISCO. The total number of profitable transactions was greater than that of loss transaction. The average profitable transaction for these NASDAQ stocks was 75.8%.

Although the above results are encouraging, it can only reflect the performance of GFETS under perfect information because it was trained and optimized for all testing data. To evaluate the actual performance of the system, a blind test was conducted. The data was divided into two half with the first half for in-sample training and the second half for

Table 9.1. Performance of GFETS for the Testing Data

	Investment			*Buy-Sell*		
Stock Name	*Gain*	*Capital*	*Profit%*	*Total*	*+ve*	*+ve%*
CISCO	156.7	72.2	217.0	76	60	79.0
Microsoft	145.1	112.6	128.8	55	39	70.9
IBM	155.1	121.0	128.2	53	41	77.4
Starbuck	37.7	24.2	156.1	72	59	81.9
Oracle	59.1	43.0	137.3	40	28	70.0
		Average :	153.5		*Average :*	75.8

out-sample testing. GFETS was first used to find the optimized fuzzy trading rules for the in-sample training data. The selected rules were then tested by applying them to perform the trading with the out-sample data. The results of the in-sample training were tabulated in Table 9.2. The highest and the lowest return were 125.6% and 54.0% for CISCO and IBM, respectively, and the average return was maintained at 90.8%. The highest percentage of profitable transaction was 86.7% for CISCO and the lowest was 72.2% for Microsoft. On average, over 80% of the transaction was profitable.

Table 9.2. Performance of GFETS for In-sample Training

	Investment			*Buy-Sell*		
Stock Name	*Gain*	*Capital*	*Profit%*	*Total*	*+ve*	*+ve%*
CISCO	68.1	54.2	125.6	30	26	86.7
Microsoft	50.1	72.1	69.6	18	13	72.2
IBM	63.4	117.5	54.0	27	20	74.1
Starbuck	18.2	24.2	75.2	31	27	87.1
Oracle	42.8	33.1	129.6	24	20	83.3
		Average :	90.8		*Average :*	80.7

The results of the out-sample testing were tabulated in Table 9.3. Using the selected rules to perform buy-sell of the stocks gave significant profit except IBM. The profit ranged from -10.5% to 35.6% and the average returns were equal to 11.1%, with the highest and the lowest returns for CISCO and IBM again. It is interested to note that the performance ranking of the stocks are the same for both the in-sample and out-sample data. On average, 68.3% of the transaction resulted in a profit. The highest percentage of profitable transaction was 76.0% for Starbuck and the lowest was 50.0% for IBM. Figure 9.4 shows the buy/sell point for the CISCO stock. From the graph, we find that GFETS in many cases were "buy low and then sell high", and therefore obtained in a significant gain from the transaction.

Table 9.3. Performance of GFETS for Out-sample Testing

	Investment			*Buy-Sell*		
Stock Name	*Gain*	*Capital*	*Profit%*	*Total*	*+ve*	*+ve%*
CISCO	26.5	74.4	35.6	25	18	72.0
Microsoft	11.9	117.9	10.1	16	12	75.0
IBM	-13.7	131.5	-10.5	16	8	50.0
Starbuck	1.4	20.7	7.0	25	19	76.0
Oracle	5.8	44.2	13.1	19	13	68.4
		Average :	11.1		*Average :*	68.3

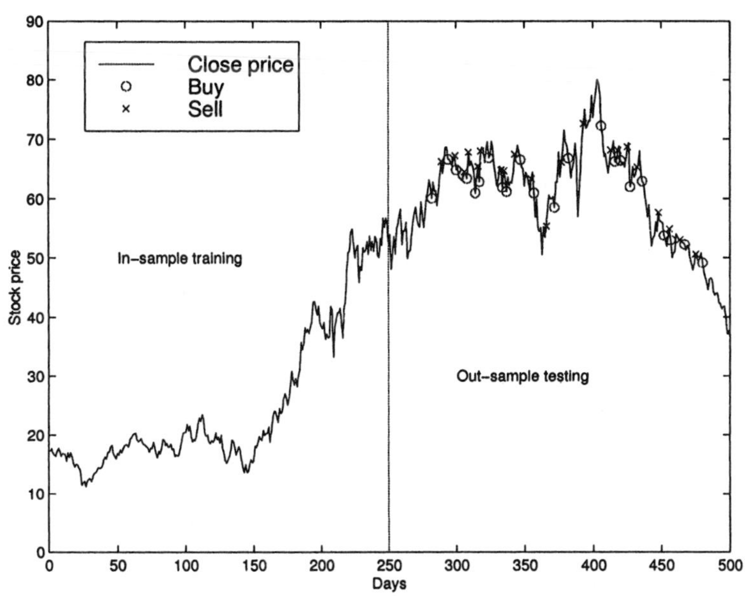

Figure 9.4. Buy/sell CISCO stock using simple GFETS.

Compared the results of the in-sample training with that of the out-sample testing, it is easily to note that there is significant difference in the performance of GFETS. Overall, although the average returns were dropped dramatically from 90.8% to 11.1%, the average profitable transaction was maintained at a reasonably high level (over 68%). In fact, this is an expected outcome because in this experiment GFETS only adopted a single set of fuzzy trading rules in the whole out-sample testing period. Since market price is volatile and the factors determining the stock price vary from time to time, fuzzy trading rules to be used for market timing should be adapted to the changes in the market. It is believed that the performance of the system can be improved through re-training and the frequency of re-training will directly affect its overall performance. In this paper, we proposed two training approaches: incremental and dynamic approaches. The details and the performance of these training approaches will be given in the coming sections.

2. Incremental Training Approach

In the incremental approach, GFETS is trained with m-day daily data and then re-trained after every n days. It will be initialized by training with the first m-day preceding daily data. GA will be used to select the good fuzzy trading rule combination from the fuzzy trading rules knowledge base. The selected fuzzy trading rules will then be used to buy/sell for the subsequent n days. After the $1st$ n-day trading, the system will be re-trained with the $2nd$ m-day preceding daily data, which will include the daily data of this first n trading days and the last $m-n$ days of the preceding training data. The good fuzzy trading rule obtained by GA will be used to buy/sell for another n days before it will be re-trained again with m-day preceding daily data. This procedure will be repeated as illustrated in Figures 9.5 and 9.6.

The period of the training data and the interval to re-train the system are determined by the parameters (m, n). The value of the (m, n) pair directly affects the sensitivity and the responsiveness of GFETS to the market fluctuation. In general, the larger the m value, the more the historical data will be used for selecting the fuzzy trading rules. This will reduce the effect of short-term market fluctuation on the selection process and thus makes GFETS less sensitive to the short-term fluctuation. Similarly, increase the n will increase the time between the re-training of GFETS and thus reduce the chance of making appropriate adjustment in the system to sudden change in market situation. Consequently, the responsiveness of GFETS will be reduced.

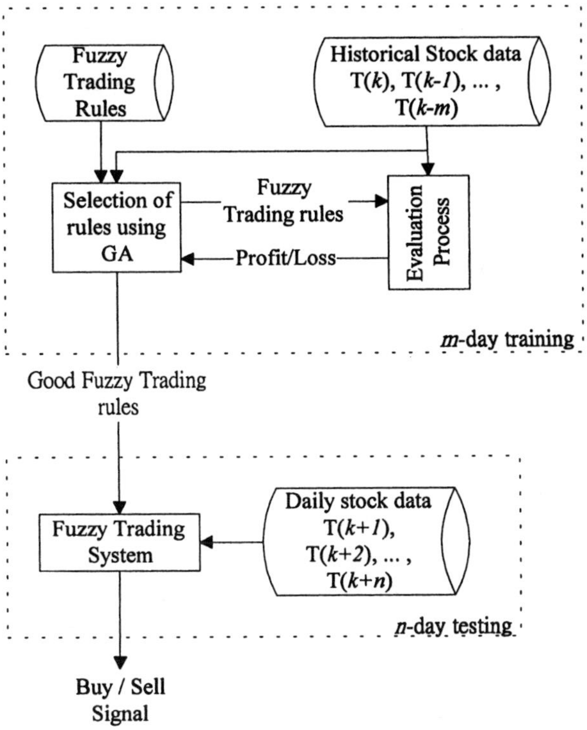

Figure 9.5. Incremental training approach for m-day training and n-day testing.

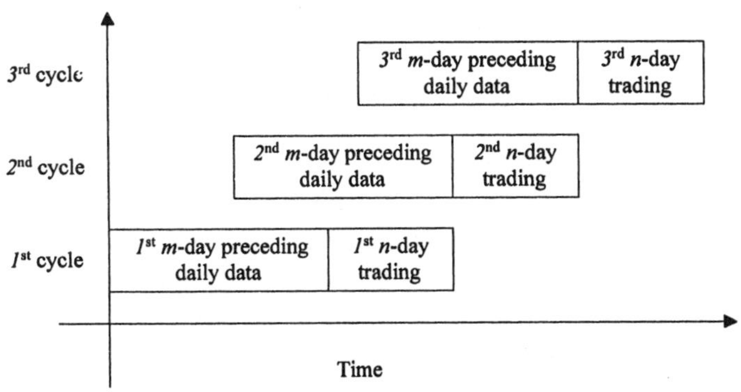

Figure 9.6. Time frame of Incremental Training approach from the $1st$ to $3rd$ m-day training and n-day testing.

2.1. Performance Evaluation of GFETS under Incremental Training

To evaluate the performance of the incremental approach, we tested the system using the daily data of the five stocks for the following pairs of (m, n) value:

1. $m = 250$ and $n = 100$
2. $m = 90$ and $n = 60$
3. $m = 60$ and $n = 30$
4. $m = 30$ and $n = 15$

The values are typical day used in technical analysis. The training period and re-training interval of the system were decreasing simultaneously from 250 days to 30 days and from 100 days to 15 days, respectively. Decreasing these two values changes the system from low to highly sensitive and responsive to market variations. The performance of GFETS for each stock and for each (m, n) pair were tabulated in Table 9.4. For each stock, we derived the average performance of GFETS by taking the average of the performance of GFETS for each (m, n) pair. The results were tabulated in Table 9.5.

The results indicated that for individual stock, the performance of the system would vary for different (m, n) pairs. For CISCO, Microsoft and ORACLE, the best returns were obtained at $(60, 30)$. For IBM and Starbuck, the best returns were obtained at $(90, 60)$ and $(30, 15)$, respectively. GFETS could make profit for more than half of the transactions for all case except the $(250, 100)$ for Oracle. The highest percentage of profitable transaction was 78.6%. Although the performance of GFETS for individual stocks may not be very good e.g. Oracle, the average performance of GFETS was very good. The average profit for individual stocks ranged from 10.0% to 89.7% and the overall average was 42.5%. On average, 62.3% of the transaction resulted in a profit. Compared with the simple GFETS, the performance of GFETS under the incremental training approach is much better. Although the average percentage of profitable transaction dropped slightly by 6% (from 68.3% to 62.3%), the average profit was greatly increased by 31.4% (from 11.1% to 42.5%). Figure 9.7 shows the buy/sell point for the CISCO stock under $(60, 30)$ incremental training. Once again, GFETS in many cases brought at low and then sold at high.

Although GFETS under incremental training approach gave better returns, the returns would depend on the setting of the parameter (m, n),

Table 9.4. Performance of GFETS Using Incremental Training Approach

Stock Name	(m, n)	Investment			Buy-sell		
		Gain	*Capital*	*Profit%*	*Total*	*+ve*	*+ve%*
CISCO	(250,100)	45.1	76.1	56.7	34	21	61.8
	(90,60)	69.8	76.1	97.7	50	34	68.0
	(60,30)	91.5	76.1	120.2	47	33	70.2
	(30,15)	66.7	76.1	87.8	49	32	65.3
Microsoft	(250,100)	30.1	111.7	27.0	14	7	50.0
	(90,60)	15.7	115.9	13.6	25	15	60.0
	(60,30)	41.6	112.8	36.9	32	19	59.4
	(30,15)	28.1	112.8	25.0	35	23	65.7
IBM	(250,100)	18.3	125.1	14.7	20	11	55.0
	(90,60)	58.1	124.4	46.7	44	31	70.5
	(60,30)	12.0	122.5	9.8	31	18	58.1
	(30,15)	46.0	129.5	35.5	39	27	69.2
Starbuck	(250,100)	15.0	21.4	70.0	28	22	78.6
	(90,60)	16.4	23.7	69.4	34	23	67.7
	(60,30)	6.9	24.0	29.0	39	23	59.0
	(30,15)	18.1	24.3	74.4	45	32	71.1
Oracle	(250,100)	-8.7	41.8	-20.8	15	6	40.0
	(90,60)	15.8	46.1	34.3	42	23	54.8
	(60,30)	18.6	44.2	42.0	41	21	51.2
	(30,15)	-8.0	44.2	-18.2	41	21	51.2

Table 9.5. Average Performance of GFETS Using Incremental Training Approach

	Investment			Buy-Sell		
Stock Name	*Gain*	*Capital*	*Profit%*	*Total*	*+ve*	*+ve%*
CISCO	68.3	76.1	89.7	45.0	30.0	66.7
Microsoft	28.9	113.3	25.5	26.5	16.0	60.4
IBM	33.6	125.4	26.8	33.5	21.8	64.9
Starbuck	14.1	23.4	60.4	36.5	25.0	68.5
Oracle	4.4	44.1	10.0	34.8	17.8	51.1
		Average :	42.5		*Average :*	62.3

which is not straightforward as shown in the experiment. The main shortfall of this approach is the time to re-train the system must be

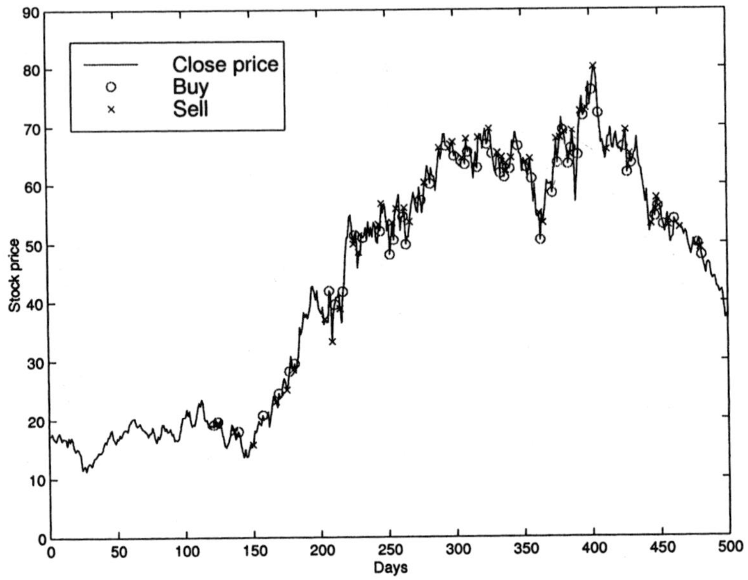

Figure 9.7. Buy/sell CISCO stock using GFETS under (60, 30) incremental training.

fixed in advance. This makes the system inflexible and too rigid for the stock market, which is dynamic by nature. To improve the flexibility of the system, we introduce a mechanism to determine the need and the time to re-train the system. The modified approach is called dynamic training approach. It will be introduced in the next section.

3. Dynamic Training Approach

Unlike the incremental training approach, the time to re-train the system under the dynamic training approach will not be fixed. Instead, GFETS will be re-trained whenever its performance drops below a predefined threshold level as shown in Figure 9.8. In this approach, the system will be first trained using m preceding daily data. The selected good fuzzy trading rules will then be used for 1-day trading. At the end of the first trading day, the system will be assessed and checked whether or not it has to be re-trained. The assessment is performed in the following way. The current daily data will be combined with the last $(m-1)$ preceding daily data to form an m-day testing data. This m-day testing data will then be used for the current rules to perform simulated trading. If the earning of the simulated trading is greater than the threshold level, the current rules will be kept for another trading day. At the end of the second trading day, the system will be assessed again on the re-training need. But in this time the simulated trading will be conducted with the testing data that is made up of the two current daily data and the last $(m - 2)$ preceding daily data. Once again, if the earning is greater than the threshold level, it will be kept for another trading and the assessment will be repeated at the end of the trading day as illustrated in Figure 9.9.

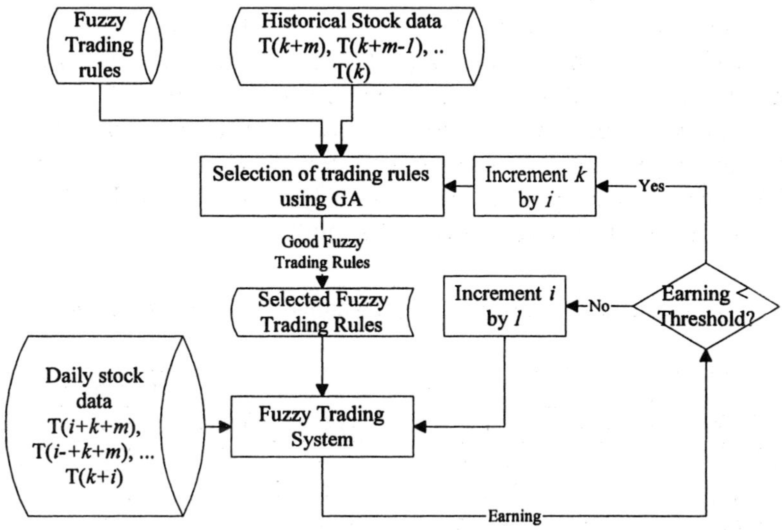

Figure 9.8. Dynamic training approach.

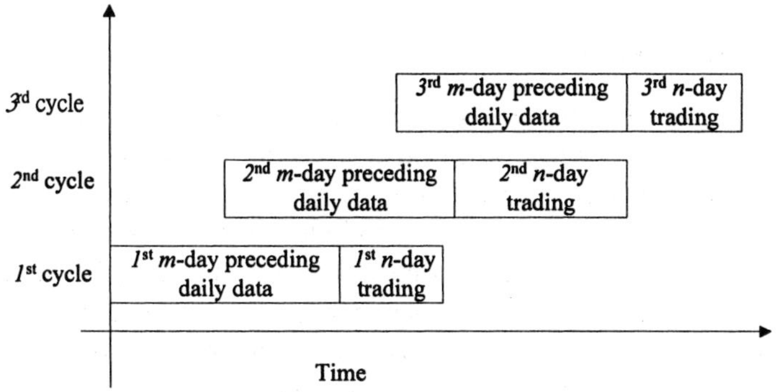

Figure 9.9. Time frame of dynamic training approach.

3.1. Performance Evaluation of GFETS under Dynamic Training

We used the same data set to test the performance of GFETS under dynamic training. The results were tabulated in Table 9.6. The results are much better than that of the simple GFETS. Compared to incremental training, dynamic training gave significantly better results. It gave very high returns for all stocks except Microsoft. The highest returns were 76.4% for CISCO and the lowest returns were 6.0% for Microsoft. The average returns were 51.5% and were 9.0% higher than that of obtained in the incremental training. The percentage of profitable transaction ranged from 60.8% to 71.1% with the average of 67.8%, which is 5.5% higher than that of the incremental training and compatible with that of the simple GFETS. Figure 9.10 depicts the buy/sell point for the CISCO stock under dynamic training. The figure clearly showed that GFETS gave highly reliable buy-sell signals in the experiment.

Table 9.6. Performance of GFETS Using Dynamic Training Approach

	Investment			*Buy-Sell*		
Stock Name	*Gain*	*Capital*	*Profit%*	*Total*	*+ve*	*+ve%*
CISCO	59.1	77.3	76.4	36	25	69.4
Microsoft	6.7	112.8	6.0	38	26	68.4
IBM	46.0	129.5	35.5	39	27	69.2
Starbuck	18.1	24.3	74.4	45	32	71.1
Oracle	29.9	46.1	65.0	46	28	60.8
		Average :	51.5		*Average :*	67.8

4. Conclusion

In this paper, we have proposed a Genetic Fuzzy Expert Trading System (GFETS) to solve the market timing problem. We have formulated commonly used trading rules in technical analysis as fuzzy rules and incorporated them into the knowledge base. Good fuzzy rules were selected from the knowledge based using genetic algorithm. The performance of the system was evaluated using the stock data of NASDAQ. It gave an average of 11.1% returns and 68.3% profitable transactions in the out-sample testing. To improve the performance of the system, we have proposed two training approaches including an incremental approach and a dynamic approach. The performance of these training approaches was studied and reported in the paper. The results showed

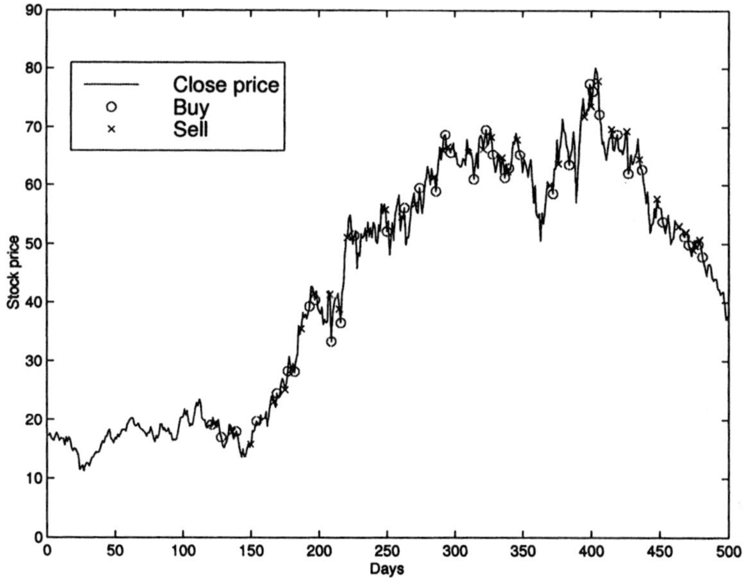

Figure 9.10. Buy/sell CISCO stock using GFETS under dynamic training.

that both approach outperformed the original GFETS. The incremental training approach generated an average of 42.5% returns and 62.3% profitable transaction while the dynamic training approach generated an average of 51.5% returns and 67.8% profitable transaction. Overall, GFETS with dynamic training approach is the most reliable.

Some future work includes expanding the knowledge base with more trading rules from technical analysis to provide evidential information for the fuzzy expert system to make better market timing. The impact of using different membership function in modelling the fuzzy trading rules should also be investigated. So far, GFETS only relies on the use of information provided by historical data. How to integrate and apply the neuro-forecasting technique in market timing is another important area to be explored.

Appendix: Fuzzy Trading Rules of GFETS

Rule	*IF*	*THEN*
1.	DMA(5) is upcross	long
2.	DMA(5) is downcross	short
3.	DMA(5)-DMA(20) is upcross	long
4.	DMA(5)-DMA(20) is downcross	short
5.	WMA(5) is upcross	long
6.	WMA(5) is downcross	short
7.	WMA(5)-WMA(20) is upcross	long
8.	WMA(5)-WMA(20) is downcross	short
9.	EMA(5) is upcross	long
10.	EMA(5) is downcross	short
11.	EMA(5)-EMA(20) is upcross	long
12.	EMA(5)-EMA(20) is downcross	short
13.	RSI is small	long
14.	RSI is moderate	hold
15.	RSI is large	short
16.	RSI(2) is less than 35 and RSI-RSI(MA) is greater than 65	long
17.	RSI(2) is greater than 65 and RSI-RSI(MA) is less than 65	short
18.	MACD is negative	short
19.	MACD is zero	hold
20.	MACD is positive	long
21.	K-line is small and D-line is small	long
22.	K-line is large and D-line is large	short
23.	KD is negative	short
24.	KD is positive	long
25.	OSC is negative	long
26.	OSC is positive	short
27.	ROC is high	short
28.	ROC is low	long
29.	MOM is decline	long
30.	MOM is raise	short
31.	P-R is low	short
32.	P-R is high	long
33.	PDI-NDI is downcross	short
34.	PDI-NDI is upcross	long
35.	VI is long	long
36.	VI is short	short

References

Allen, F. and R. Karjalainen (1999). "Using Genetic Algorithm to Find Technical Trading Rules," *Journal of Financial Economics*, 51, 245–271.

Dawid, H. (1996). *Adaptive Learning by Genetic Algorithms: Analytical Results and Applications to Economical Models.* Springer.

Driankov, D., H. Hellendoorn, and M. Reinfrank (1993). *An Introduction to Fuzzy Control.* Springer.

Edmonds, A. N., D. Burkhardt, and O. Adjei (1996). "Genetic Programming of Fuzzy Logic Production Rules with Application to Financial Trading," *Neural Networks in Financial Engineering*, 179–188. World Scientific Publishers.

Holland, J. H. (1975). *Adaption in Natural and Artifical Systems.* MIT Press.

Houck, C. R., J. A. Joines, and M. G. Kay (1995). "A Genetic Algorithm for Function Optimization: A Matlab Implementation".

Klir, G. J. and B. Yuan (1995). *Fuzzy Sets and Fuzzy Logic: Theory and Applications.* Prentice Hall.

Lynch, P. and J. Rothchild (2000). *One Up on Wall Street*, 1st ed. Fireside.

NASDAQ website, "The NASDAQ Stock Market," <http://www.nasdaq.com/about/about_nasdaq.stm>.

Richard, J., Jr. Bauer and D. R. Julie (1999). "Technical Market Indicators: Analysis and Performance". John Wiley and Sons.

Surgeno, M. and K. Murakami (1985). "An Experimental Study on Fuzzy Parking Control Using a Model Car," in M. Sugeno (ed.), *Industrial Applications of Fuzzy Control.* North-Holland.

IV

MISCELLANEOUS APPLICATIONS DOMAINS

Chapter 10

PORTFOLIO SELECTION AND MANAGEMENT USING A HYBRID INTELLIGENT AND STATISTICAL SYSTEM

Juan G. Lazo Lazo

ICA : Applied Computational Intelligence Laboratory
Department of Electrical Engineering
Pontifical Catholic University of Rio de Janeiro, PUC-Rio
Rua Marquês de S. Vicente 225, Gávea, Rio de Janeiro, CEP 22453-900, RJ, Brazil
juan@ele.puc-rio.br

Marco Aurélio C. Pacheco

ICA : Applied Computational Intelligence Laboratory
Department of Electrical Engineering
Pontifical Catholic University of Rio de Janeiro, PUC-Rio
Rua Marquês de S. Vicente 225, Gávea, Rio de Janeiro, CEP 22453-900, RJ, Brazil
Department of Computer & Systems Engineering
State University of Rio de Janeiro, UERJ, RJ, Brazil
marco@ele.puc-rio.br

Marley Maria R. Vellasco

ICA : Applied Computational Intelligence Laboratory
Department of Electrical Engineering
Pontifical Catholic University of Rio de Janeiro, PUC-Rio
Rua Marquês de S. Vicente 225, Gávea, Rio de Janeiro, CEP 22453-900, RJ, Brazil
Department of Computer & Systems Engineering
State University of Rio de Janeiro, UERJ, RJ, Brazil
marley@ele.puc-rio.br

Abstract This paper presents the development of a hybrid system based on Genetic Algorithms, Neural Networks and the GARCH model for the selection of stocks and the management of investment portfolios. The hybrid system comprises four modules: a genetic algorithm for selecting the assets that will form the investment portfolio, the GARCH model for forecasting stock volatility, a neural networks for predicting asset returns for the portfolio, and another genetic algorithm for determining the optimal weights for each asset. Portfolio management has consisted of weekly updates over a period of 49 weeks.

Keywords: Genetic Algorithms, Neural Networks, GARCH, VaR, Volatility

Introduction

The construction of investment portfolios poses a problem with multiple objectives. In order to form an investment portfolio, a set of stocks which are expected to yield profits must be selected, a difficult task due to the great number of possibilities and parameters to be considered. Once the stocks in which one wishes to invest have been chosen, there is the problem of defining the percentage of the fund to be invested in each asset, which is also called the weight of the asset in the portfolio. The vector of weights is defined by considering the expected returns on each asset and the risk each investor is willing to take on the investment. Portfolio management consists of periodic evaluations of the portfolio's performance and of the modification of the weight values.

This paper makes use of a Genetic Algorithm (Goldberg 1989; Michalewicz 1996) to select the assets that will comprise the portfolio, based on a subset of assets that are traded on the São Paulo Stock Exchange — Brazil (BOVESPA). A Neural Network (Haykin 1998; Zurada 1995) contributes to portfolio management by predicting the asset returns for the next period of portfolio evaluation (Lazo 2000). The GARCH model (Engle 1982; Bollerslev 1986; Engle 1995) is employed for predicting the volatility of each asset. Another Genetic Algorithm is used for optimizing the assets within the portfolio by estimating the Value at Risk (VaR) (Jorion 1998; Down 1998; Hull 1999) on a weekly basis. The portfolio is managed for a period of 49 weeks and its evaluation consists of comparing its behavior with that of the market — the BOVESPA Index. In this paper, Markowitz's model and the Efficient Frontier model (Markowitz 1959; Levy 1984; Elton 1995) have been used to perform the selection and to determine the percentage represented by each asset in the portfolio.

The system is evaluated with the use of the return series of 137 Brazilian assets traded on the BOVESPA between July 1994 and December

1998. One part of the data was used for training the model and the other part (January 1998 to December 1998) was used for testing, i.e., for portfolio management.

The system that has been developed is of an academic nature and does not purport to be a tool to be used by individual investors. Its purpose is to evaluate the performance of Genetic Algorithms and of Neural Networks in portfolio construction and management in emerging markets (the Brazilian Market).

This paper assumes that there are no fees or transaction costs, and the bankruptcy cost is disregarded. In addition, it presupposes that all the assets may be split and liquidated. It also considers that information is free and is available to all the investors. In order to avoid the possibility of premature convergence, risk-free assets have not been considered for the construction of the portfolio. These assets could dominate the portfolio when the risk of other assets is high or when the market is going down.

Section 1 describes the modeling of the genetic algorithm that has been employed for selecting the assets that make up the portfolio. Section 2 presents the volatility forecasts performed by the GARCH model. In section 3, the returns are predicted by the neural networks. Section 4 describes the modeling of the genetic algorithm that has been developed for weight optimization and for portfolio management via the results obtained by the neural nets, while section 5 presents the calculation of the VaR for the portfolio. The results obtained with the proposed hybrid model are analyzed in section 6, and finally, section 7 presents the conclusions that have been drawn from this work.

1. Construction of the Investment Portfolio by Genetic Algorithms

In this model, it is the genetic algorithm that, by means of the Efficient Frontier Criterion, selects and determines which and how many of the 137 assets traded on the BOVESPA will form the investment portfolio.

To this end, the monthly return rates and risk are calculated for each one of the 137 assets in accordance with the Mean-Variance criterion proposed by Markowitz (Markowitz 1959), where the return estimate is represented by the mean and asset risk is represented by the variance. The covariance and correlation matrices of these assets, which represent the portfolio's systematic and non-systematic risk, respectively, are also calculated. Systematic risk occurs by force of circumstances that affect companies, such as inflation, political events, etc.; non-systematic risk, however, may be eliminated by means of portfolio diversification.

Basically, the problem contemplates an initial portfolio comprised of 137 assets and the GA must determine the percentage to be invested in each asset, which is also called asset weight. The GA must confer a significant weight to all the assets that are expected to yield profits and will confer zero weight to assets that do not comply with this condition. In other words, it will attempt to obtain the optimal portfolio according to the Efficient Frontier Criterion in order to minimize the risk for the portfolio. The following constraints must be met: the sum of all the weights must be equal to 1, and the weight attributed to an asset must be greater than or equal to zero.

The Genetic Algorithm generates the possible solutions, evaluates them and supplies the best set of assets. The part below provides a detailed description of the basic components of the GA: representation, evaluation and operators.

1.1. Representation and Decoding of the Chromosome

The representation of the chromosome comprises 137 genes, (Figure 10.1), where each gene represents the weight of the asset in the portfolio.

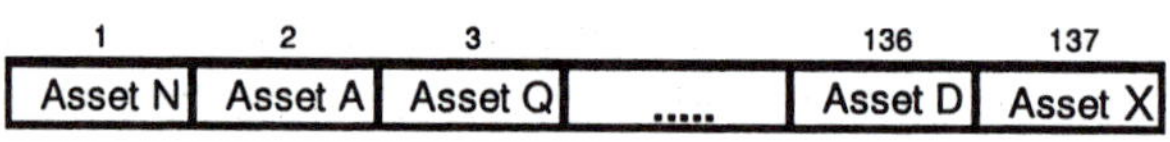

Figure 10.1. Chromosome for asset selection.

1.2. Evaluation of the Chromosome

When the chromosome is evaluated, an attempt is made to minimize the portfolio's risk defined by Equation (10.1), which represents the standard deviation of the portfolio's returns.

$$Min \quad \sum_{i=1}^{N} {X_i}^2 {\sigma_i}^2 + \sum_{i=1}^{N} \sum_{\substack{j=1 \\ j \neq i}}^{N} X_i X_j \sigma_{ij} \tag{10.1}$$

where X_i is the weight for the asset that corresponds to gene i of the chromosome; X_j is the weight for the asset that corresponds to gene j of the chromosome; σ_i is the risk that pertains to the asset that corresponds to gene i of the chromosome, i.e., the standard deviation of asset i; and

σ_{ij} is the covariance of the asset that corresponds to gene i with the asset that corresponds to gene j of the chromosome.

1.3. Genetic Operators and Evolution Parameters

The operators employed in this paper were:

- Uniform crossover (Michalewicz 1996)
- Mutation

The algorithm was tested in several experiments, where the following parameter values were modified:

Population Size	: 100, 500, 3500 individuals.
Crossover Rate	: 0.5, 0.1, 0.25, 0.55, 0.68.
Mutation Rate	: 0.06, 0.1, 0.25.
Number of Generations	: 100 generations.

1.4. Results Obtained

In every run, the genetic algorithm converges to the same result, giving significant weights to the same 13 assets listed below:

Aços Vilares Pn, Albarus On, Bemge On, Brahma On, Brahma Pn, Cemig Pn, Ciquine Pna, Docas Pn, Electrolux Pn, FrasLe Pna, Light On, Telerj On, Unibanco On

In order to evaluate the results of this genetic algorithm, the selected assets were used for setting up and managing a portfolio with a view to the maximization of portfolio returns. This portfolio was managed over a period of 49 weeks, from February 1998 to March 1999, and the asset weights were updated each week with the use of the Mean-Variance Criterion and the Efficient Frontier (Markowitz 1959; Levy 1984). In other words, the estimated returns on each asset for the next period is given by the mean, and the risk is the variance-covariance matrix of the returns. The sample window for the re-estimations was of 6 months. The performance of the portfolios over the managed period is compared with the BOVESPA market index. Figure 10.2 shows the result obtained after managing the portfolio that maximizes the returns.

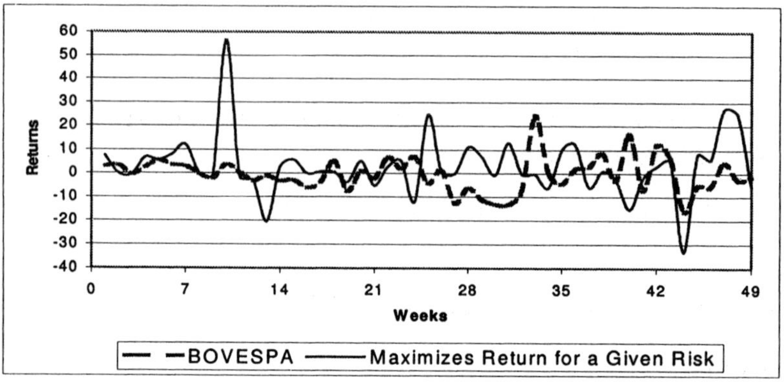

Figure 10.2. Comparison between the performance of the portfolio managed by the Genetic Algorithm and the market portfolio (BOVESPA).

Table 10.1 presents a few comparative measurements of the performance of the managed portfolios, such as the Mean Return (mean weekly portfolio returns) and the variance (risk for the portfolio).

Table 10.1. Comparison between the Performance of the Managed Portfolio and the Market Portfolio

	Market Portfolio	Portfolio that Maximizes Return
Mean Return	-0.460	3.120
Variance(%)	56.302	165.233

Figure 10.2 and Table 10.1 demonstrate that the managed portfolio has obtained better returns than the market portfolio, although at a much higher risk. However, this was expected because, once the objective was to maximize returns, there were no restrictions on risk. Besides, the market in this period presented a negative return and a high risk. This all suggests that the selected assets may generate profits.

2. Volatility Forecasting by the GARCH Model

The term volatility denotes the temporal variability of the degree in which the data scatter around their central trend. Considering the mean as the central trend, what the variance (or standard deviation) supplies is precisely this degree of dispersion. Hence, volatility is the variation, along the time horizon, of conditional mean. Volatility measures the risk that is associated with a specific series. Intuitively, the more volatile a price (or return) series associated with a specific asset, the greater the degree of uncertainty with regard to its behavior and therefore, the greater the risk incurred when trading such an asset.

In the Brazilian market, it is particularly important to predict volatility because according to Duarte, Pinheiro and Heil (1997), there is empirical evidence that Brazilian assets and indexes are considerably more volatile than their North-American, European and Japanese counterparts.

This paper has made use of the GARCH model (Generalized Autoregressive Conditional Heteroscedasticity Model) (Engle 1982; Bollerslev 1986) to perform the volatility forecasts. The GARCH(p,q) model has been selected for the purposes of this paper on account of its efficiency and broad applicability in the financial market (Engle 1995; Jorion 1998; Drost 1992). The return series of the assets that form the portfolio were used for testing different representations for the GARCH(p,q) model. The Akaike criterion (AIC) and the Schwartz criterion (SBC) (1995) were used as the criteria for evaluating the best GARCH(p,q) model. On account of the results obtained, the representation of the GARCH(1,1) was elected, in agreement with the type of representation that is most commonly suggested for financial series in the bibliography (Engle 1995; Jorion 1998; Campos 1998; Nelso 1990). The said criteria consist of choosing the one model that minimizes the AIC or SBC values expressed by Equations (10.2) and (10.3).

$$AIC(k) = \frac{1}{N}\{-2\ln(L) + 2k\} \tag{10.2}$$

$$SBC(k) = \frac{1}{N}\{-2\ln(L) + k\ln(n)\} \tag{10.3}$$

where L is the likelihood function, k is the number of independent parameters in the model, and n is the number of observations.

For the purpose of determining the parameters (a0, a1, and b1) of the GARCH(1,1) model employed for the volatility forecast, the ARMA(1,1) was used initially, in accordance with the suggestion in Bollerslev (1986), in order to reduce the search space and to have an initial approximation for these parameters. Given the fact that parameters that are calculated in this manner present an embedded error, and with a view to optimizing the parameters of the GARCH (1,1) model, the values of the parameters (a0, a1, and b1) obtained by the ARMA model were used as initial points in another algorithm for parameter optimization (mixed gradient algorithm). The result of this optimization provided the parameters that were employed in the GARCH(1,1) model for the calculation of the volatility forecasts.

3. Financial Asset Return Forecasting

In order to manage the portfolio, it is necessary to have estimates of the returns for the next period to be managed. This paper has made use of Neural Networks to make predictions of the returns by considering historical information on the returns of each asset and their volatility, which has been calculated by means of the GARCH model. Since the results obtained by the Backpropagation neural network (Haykin 1998) were satisfactory in terms of producing smaller prediction errors, it was decided that neural nets would be employed for obtaining return forecasts.

3.1. The Neural Networks Approach to Financial Asset Return Modeling

The architecture of the neural network is formed by 11 inputs (the 10 previous values of the return series of the asset plus a value that corresponds to the volatility value that has been calculated with the GARCH model), with one hidden layer and one output; since each series presents distinct features, the quantity of neurons in the hidden layer is different for each return series. The activation function of the neurons in the hidden layer is the hyperbolic tangent, while the output activation function is linear; the forecast is made one step ahead.

The number of inputs for the neural network was determined experimentally, based on the auto-correlation analysis performed on the series, of the square of the returns, which presented several significant lags among the first 10 lags. This auto-correlation analysis was performed because the other input of the neural network is volatility, which is the variation, along a time horizon, of conditional mean. Since correlation is a linear measurement and neural networks have a nonlinear nature, it is expected that, in the tests performed, the neural network will find some type of nonlinear relation between the past data of the return series and the past data of asset volatility.

In order to evaluate the forecasting results, the following error measurements have been employed:

MAD : Mean absolute deviation;

NRMSE : Normalized root mean square error;

MSE : Mean square error;

U-Theil : Metric that measures the extent to which a result is better than one obtained by means of naive prediction.

Table 10.2 below presents the prediction error statistics for a few of the predicted series; in Figure 10.3, the real asset returns are compared with the weekly predictions generated by the neural network.

Table 10.2. Prediction Errors

	A. Vilares Pn	Albarus On	Bemge On	Brahma On
MAD	0.11590	0.09990	0.52064	0.05637
MSE	0.02542	0.01953	0.76110	0.00697
RNMSE	1.42013	1.15564	3.86827	1.05777
UTHEIL	0.66973	0.75169	0.70053	0.81542

	Electrolux Pn	Cemig Pn	Unibanco On
MAD	0.02640	0.16256	0.07682
MSE	0.00211	0.05260	0.01011
RNMSE	0.68672	1.85865	1.21381
UTHEIL	0.98026	0.64345	0.75923

In Table 10.2, it may be observed that the prediction errors of the neural net are small, and the U-Theil statistic indicates that the forecasts obtained are much better than those of the naive prediction.

4. Portfolio Management by Genetic Algorithms

The portfolio has been managed by means of the evaluation of its performance over a period of 49 weeks with weekly updates of asset weights and of the return and risk estimates for each week with the use of the predictions made by the neural network and the GARCH model, respectively. A genetic algorithm has been used for the purpose of determining the percentage of the fund to be invested in each of the assets in the portfolio. The algorithm must meet the constraints regarding minimum risk or maximum return imposed by the investor.

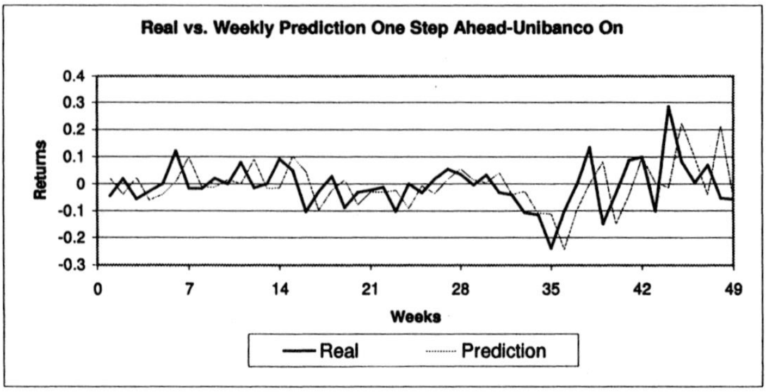

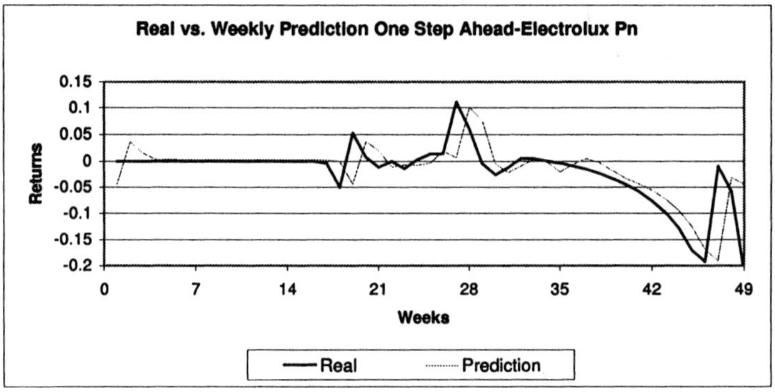

Figure 10.3. Real data vs. weekly predictions 1 step ahead.

This paper has opted for the management of two portfolios. Thus, the basic problem is to find the asset weights that will allow one portfolio to maximize the return and the other, to minimize the risk. Both portfolios must meet the following constraints:

The sum of the portfolio weights must be equal to 1;

The weight of each asset must be greater than or equal to zero.

4.1. Representation and Decoding of the Chromosome

The chromosome for managing the investment portfolio is formed by 13 genes, which represent the weights of the assets in the portfolio (Figure 10.4).

Figure 10.4. Chromosome.

Each gene of the chromosome (Figure 10.4) contains a value between 0 and 1, which indicates the percentage to be invested in the respective asset.

4.2. Evaluation of the Chromosome

Since the two portfolios to be managed have different objectives, each portfolio presents its own evaluation function.

In the first case, the evaluation function of the chromosome attempts to maximize the portfolio's returns and is defined by Equation (10.4):

$$Max\ \theta = \frac{\sum_{i=1}^{N} X_i(\bar{R}_i - R_F)}{\left[\sum_{i=1}^{N} X_i^2\sigma_i^2 + \sum_{i=1}^{N}\sum_{\substack{j=1\\ j\neq i}}^{N} X_i X_j \sigma_{ij}\right]^{\frac{1}{2}}} \tag{10.4}$$

where X_i is the weight for the asset that corresponds to gene i of the chromosome; X_j is the weight for the asset that corresponds to gene j of the chromosome; R_i represents the predicted return on the asset that corresponds to gene i of the chromosome; R_F is the return of a risk-free asset; σ_i is the risk that pertains to the asset that corresponds to gene i of the chromosome (predicted volatility for asset i); σ_{ij} is the covariance of the asset that corresponds to gene i with the asset that corresponds to gene j of the chromosome.

This function attempts to maximize portfolio returns by maximizing the Sharpe ratio (Elton 1995). The numerator of the formula expresses the portfolio return that exceeds the return of a risk-free asset, which is the return obtained from an asset that pays known interest rates. This paper has used the Brazilian CDI (Interbank Deposit Certificate) as the risk-free asset. The denominator of the formula is the risk for the portfolio, which is represented by the standard deviation of its returns.

For the second case, the evaluation function of the chromosome tries to minimize the portfolio's risk for a given return; the function is defined by Equation (10.5), which represents the standard deviation of the portfolio's returns:

$$Min\ \sigma_P = \left[\sum_{i=1}^{N} {X_i}^2 {\sigma_i}^2 + \sum_{i=1}^{N} \sum_{\substack{j=1 \\ j \neq i}}^{N} X_i X_j \sigma_{ij} \right]^{\frac{1}{2}} \tag{10.5}$$

where σ_P represents the risk for the portfolio or its volatility; X_i is the weight for the asset that corresponds to gene i of the chromosome; X_j is the weight for the asset that corresponds to gene j of the chromosome; σ_i is the risk that pertains to the asset that corresponds to gene i of the chromosome (predicted volatility for asset i); σ_{ij} is the covariance of the asset that corresponds to gene i with the asset that corresponds to gene j of the chromosome.

4.3. Genetic Operators and Evolution Parameters

The genetic operators employed were of the uniform crossover and mutation type (Michalewicz 1996). The procedure for testing the algorithm consisted of varying the parameters with the same values that were used in the previous case:

Population Size	: 40000 individuals.
Crossover Rate	: 0.5, 0.68.
Mutation Rate	: 0.06, 0.1, 0.3.
Number of Generations	: 100 generations.

5. Calculation of the VaR for the Portfolio

The VaR is a measure of exposure which attempts to quantify the maximum potential loss possible for a given portfolio (or asset) within a time horizon and with a specific confidence interval (Jorion 1998; Down 1998; Hull 1999).

The time horizon is determined by the period during which the weights are updated in the course of portfolio management (one week). Therefore, the portfolio's variance or risk (σ_P^2) for each period depends on the weight that has been allocated to each asset and this variance is calculated each week when the it is time to optimize the weights, Equation (10.6). The correlation matrix needed for calculating the VaR was calculated for each week, based on the return forecasts.

$$\sigma_P^2 = \sum_{i=1}^{N} {X_i}^2 {\sigma_i}^2 + \sum_{i=1}^{N} \sum_{\substack{j=1 \\ j \neq i}}^{N} X_i X_j \sigma_{ij} \tag{10.6}$$

The VaR for the portfolio is calculated with the use of Equation (10.7) for a confidence interval of 95% (a = 1.65) and for a portfolio value (V_P) of 100 thousand US dollars.

$$VaR_P = -\alpha \sigma_P V_P \tag{10.7}$$

In this manner, the VaR for the portfolio is recalculated each week according to the updates of the portfolio's weights.

6. Results

Both of the portfolios created were managed over a period of 49 weeks with weekly updates of the asset weights, and with the use of the weekly return forecasts performed by neural networks and the volatility forecasts provided by the GARCH model. The results obtained by portfolio management with the proposed model are shown in Table 10.3.

Table 10.3. Comparison of the Results for the two Managed Portfolios

	Market Portfolio	Portfolio that Maximizes Return	Portfolio that Minimizes Risk for Given Return $R > 5\%$
Mean Return	-0.221	5.227	0.540
Variance(%)	59.395	142.391	12.2094
Beta	1.00	0.549	0.203

Figures 10.5 and 10.6 present the performance of each portfolio comparing with the returns of the market portfolio (BOVESPA Index).

It may be observed that on the average, the returns produced by the managed portfolios are higher than the market returns and that the portfolio's risk (variance) is lower than the market's risk for the portfolio that minimizes risk, in contrast with the portfolio that maximizes return which presents a higher risk. However, this was expected once there were no restrictions on risk and also considering that the return achied was 24 times the market return.

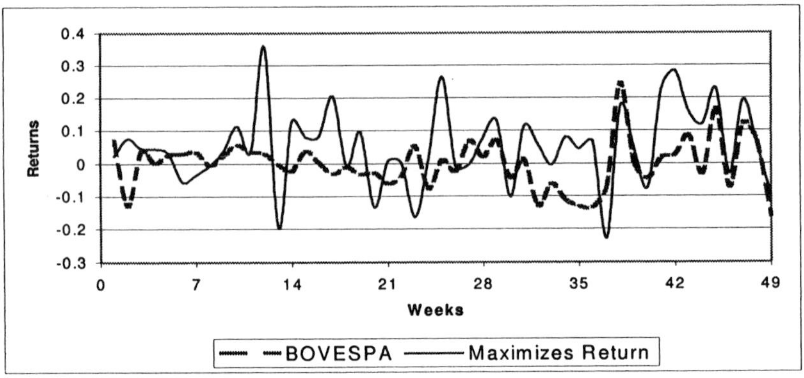

Figure 10.5. Comparison between the performance of the portfolio that maximizes the return and the market portfolio.

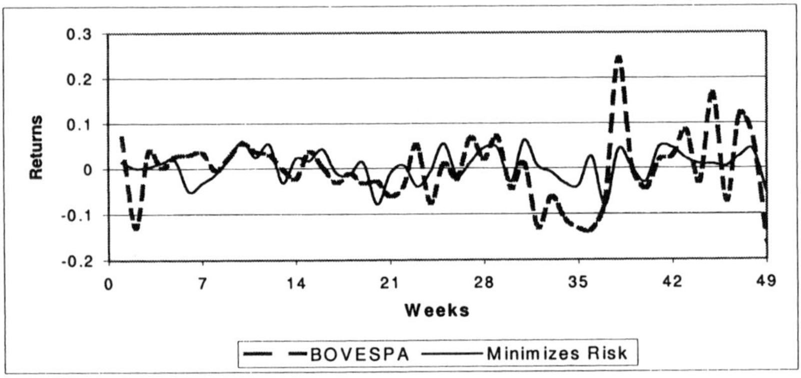

Figure 10.6. Comparison between the performance of the portfolio that minimizes the risk for a given return of more than 95% and the market portfolio.

The Beta values of the portfolios reveal that both portfolios are of a defensive nature. It may also be observed that the managed portfolios perform well during the deepest market dips. Therefore, given its constraints, this model may generate gains for the investor and possibly increase such gains by allowing a risk-free asset to be incorporated into the portfolio.

In Figure 10.7 below, the weekly forecasts of the VaR and the value of the portfolio after the management period are compared.

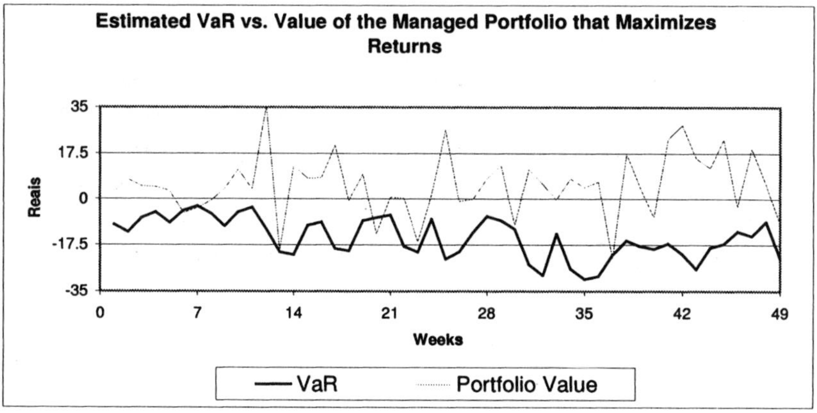

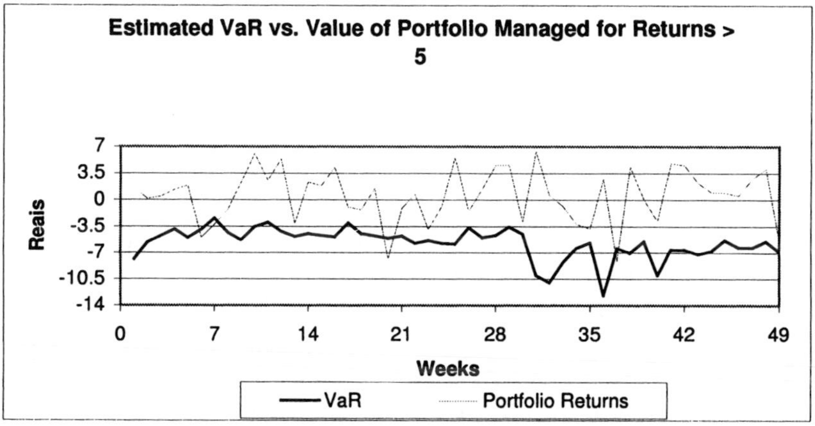

Figure 10.7. Comparison between the predicted VaR for the portfolio and the value of the portfolio at the end of the management period, for each of the two managed portfolios.

It may be observed in the graphs above that the predicted value of the potential loss for the portfolio (VaR) for a confidence interval of 95% was surpassed by the portfolio, in the worst case, only in 3 weeks throughout the entire management period, in other words, the model is able to make good predictions of the VaR for the portfolio.

7. Conclusions

This paper has proposed a Hybrid System for the selection of stocks and the management of investment portfolios with the use of genetic algorithms for portfolio selection and management, the GARCH model for volatility forecasts, neural networks for predicting asset returns, and the methodology for calculating the VaR as a measure of risk.

The assets selected to form the portfolio contemplated in this study were not modified throughout the entire management period; only portfolio weights were changed.

In general, the results of the tests performed with the proposed model that uses Genetic Algorithms for selecting the assets in the portfolio proved to be satisfactory (Table 10.1) since the selected assets were capable of generating profit for the investor. It has been observed that the selected assets are more sensitive to market variations.

It should be pointed out that, in the course of the 49 weeks that were selected for portfolio management, the Brazilian market presented periods of greater instability (with greater volatility). It is more difficult to make forecasts during such periods and they account for the largest errors in the forecasts provided by the neural network.

In order to obtain good volatility forecasts, it is essential to estimate optimal parameters (a0, a1, and b1) for the GARCH(1,1) model employed, and for this reason special care must be taken with regard to the optimization algorithm employed and also to the initial point ascribed to it.

The GARCH model once more proved to be efficient in volatility forecasting since this type of forecast plays an important part in the modeling of the neural nets that provide return forecasts and of the genetic algorithm for weight optimization, both of which are employed when calculating the VaR for the portfolio.

Because the neural network represents the market model, it is necessary to supply the network with the greatest possible amount of information on the variables that influence the market.

It has been demonstrated that the return forecasts were considerably improved by the fact that the asset volatility forecasts produced by the GARCH model were introduced into the modeling of the neural network for return forecasts.

This study has shown how the proposed system is able to perform good forecasts of the VaR for the portfolio over its management period.

As a rule, the managed portfolios obtained a better performance during the deepest dips in the market.

It has been observed that on the average, the managed portfolio obtains higher returns at a lower risk than the market portfolio.

One of the possible ways by which to improve the modeling of the algorithm for asset selection is to introduce other variables, such as a company's liquidity or volatility.

In this study, the assets were not modified throughout the entire portfolio management period; the possibility of updating the assets that comprise the portfolio on a weekly basis may also contribute to the improvement of the results obtained.

References

Bollerslev, T. (1986). "Generalized Autoregressive Conditional Heteroscedasticity," *Journal of Econometrics*, 31, 307–327.

Campos L., E. (1998). "Modelo de Escala Local: Uma Alternativa de Especificação Multiplicativa para Estimação e Previsão de Volatilidade de Séries Financeiras," Dissertação de Mestrado Dee Puc-Rio, Rio de Janeiro, Brazil.

Down, K. (1998). *Beyond Value at Risk — The New Science of Risk Management.* John Wiley & Sons.

Drost, F. C. and T. N. E. (1992). "Temporal Aggregation of GARCH Processes," Center Discussion Papers 9066 e 9240. Tilburg University.

Duarte Jr., Antonio M., M. A. Pinheiro, and T. B. B. Heil (1997). *Previsão da Volatilidade de Ativos e Índices Brasileiros.* Resenha BM&F, Number 112.

Elton, E. J. and M. J. Gruber (1995). *Modern Portfolio Theory and Investment Analysis*, 5th ed. John Wiley & Sons.

Engle, R. F. (1982). "Autoregressive Conditional Heteroscedasticity with Estimates of the Variance of the United Kingdom Inflation," *Econometrica*, 50(4), 987–1007.

Engle, R. F. (1995). *ARCH Selected Readings — Advanced Texts in Econometrics.* Oxford University Press.

Goldberg, D. E. (1989). *Genetic Algorithms in Search, Optimisation, and Machine Learning.* Addison-Wesley.

Haykin, S. (1998). *Neural Networks — A Comprehensive Foundation.* McMillan College Publishing Co.

Hull, J. C. (1999). *Options, Futures & other Derivatives*, 4th ed. Prentice Hall.

Jorion, P. (1997). *Value at Risk: The New Benchmark for Controlling Market Risk.* McGraw-Hill.

Lazo, J. G. L., Marley M.B.R. V., and M. A. C. Pacheco (2000). "A Hybrid Genetic-Neural System for Portfolio Selection and Management," in Dimitris Tsaptsinos (ed.), *Proceedings of the Sixth International Conference on Engineering Applications of Neural Networks (EANN 2000)*, 147–154.

Levy, H. and M. Sarnat (1984). *Portfolio and Investment Selection: Theory and Practice.* Prentice-Hall.

Markowitz, H. M. (1959). *Portfolio Selection.* Cambridge, MA: Black-Well.

Michalewicz, Z. (1996). *Genetic Algorithms + Data Structures = Evolution Programs.* Springer-Verlag.

Nelso, D. B. (1990). "Stationarity and Persistence in the GARCH(1,1) Model," *Econometric Theory*, 6, 318–334.

Zurada, J. M. (1995). *Introduction to Artificial Neural System.* Boston: Mass.

Chapter 11

INTELLIGENT CASH FLOW: PLANNING AND OPTIMIZATION USING GENETIC ALGORITHMS

Marco Aurélio C. Pacheco
ICA : Applied Computational Intelligence Laboratory
Department of Electrical Engineering
PUC-Rio, Rua Marquês de São Vicente 225, Gávea
Rio de Janeiro, RJ, Brazil, CEP 22453-900
marco@ele.puc-rio.br

Marley Maria R. Vellasco
ICA : Applied Computational Intelligence Laboratory
Department of Electrical Engineering
PUC-Rio, Rua Marquês de São Vicente 225, Gávea
Rio de Janeiro, RJ, Brazil, CEP 22453-900
marley@ele.puc-rio.br

Maíra F. de Noronha
ICA : Applied Computational Intelligence Laboratory
Department of Electrical Engineering
PUC-Rio, Rua Marquês de São Vicente 225, Gávea
Rio de Janeiro, RJ, Brazil, CEP 22453-900
maira@ele.puc-rio.br

Carlos Henrique P. Lopes
ICA : Applied Computational Intelligence Laboratory
Department of Electrical Engineering
PUC-Rio, Rua Marquês de São Vicente 225, Gávea
Rio de Janeiro, RJ, Brazil, CEP 22453-900

ch@ele.puc-rio.br

Abstract This article describes an intelligent system for financial planning and cash flow optimization, designed and developed by PUC-Rio (Pontifícia Universidade Católica do Rio de Janeiro) for the Souza Cruz, Brazilian tobacco company. The system, named ICF: Intelligent Cash Flow, is a computational tool for decision making support which provides short-term and long-term financial managing strategies, considering financial products of the market. The ICF makes use of Genetic Algorithms, a search and optimization technique inspired in natural evolution and the genetics, in order to elaborate plannings and cash flow projections which offer more profitability and liquidity for the considered period. From the planning alternatives offered by the system, the user can decide each day the investment and resource allocation options that suit better the demands and policy of the firm. The ICF analyses 500.000 different planning options in 20 minutes in order to identify the most profitable cash flows.

Keywords: Genetic Algorithms, Cash Flow Planning, Decision Making Support, Investments

Introduction

The Souza Cruz group deals everyday with the planning and execution of its cash flow, which operates millions of dollars daily. The aim of this activity is to promote larger profitability and liquidity to the company. Thus, the company identifies daily the best investments for the available operational cashbox balance or the best options of resource allocation to cover the deficit (Ross 1995). In simple terms, the problem consists of deciding, each day, what to do (investment, bank account etc) with the money available and where to borrow money from to pay the bills when cash is not available.

There are many variables that have influence in the cash flow: the types of investments or loans, terms, return rates, rules of redemption, risks, taxes, involved institutions, etc. The cash flow planning for a period of 90 days, for example, turns out to be computationally untreatable if it is to be done by exhaustively evaluating all the search space of possible plannings (approximately 68^{90} if we considered 68 options of different investments products for each day of 90 days period), in search of the combinations of investments and loans for 90 days which result in bigger profitability and liquidity.

In order to solve this problem, the ICF system—Intelligent Cash Flow—was developed. This system uses two models: the financial and the genetic model. The financial model is used to calculate the cash flow profitability, based on the IDC (Interbankary Deposit Certificate), by projecting profits and taxes for each kind of investment, for any term in the considered period. The genetic model, on the other hand, is used to search for a cash flow planning that promotes both, highest profitability and liquidity, simultaneously (Ross 1995; Goldberg 1989; Dasgupta 1997).

1. The ICF System

1.1. The Financial Model

The financial markets typically offer many types of products, for example fixed and floating rates investment and resource allocation (money borrow). The products considered by the model of the cash flow optimization were typical Brazilian ones.

The reckoning of profits and taxes for each product is based on the IDC (Ross 1995), which are papers of the monetary and non-monetary financial institutions that ballast the operations of the interbankary market, having the task of transferring resources from one financial institution to another.

The purpose is to re-negotiate the effective tax rate of the IDC, defined as the accumulated average daily rates of IDC, calculated by the Central Bank (Fortuna 1997) for the period between the date of operation in the market and the last negotiation date (redemption date). The quotation of the tax rate is made by the purchase or selling of a Unitary Price, and calculated through the division of the redemption value (defined by the BM&F as R$100,00) (Fortuna 1997) by the accumulation factor of the daily average rates for 30 days, until the redemption date (Equation 11.1). The UP therefore represents the discount of the redemption value by the tax rate object of the contracts IDC-1 day. These contracts make reference to the average rates calculated by the Central Bank, which reflect the average cost of the available resource exchanging operations between the financial institutions in the overnight market.[1]

$$UP = \frac{100.000}{\left(1 + \frac{TEO}{30 \times 100}\right)^n} \tag{11.1}$$

In Equation 11.1, TEO is the monthly rate equivalent to the *over*, n is the number of weekdays from the operation date to the redemption date, and 100 is the factor used to calculate the percentage value.

From the manipulation of UPs and the projection of forthcoming UPs, the financial model of the ICF calculates the liquid profitability of a cash flow, projecting all the profits and tributes to the last day of the cash flow, according to the current interest rates, so as not to exist discrepancy between the present and the forthcoming values.

For the reckoning of profits and taxes, the model includes its own calendar, which gives the system information such as: number of weekdays and total of days between two dates; non-fixed holidays, day of the week of a certain date; and redemption date corresponding to the date and term of the investment.

1.2. The Genetic Model

1.2.1 Representation. The chromosome of the ICF genetic model consists of n genes, represented by the data structure in Figure 11.1a. Each gene stands for a day in the considered period and has four fields. The first two identify an investment option (IdAplic) and its term (Pr), which are used when the cash operational balance is positive (cash available); the last two identify a resource allocation option (IdTRec) and its term (Pr), are used when cash operational balance is negative and money needs to be borrowed. For each analyzed day, only two of these fields are used, which depends on whether the cash operational balance is positive or negative that day.

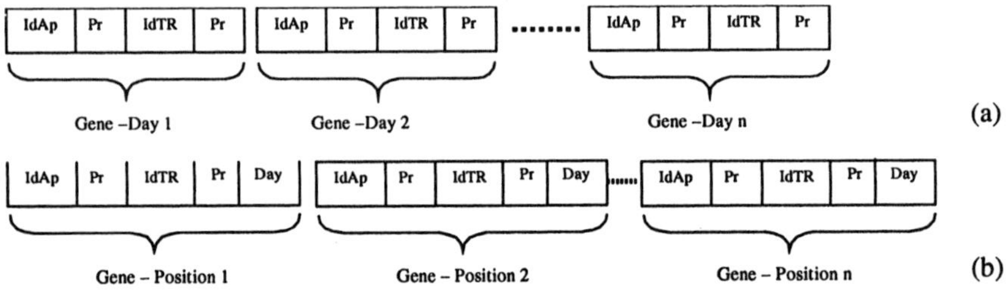

Figure 11.1. Chromosome representation (a) first model (b) positional rigidity relaxing model for epistatic problems.

According to the Evolutionary Computation theory, problems such as the optimization of the cash flow are highly epistatic (Dasgupta 1997), which means that there is strong interdependency between genes of the respective representation (for example, the investment on day d depends on the availability of financial resources that day, which can be due to the redemption made on day $d - n$). Such genes consist of genetic patterns that can be separated by the crossover operator.

In order to deal with the epistasy in this problem, a second model was created in which each gene is represented by its allele (Figure 11.1b) and by its locus (position in the chromosome). This kind of representation has the objective to decrease the original positional rigidity (Figure 11.1a), increasing the chances of distant interdependent genes to come closer to each other. Thus, genetic patterns with high fitness have more chances to proliferate in forthcoming generations.

1.2.2 Genetic Operators. In this project, two new crossover operators were created for this representation (allele, locus). These operators are an extension of the partially-mapped crossover operator proposed by Goldberg (Goldberg 1989), and explore important similarities of value and order simultaneously.

The first operator generates the offspring by choosing two random cut points on the chromosomes of the parents. These cut points define a subsequence of the chromosome, and the offspring receives the subsequence of one parent and has preserved the order and position of as many genes as possible from the other parent. As an example, the following discussion is focused only on the day field: If the chromosome has 9 genes, corresponding to the options and terms for 9 days, the two parents could be:

$p_1 = (\ 8\ 9\ 6 \mid 7\ 4\ 5\ 2 \mid 3\ 1\ )$ and
$p_2 = (\ 1\ 4\ 2 \mid 3\ 5\ 8\ 7 \mid 6\ 9\ )$,
where the numbers represent the gene of day n.

First, the offspring receives the swapped subsequences:

$o_1 = (\ \mathrm{x}\ \mathrm{x}\ \mathrm{x} \mid 3\ 5\ 8\ 7 \mid \mathrm{x}\ \mathrm{x}\ )$ and
$o_2 = (\ \mathrm{x}\ \mathrm{x}\ \mathrm{x} \mid 7\ 4\ 5\ 2 \mid \mathrm{x}\ \mathrm{x}\ )$,
where the x means that the gene is at present unknown.

The other positions receive the same genes as the original parents where there is no conflict (repetition):

$o_1 = (\ \mathrm{x}\ 9\ 6 \mid 3\ 5\ 8\ 7 \mid \mathrm{x}\ 1\ )$
$o_2 = (\ 1\ \mathrm{x}\ \mathrm{x} \mid 7\ 4\ 5\ 2 \mid 6\ 9\ )$

Finally, the first empty position of the offspring receives from the other parent the gene of the first missing day and so on, until all the positions are filled:

$o_1 = (\ 2\ 9\ 6 \mid 3\ 5\ 8\ 7 \mid 4\ 1\ )$
$o_2 = (\ 1\ 3\ 8 \mid 7\ 4\ 5\ 2 \mid 6\ 9\ )$

The mutation operator applied in the ICF implements a random choice of a gene (day) and the random assignment of a new term and a new type of financial application (investment or resource allocation).

1.2.3 The Fitness Function. The fitness function evaluates the fitness of a cash flow and gives the probability ($p_i = f_i / \sum^N f_j$) with which the flow will be selected for reproduction. In this case, the fitness function calculates the liquid returns (profit or tax) of each suggested application for each day in the considered period, projected to the last day of the same. A more satisfactory planning is obtained by finding the maximum return value to this function. The following procedure exemplifies the reckoning of the liquid return of the operations for the cash flow.

```
For day = 1 to day = last Day of Cash flow
{
        If Balance[day] different from 0
      {
            If Balance[day] > 0 /* An application is made */
            {
                  Fitness = Fitness +
                  ProjectedLiquidReturn(IdAplic, Term, day, Value)
            }
            Else /* A loan is made */
            {
                  Fitness = Fitness −
                  ProjectedLiquidTribute(IdTRec, Term, day, Value)
            }
      }
}
```

In the expression above, the terms ProjectedLiquidReturn and ProjectedLiquidTribut refer to the liquid return and tribute, respectively, projected to the last day of the period according to the current interest rate. The above procedure is executed for all the plannings created during the evolving process. The return value is a number that measures the fitness (profitability) of each planning. The selection of genitors for the descendents generation is probabilistic and based on the fitness.

The genetic model of the ICF was tested with real cashbox and market data. The results are shown in the next section.

2. Results

The ICF presents in an Excel like screen (Figure 11.2) the most profitable cash flow optimization found during evolution. (This example is based on fictitious data). This table shows the suggestions of applications (row) or resource allocation for each day (column) in a specified period (only the first 6 days can be seen in the picture).

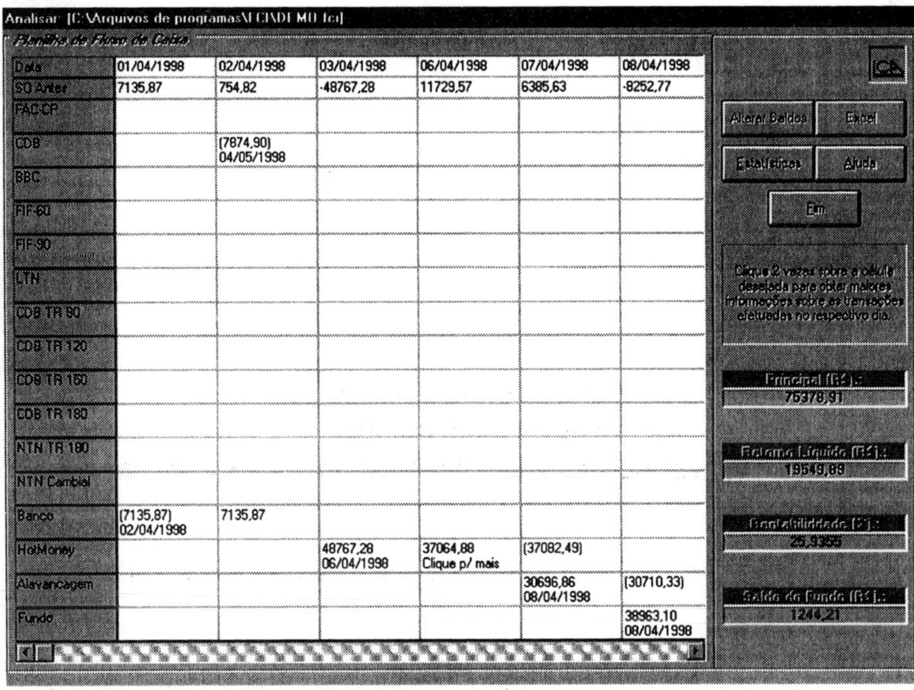

Figure 11.2. A table of a cash flow suggested by the ICF.

Figure 11.3 shows the performance of the Genetic Algorithm in the optimization of the profitability of the cash flow. This Genetic Algorithm evaluated 500000 cash flows in 1000 generations of 500 individuals each. The general characteristics of the GA is as follows: roulette wheel selection; linear normalization of the chromosome fitness; crossover rate 0.8; and mutation rate 0.1. At the beginning of the evolution, the cash flow found by the ICF has a profitability of approximately 2100. At the end, The model manages to identify plannings with 38.1% higher profitabilities (2900), making evident the importance of a decision-supporting system when it comes to optimizing the application of financial resources.

These results have been compared to random search, which is commonly used when other practical solution do not exist. In average, the ICF generates results with 50% more profitability than the random search.

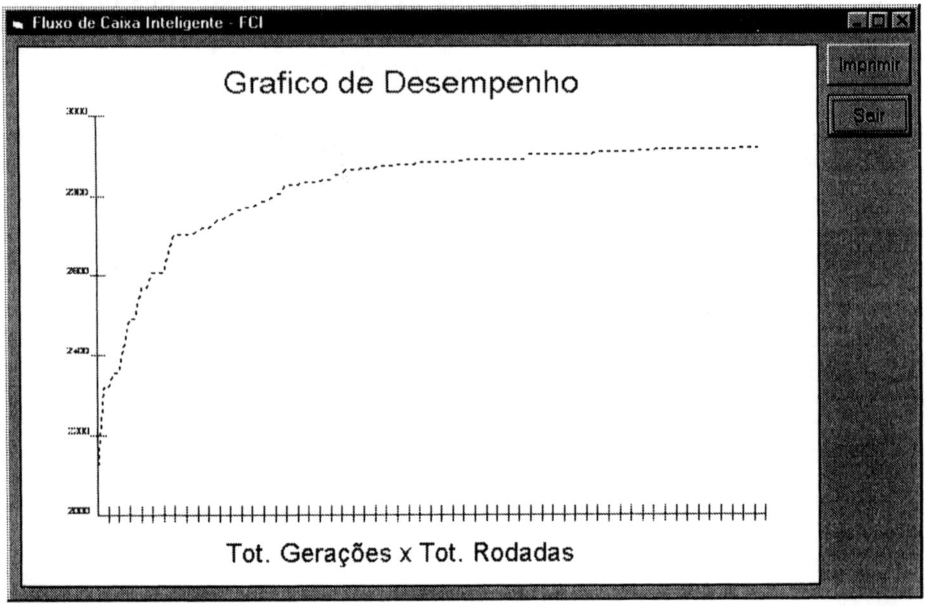

Figure 11.3. ICF performance graph of the cash flow example.

Many experiments were made for different periods of the year. The results show that the profitability is affected by the availability of cash operational balance, but it is also strongly influenced by the planning strategy.

3. Conclusions

The ICF is a result of a partnership between PUC-Rio University and Souza Cruz company. For Souza Cruz, the complexity of the problem made it necessary to search for more scientific ways to project the financial cash flow. A research group, including professors and students of PUC-Rio, used the state of art in the area of Genetic Algorithms in order to offer the performance demanded by the company.

The ICF was incorporated to the routine of Souza Cruz and made possible the automation of the planning of the company's cash flow for horizons of 4 to 90 days, suggesting operations that help the operator to search the highest profitability and liquidity of the company's capital. The ICF analyses 500.000 different cash flow in approximately 20 minutes (Pentium II, 450 Mhz), in order to find a (sub)optimal planning.

Through the daily use of the program, it is possible to interfere in order to adjust it to the new conditions of the market, of the legislation and of the national financial cenario.

Notes

1. The Central Bank sells public titles to the financial institutions, with certain tax rates; the banks pay in cash and go daily to the market in order to obtain the resources to finance these positions, re-passing these titles to the investors, with the compromise to re-buy them the following day, paying a daily rate. Therefore, it is called an overnight operation.

References

Ross, S. A., R. W. Westerfield, and J. F. Jaffe (1995). *Administração Financeira — Corporate Finance*. Editora Atlas.

Goldberg, D. E. (1989). *Genetic Algorithms in Search, Optimization, and Machine Learning*. Reading, MA: Addison-Wesley.

Dasgupta, D. and Z. Michalewicz (1997). *Evolutionary Algorihtms in Engineering Applications*. Springer.

Davis, L. (1996). *Handbook of Genetic Algorithms*. Int. Thomson Comp. Press.

Fortuna, E. (1997). *Mercado Financeiro - Produtos e Serviços*, 10th ed., Editora Qualitymark.

Chapter 12

THE SELF-EVOLVING LOGIC OF FINANCIAL CLAIM PRICES

Thomas H. Noe
A. B. Freeman School of Business, Tulane University
New Orleans, LA 70118-5669
tnoe@mailhost.tcs.tulane.edu

Jun Wang
SAS Institute Inc.
SAS Campus Drive, Room R5217
Cary, NC 27513
Jonathan.Wang@sas.com

Abstract In this paper, we use Genetic Programming, an optimization technique based on the principles of natural selection, to price financial contingent claims. Compared to the traditional arbitrage-based approach, this technique is useful when the underlying asset dynamics are unknown or when the pricing equations are too complicated to solve analytically. Comparing to other established data-driven option pricing techniques such as neural networks, implied binomial trees, etc., genetic programming has the advantage of not restricting the structure of the pricing formulas. In addition, because it is very easy to incorporate existing analytical pricing formulas into the evolutionary process, genetic programming can be applied in combination with existing pricing methods. In this paper, we show that genetic programming can recover Black-Sholes formula from a simulated data sample of fairly small size. The application to S&P 500 futures options also show promising results.

Keywords: Contingent Claims, Derivative Pricing, Genetic Programming

Introduction

One of the major challenges facing finance researchers and practitioners is to price financial contingent claims correctly. The traditional approach to pricing a derivative security, pioneered by Black and Scholes (1973) and Merton (1973), is to specify a stochastic process for the underlying asset, derive the pricing equation for the derivative using the no-arbitrage argument, and solve the pricing equation to find the derivative price. Although this approach has enjoyed tremendous success both in practice and in research, this approach may produce biased prices estimates in some cases. For example, the stochastic process for the underlying asset could be misspecified, the parameters for the process could be mis-estimated, or the pricing equation could be too complicated to solve analytically. The well-known volatility smile phenomenon illustrates the difficulty of fitting actual prices with simple pricing models. On the other hand, the booming derivative markets generate a large amount of data. This has lead a number of researchers to suggest data-driven approaches to pricing problems which extract pricing measures directly from observed data. In fixed income research, this is called the no-arbitrage approach.[1] For equity options, Rubinstein (1994), Dupire (1994), Derman and Kani (1994), Jackwerth and Rubinstein (1996) have provided methods to construct a discrete representation of the risk neutral density that is exactly consistent with a set of observed option prices. A different data-driven approach is taken by Hutchinson, Lo and Poggio (1994). They utilize a radial-bias neural network algorithm to price and hedge options.

These data-driven techniques have one common characteristic. They require the researcher to pre-specify the underlying model structure and then estimate the models' parameters. One natural question is whether there is a way of pricing assets without pre-specifying the pricing structure or the form of the relation between the underlying asset's and the derivative's price. In this paper, we investigate such a method, genetic programming (GP), for searching for the optimal pricing structure.

In genetic programming, pricing formulas are represented as tree-like computer programs. This provides great flexibility in the form of the pricing formulas. Further more, it is quite simple to represent known analytic formulas as tree-like programs so that these formulas can be incorporated in the evolution process. This makes it easy to combine the power of genetic search method with the current knowledge base on derivative pricing. The end result can be a more accurate pricing formula and a better understanding of the relationship between underlying and derivative securities.

Using a small sample of simulated data, we show that genetic programming can find a pricing formula with small pricing errors. In addition, by incorporating Black-Sholes formulas (with incorrect volatilities) in the initial population, the evolution process can be shortened substantially to achieve the same accuracy in pricing. This combination of existing knowledge base with genetic programming should be useful for derivative pricing.

This paper is organized as follows. Section 1 contains a brief description of genetic programming and our application to the derivative pricing problem. Section 2 shows the results when we apply to a small sample of simulated option prices. Section 3 shows the results of applying GP to S&P 500 futures options. Section 4 concludes the paper.

1. Method

1.1. Genetic Programming

Koza (1992) developed genetic programming, which shares many of the same features as genetic algorithm, developed by Holland (1962,1975). Both are optimization techniques based on the principles of natural evolution. Borrowing terminology from biology, there is an environment in which evolution occurs. This environment consists of a population of solution candidates, each of which is valuated according to a problem-specific function, "fitness," which measures how good the solution is to the problem. Then the solution candidates are recombined through crossover and mutation to generate new solution candidates, or "offspring." Crossover "marries" two computer programs to generate new programs. Mutation involves random modification of the code of an existing program. Similar to natural selection, the candidates with higher fitness measures are more likely to reproduce offspring and to keep their "genes" alive. The new generation of solution candidates are evaluated again and reproduce their offspring following the same selection process. This evolution process is stopped whenever certain terminating criteria are met.

The major difference between genetic algorithms and genetic programming is the representation of the solution candidates. In genetic algorithms, a solution candidate is called a "chromosome," which is a string of characters with a fixed length. For a specific problem, chromosomes are mapped to the specific solution domain of the problem. In genetic programming, a solution candidate is represented as a tree-like hierarchical structured computer program. Each node of the tree is a function, or "building block," which takes the output of its branch nodes as inputs and produces an output. If a function requires no input, it is a terminal function and becomes a terminal node of the tree. The whole program is computed recursively from the terminal nodes to the root of the tree. The size, shape and content of the tree can change dynamically during the evolution process. The dynamic nature of the solution structure makes genetic programming a flexible optimizing tool.

The search space of genetic programming is huge: the space of all possible tree-like programs. To be effective in searching such a large and complex space, the algorithm not only conducts a parallel search of many points, through fitness evaluation of an entire pool of programs, but also implicitly processes, in parallel, a large amount of useful information regarding the attributes in each solution program. The Darwinian fitness-proportionate reproduction allocates an exponentially increasing weight to those better attributes. This selection routine, when combined with spontaneous and dynamic emergence of new programs through crossover and mutation, leads to an optimal search in the stochastic environment by trading off, in a judicious fashion, the benefits of extensive search in a promising direction against the costs of "overfocusing," and thus ignoring large regions of the search space.

Economics researchers have applied genetic algorithms to study the learning dynamics of genetic agents in different game settings.[2] It is well established that agents using genetic algorithms are able to find the optimal strategies in these games. The application of genetic programming in economics is mainly in the area of generating smart computer traders since the tree structure of the solution candidates can be easily interpreted as the decision tree of human traders.[3]

In the context of derivative pricing, the tree structured computer programs can be naturally linked to pricing formulas. The attributes of the derivative contract (e.g. time to maturity and strike price) and the market price of the underlying asset are the inputs of the program and the output is the price of the derivative. The error between the market option price and the price from the program measures the performance of the program. The smaller the error, the higher the fitness of the program. The evolution terminates when one (best) program can price almost all the options correctly or when a predetermined number of evolution generation is reached.

1.2. Setup

Following Hutchinson, Lo and Poggio (1994) and other works on non-parametric pricing methods,[4] we first apply genetic programming in a simulated market governed by the Black-Sholes pricing formula. In our application, three terminal functions are specified, the ratio of spot stock

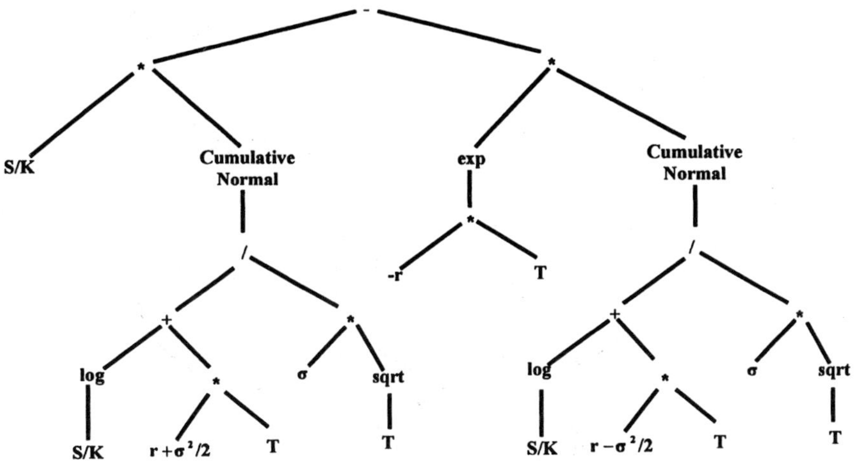

Figure 12.1. Black-Sholes option pricing formula in tree representation.

price and strike price of the option, time to maturity, and a real number between -1 and 1. Non-terminal functions are operators such as plus, minus, multiplication, division, logarithm, exponential, minimum, maximum, square root, and cumulative normal.[5] The output of the tree structure is the option price normalized by the strike price. We can obtain the option price by multiplying the output with the strike. Given these building blocks, the Black-Sholes option pricing formula can be easily represented in the tree form, as illustrated in Figure 12.1.

The initial population of the evolution consists of N randomly generated programs. The fitnesses of these N programs are then evaluated and the evolution of option pricing formulas starts. We adopt a "steady-state" evolution process and "tournament" selection. In particular, three different programs are selected randomly from the population. Of the three programs, the two programs with the highest and second highest fitness crossover to produce their "child" program. Figure 12.2 is an illustration of the crossover operation. In crossover, first one node is randomly selected from each program tree; then the node and its branches in one tree is replaced by the node and its branches from the other tree to form the child program. With a small probability p_m, mutation occurs, where the child program is substituted by a randomly generated one. Finally, the least fit program of the three is replaced by the child program from crossover and possibly mutation. This sequence of selection-crossover-replacement is repeated N times to generate a new generation of pricing programs. The best program in this generation is then compared to the best program so far. If the best program in this

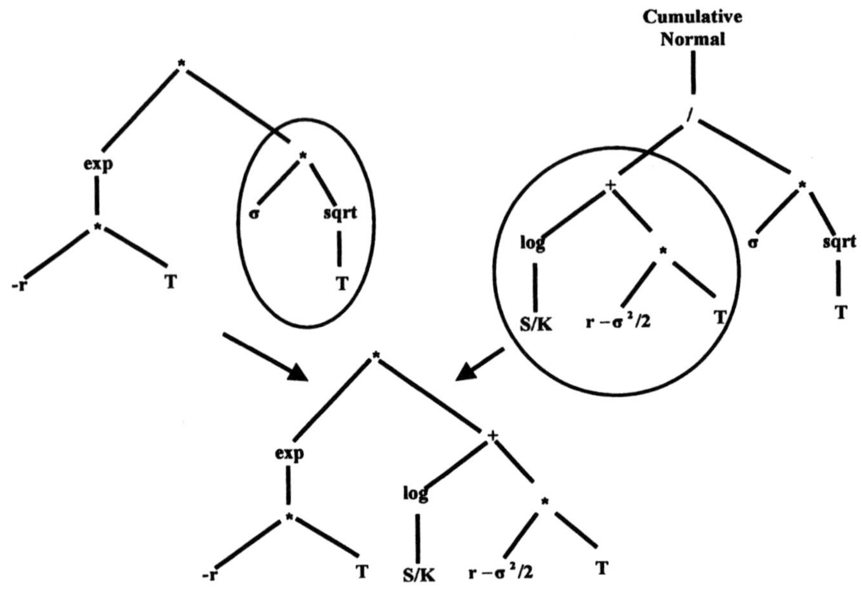

Figure 12.2. Crossover operation.

generation performs better, it becomes the best program of the evolution up to this generation. The whole evolution stops if the best program correctly prices α percent of the options, or there is no improvement for n_{NI} generations, or a total of n_T generations are reached.

Let M be the number of training options. Then we define the raw fitness of a trading program as

$$FIT_{raw} = \sum_{i=1}^{M}\{|p_i - p_i^0| + f(\frac{|p_i - p_i^0|}{p_i^0})\} + g(S) \tag{12.1}$$

where p_i is the option price generated by the genetic program; p_i^0 is the option price from market, or in simulation, from Black-Sholes option pricing formula. S is the size of the program represented by the number of nodes in the tree. f and g are monotone increasing function. The first term in the raw fitness equation is the sum of absolute errors. The second term is an increasing function of relative pricing errors. When the option is deep out of the money, its price will be very small and absolute error will not be enough to indicate the magnitude of the pricing error. The relative pricing error term is then useful to provide a more accurate picture of errors. The third term is to control the size of genetic programs because we hope to find formulas that are accurate and simple. By incorporating the size in the fitness, we can prevent the tree size to grow too large. The smaller the raw fitness, the better the program performs. We can get the fitness by applying a monotone decreasing function on the raw fitness. On the other hand, since we use tournament selection, only the rank of programs is relevant. The raw fitness is enough to provide such a rank.

2. Simulation Results

We use Black-Sholes formula to generate 120 call prices with different price/strike ratio and time to maturity. In particular, we fix the strike price at 50, annual volatility at 20% and risk-free rate at 5%. The spot price is set at 40, 45, 50, 55, 60 respectively and the time to maturity starts from 2 weeks to 26 weeks with one week increment. In this way, 125 call prices are generated. Calls with stock price 40 and time to maturity from 2 to 6 weeks have prices less than 0.1 cent, so we drop these five from the sample. Hence the final training sample consists of 120 observations.

In this application, relative pricing errors contribute to the raw fitness through a simple linear function,[6] namely,

$$f(\frac{|p_i - p_i^0|}{p_i^0}) = 0.2\frac{|p_i - p_i^0|}{p_i^0} \tag{12.2}$$

The size effect is specified by a quadratic function,[7]

$$g(S) = 0.00001S^2 \tag{12.3}$$

The number of programs in the evolution population (N) is 2000, and the mutation probability p_m is 0.001. The termination criteria is set to 95 or 50 generations without improvement, or a total of 400 generations.[8]

The evolution process is run ten times and ten genetic programs are generated. All ten runs are terminated because 400 generation is reached. Table 12.1 summarizes the performance of these ten programs. We report the mean absolute pricing errors of these pricing programs in the training sample and in an enlarged sample. The enlarged sample consists of 525 options prices and is generated when we allow the spot price to vary from 40 to 60 with increment 1 and keep others the same as we generate the training sample. As shown in Table 12.1, the performance of these ten programs are not uniform due to the stochastic nature of the evolution process. The best program, program No.2, has a mean pricing error of 1.2 cents in the training set and 2.7 cents in the enlarged set, while program No.7 has a pricing error of 55 cents per option in the enlarged set. However, since one program is sufficient for pricing purposes, the performance of program No.2 is encouraging. Figure 12.3 graphs its pricing errors. For most parts, the errors are small. Large pricing errors appear when the option is close to expiration date and when the spot/strike ratio is 0.84 and time to maturity is longer than 3 months. The large pricing errors for options near expiration are easy to understand as the pricing formula is highly nonlinear there. Hutchinson, Lo and Poggio (1994) found the same effect using neural networks. The pricing error at spot/strike ratio 0.84 is probably caused by some errors in the program that is preserved in the evolution because the ratio of 0.84 is not in the training sample. Finally, note that the errors of this program are smaller than the pricing errors of the neural networks reported by Hutchinson, Lo and Poggio (1994). In Figure 5 of Hutchinson, Lo and Poggio (1994), the call price errors normalized by the strike price of typical Radial Basis Functions are of the magnitude of 0.01. In Figure 12.3, if we divide the pricing errors by the strike price, 50, the largest error is less than 0.005.

Table 12.1. Performance Summary of Genetic Option Pricing Programs

Program	Raw Fitness	Size	Mean Absolute Error in Training Sample	Mean Absolute Error in Enlarged Sample
1	10.9504	179	0.0435125	0.234374
2	2.39267	229	0.0119801	0.0268274
3	10.9251	256	0.0552858	0.321717
4	15.8456	146	0.113564	0.129961
5	4.14133	266	0.0244115	0.186542
6	4.98215	306	0.0227589	0.239008
7	14.9628	213	0.106428	0.556377
8	3.91602	307	0.0200092	0.0795417
9	6.08154	273	0.0336438	0.0626854
10	5.54673	296	0.0256067	0.13877

One big advantage of genetic programming is the ability to incorporate known pricing formulas easily. To illustrate this, we put two Black-Sholes formulas with wrong volatilities in our initial population. One has a volatility of 30% and the other has a volatility of 10%. We also reduce the total generations of evolution from 400 to 100. Again, we run the evolution process ten times and generate ten pricing programs. A summary of the performance is in Table 12.2. As can be seen from Table 12.2, the errors of these programs are smaller than those of the

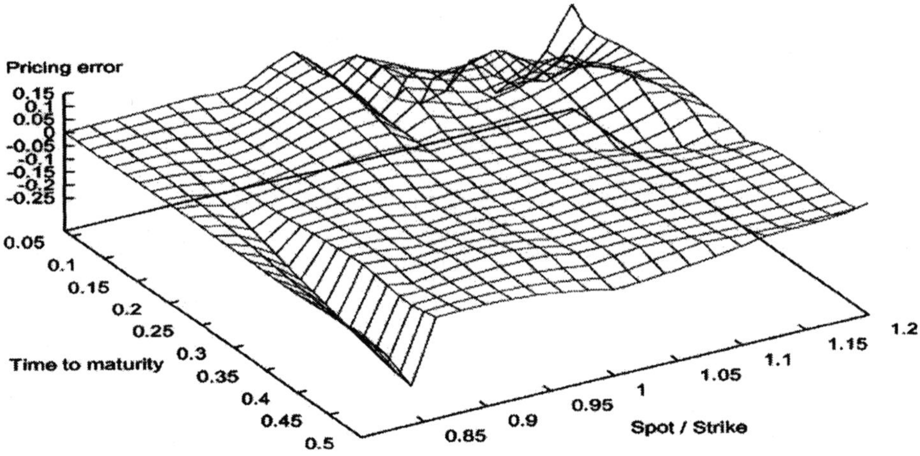

Figure 12.3. Program 2 from Table 12.1.

programs in Table 12.1. Figure 12.4 illustrates the pricing errors of the best program, No.2, from Table 12.2. Comparing to Figure 12.3, its pricing error is much smaller. In addition, the computation time to generate these pricing programs is much less than it is in previous unguided evolution.[9]

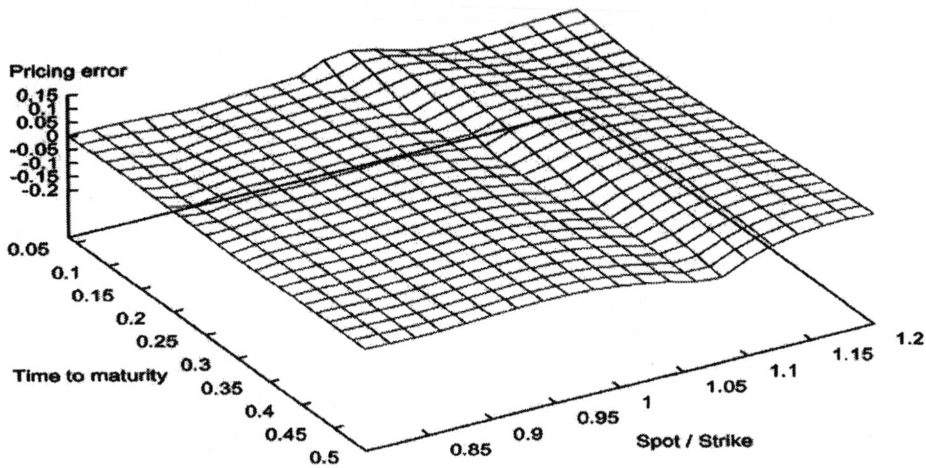

Figure 12.4. Program 2 from Table 12.2.

Table 12.2. Performance summary of genetic option pricing programs generated when two programs in the initial population are Black-Sholes pricing formulas with incorrect volatilities.

Program	Raw Fitness	Size	Mean Absolute Error in Training Sample	Mean Absolute Error in Enlarged Sample
1	7.6342	82	0.0318373	0.0797962
2	1.79956	133	0.00873717	0.0130343
3	5.14277	219	0.0310684	0.0401444
4	7.211	139	0.0290486	0.374124
5	8.48708	83	0.0377613	0.0649374
6	6.91098	239	0.0362878	0.160665
7	6.13789	124	0.0210355	0.0323546
8	5.58583	188	0.0359399	0.0402882
9	6.02011	123	0.0277535	0.047442
10	4.30658	178	0.0169761	0.038864

3. Application to S&P 500 Futures Options

The simulation shows that GP has the potential to recover the pricing formula in a Black-Sholes world. In this section, we apply GP to the S&P 500 futures options. The daily option data and futures prices from January, 1997 to June, 1997 are retrieved from Chicago Mercantile Exchange. We separate the data into six monthly subsets.[10] In each month, we run genetic programming ten times and generate 10 pricing programs. To utilize the existing knowledge in pricing, we include the 10 pricing programs generated from the previous month in the initial population of GP runs. For the first month in our sample, January, 1997, we only include the two Black-Sholes programs with volatilities of 30% and 10%. The performance measures of ten programs in the training month (In-Sample performance) and in the following month (Out-Sample performance) are reported in Table 12.3.

As shown in Table 12.3, the mean absolute errors of the genetically generated pricing programs are around 1 dollar in-sample and less than 2 dollars out-sample. This may seem a quite big pricing error. However, note that the strike prices for these options are over 700 and the absolute error normalized by the strike price is much smaller at less than 0.2%. This is comparable to if not better than the results of neural networks in Hutchinson, Lo and Poggio (1994). The other thing to note is that the performance measure standard errors of the ten pricing programs in

Table 12.3. Performance summary of genetic option pricing programs in S&P 500 futures options. Ten genetic programs are generated from the options data in the training month. The average of the mean absolute errors of ten program in the training month (In Sample) and the average of the mean normalized absolute errors are reported. The same two measures are reported using the options data in the following month (Out Sample). Normalized absolute error is the absolute error normalized by the strike price of the option. The numbers in parentheses are standard errors of the corresponding measures of the ten pricing programs.

Training Month	Mean Absout Error (In Sample)	Mean Normalized Absolute Error (In Sample)	Mean Absolute Error (Out Sample)	Mean Normalized Absolute Error (Out Sample)
1/97	1.088 (0.119)	.001409 (.000155)	1.778 (0.214)	.002219 (.000266)
2/97	1.330 (0.0370)	.001663 (.000044)	1.306 (0.070)	.001614 (.000082)
3/97	1.121 (0.008)	.001391 (.000010)	1.284 (0.025)	.001661 (.000031)
4/97	0.981 (0.024)	.001275 (.000033)	1.994 (0.381)	.002444 (.000491)
5/97	1.268 (0.020)	.001537 (.000027)	1.713 (0.056)	.001944 (.000064)

each month are about 10% of the average, indicating that the variation of GP is not very large. This bodes well for the application of GP in the real world. Although the internal randomness of the algorithm is inevitable, the end results of the algorithm do not differ from each other materially.

4. Conclusion

In this paper, we use genetic programming to search for pricing formulas of financial derivatives. Using a small sample of option prices simulated from Black-Sholes formula, we show that genetic programming can recover Black-Sholes formula from the data. In addition, when we include analytical formulas in the initial population, the evolution requires substantially fewer generations to reach the same, if not better, pricing performance. Structural flexibility is one appealing feature of genetic programming. This becomes even more so when we can utility this structural flexibility to incorporate known analytic formulas in the population of evolution. In this way, we can use genetic programming technique to improve the pricing accuracy of our current pricing models. Finally, we apply GP to S&P 500 futures options and our results indicate that the performance of GP in pricing options is at least comparable to, if not better than, the performance of neural networks in Hutchinson, Lo and Poggio (1994). This suggests that GP can be a promising way to uncover the complex pricing functions of the financial derivatives in the real world.

Notes

1. See, e.g. Black, Derman and Troy (1990), Heath, Jarrow and Morton (1992), Ho and Lee (1986), Hull and White (1993).

2. Examples include Arifovic (1994, 1996), Ho (1996), and No and Pi (2000).

3. See, e.g., Allen and Karjalainen (2000), Andrews and Prager (1994), Neely, Weller and Dittmar (1997), Wang (2000).

4. For example, Ait-Sahalia and Lo (1995), Stutzer (1996)

5. Division, logarithm, exponential and square root are altered slightly to prevent numerical overflow or illegal operation. The division operator returns zero if the denominator is zero. Logarithm returns zero if input is zero and returns natural log of the absolute value of input otherwise. Exponential returns zero if the input is too big or too small. Square root is applied to the absolute value of the input.

6. This part is not added in the raw fitness if the option price is less than 1 cent.

7. We also impose an upper limit of 500 nodes whenever a new tree is created.

8. Correct pricing is defined as the absolute pricing error is less than 2 cents.

9. The computation time to run one evolution with a total random initial population takes about 3 hours on a PC with a Pentium 200 Pro CPU. The computation time to run one evolution with two pricing programs inserted in the initial population takes about 30 minutes.

10. The options data is checked for obvious arbitrage opportunities and those observations are removed from our sample. This leaves us between 500 and 700 observations in each monthly data set.

References

Ait-Sahalia, Y. and A. Lo, (1995). "Nonparametric Estimation of State-Price Densities Implicit in Financial Asset Prices," *National Bureau of Economic Research Working Paper*, 5251.

Allen, F. and R. Karjalainen, (2000). "Using Genetic Algorithms to Find Technical Trading Rules," *Journal of Financial Economics*, 51, 245–271.

Andrews, M. and R. Prager, (1994). "Genetic Programming for the Acquisition of Double Auction Market Strategies," In K.E.Kinnear (Ed.), *Advances in Genetic Programming*. Cambridge, MA:MIT Press.

Arifovic, J. (1994). "Genetic Algorithm Learning and the Cobweb Model," *Journal of Economic Dynamics and Control*, 18, 3–28.

Arifovic, J. (1996). "The Behavior of the Exchange Rate in the Genetic Algorithm and Experimental Economics," *Journal of Political Economy*, 104, 510–541.

Black, F., E. Derman, and W. Troy, (1990). "A One-Factor Model of Interest Rates and Its Application to Treasury Bond Options," *Financial Analysts Journal*, 46(Jan./Feb.), 33–39.

Black, F. and M. Scholes (1973). "The Pricing of Options and Corporate Liabilities," *Journal of Political Economy*, 81, 637–659.

Derman, E. and I. Kani (1994). "Riding on the Smile," RISK 7, February, 32–39.

Dupire, B. (1994). "Pricing with S Smile," RISK 7, January, 18-20.

Heath, D., R. Jarrow, and A. Morton (1992). "Bond Pricing and the Term Structure of Interest Rates: a New Methodology for Contingent Claims Evaluation," *Econometrica*, 60, 77–105.

Ho, T.-H. (1996). "Finite Automata Play Repeated Prisoner's Dilemma with Information Processing Costs," *Journal of Economic Dynamics and Control*, 20, 173–207.

Ho, T. S. Y., and S. B. Lee (1986). "Term Structure Movements and Pricing of Interest Rate Claims," *Journal of Finance*, 41, 1011–1029.

Holland, J. H. (1962). "Outline for a Logical Theory of Adaptive Systems," *Journal of the Association for Computing Machinery*, 3, 297–314.

Holland, J. H. (1975). *Adaptation in Natural and Artificial Systems: An Introductory Analysis with Applications to Biology, Control, and Artificial Intelligence*, Ann Arbor, MI: University of Michigan Press.

Hull, J. and A. White (1993). "One-Factor Interest-Rate Models and the Valuation of Interest-Rate Derivative Securities," *Journal of Financial and Quantitative Analysis*, 28, 235–254.

Hutchinson, J. M., A. W. Lo, and T. Poggio (1994). "A Nonparametric Approach to Pricing and Hedging Derivative Securities via Learning Networks," *Journal of Finance*, 851–889.

Jackwerth, J. C. and M. Rubinstein (1996). "Recovering Probability Distributions from Option Prices," *Journal of Finance*, 51, 1611–1652.

Koza, J. R. (1992). *Genetic Programming: on the Programming of Computers by Means of Natural selection*, Cambridge, MA: MIT Press.

Koza, J. R. (1994). *Genetic Programming II: Automatic Discovery of Reusable Programs*, Cambridge, MA: MIT Press.

Merton, R. (1973). "Rational Theory of Option Pricing," *Bell Journal of Economics and Management Science*, 4, 141–183.

Neely, C., P. Weller, and R. Dittmar (1997). "Is Technical Analysis in the Foreign Exchange Market Profitable? A Genetic Programming

Approach," *Journal of Financial and Quantitative Analysis*, 32, 405–426.

Noe, T. H., and L. Pi (2000). "Genetic Algorithm, Learning, and the Dynamics of Corporate Takeovers," *Journal of Economic Dynamics and Control*, forthcoming.

Rubinstein, M. (1994). "Implied Binomial Trees," *Journal of Finance*, 49, 771–818.

Stutzer, M. (1996). "A Simple Nonparametric Approach to Derivative Security Valuation," *Journal of Finance*, 51, 1633–1652.

Wang, J. (2000). "Trading and Hedging in S&P 500 Spot and Futures Markets Using Genetic Programming," *Journal of Futures Markets*, forthcoming.

Chapter 13

USING A GENETIC PROGRAM TO PREDICT EXCHANGE RATE VOLATILITY

Christopher J. Neely
Research Department
Federal Reserve Bank of St. Louis
P.O. Box 442
St. Louis, MO 63166
USA

Paul A. Weller
Department of Finance
Henry B. Tippie College of Business Administration
University of Iowa
Iowa City, IA 52242
USA

Abstract This article illustrates the strengths and weaknesses of genetic programming in the context of forecasting out-of-sample volatility in the DEM/USD and JPY/USD markets. GARCH(1,1) models serve used as a benchmark. While the GARCH model outperforms the genetic program at short horizons using the mean-squared-error (MSE) criterion, the genetic program often outperforms the GARCH at longer horizons and consistently returns lower mean absolute forecast errors (MAE).

Keywords: Genetic Programming, GARCH, Foreign Exchange, Volatility, Forecasting, Heteroskedasticity

Introduction

It is well-known that exchange rate volatility displays considerable persistence. That is, large movements in prices tend to be followed by more large moves, producing positive serial correlation in absolute or squared returns. This has important practical implications for portfolio allocation strategies, risk management, and the pricing of derivative securities (Baillie and Bollerslev 1989 and 1991).

Engle (1982) developed Autoregressive Conditionally Heteroskedastic (ARCH) models to characterize this serial correlation. Bollerslev (1986) extended the ARCH class to produce the Generalized Autoregressive Conditionally Heteroskedastic (GARCH) model, which has been extensively used to estimate and forecast asset price volatility, and is therefore a natural benchmark against which to compare alternative models of volatility.

This chapter investigates the performance of a genetic program as a non-parametric procedure for forecasting volatility in the foreign exchange market. Applying genetic programming to this important issue in finance illustrates both its strengths and its weaknesses. Forecasting exchange rate volatility is an appropriate task in which to highlight the characteristics of genetic programming as it is an important problem with a well accepted benchmark solution, the GARCH(1,1) model. Genetic programs have the ability to detect patterns in the conditional mean of foreign exchange and equity returns that are not accounted for by standard statistical models (Neely, Weller, and Dittmar 1997; Neely and Weller 1999 and 2001; Neely 2000). This leads one to conjecture that a genetic program may also be useful in generating predictions of conditional volatility.

A disadvantage of genetic programming—as with any statistical modeling procedure—is the tendency to overfit the data in sample. Genetic programming's efficiency and ability to handle a very large set of potential functional forms for the forecast function exacerbate the usual difficulties.

1. The Genetic Program

Genetic algorithms are computer search procedures used to generate solutions to appropriately defined problems. The structure of the search procedure is based on the principles of natural selection. These procedures were developed by Holland (1975) and extended by Koza (1992). The essential features include (1) a means of representing potential solutions to a problem as character strings which can be recombined to form new potential solutions and (2) a fitness criterion which measures the "quality" of a candidate solution. In the original genetic algorithm of Holland, potential solutions were encoded as fixed length character strings. Koza's extension, referred to as genetic programming, instead employs variable-length, hierarchical strings that can be thought of as decision trees or computer programs. Here we use genetic programming to search for a forecasting function for the volatility of the exchange rate.

In the application in this paper we represent forecasting functions as trees, and make use of the following function set in constructing them: plus, minus, times, divide, norm, log, exponential, square root, and cumulative standard normal distribution function. In addition, we supply the following set of data functions: *data*, *average*, *max*, *min*, and *lag*. The data functions can operate on any of the four data series we permit as inputs to the genetic program: (1) daily returns; (2) integrated volatility (i.e. the sum of squared intraday returns); (3) the sum of the absolute value of intraday returns; and (4) number of days until the next business day. *Data*(*returns*), for example, returns the daily return at t. The other data functions operate in a similar fashion, but also take numerical arguments to specify the length of the window over which they will operate. The numerical arguments that the functions take are determined by the genetic program. Thus *average*(*returns*)(n) returns the arithmetic average of the return observations t, $t-1$, ..., $t-n+1$. Providing the number of days until the next business day permits the genetic program to vary volatility predictions for weekend and holidays.

The choice of elements to include in the function set is an important one. In principle it is desirable to concentrate on fairly simple functions, thus allowing the genetic program to build in more complexity only if the problem requires it. However, there is a tradeoff between this consideration and the potential reduction in search time that can result from a careful choice of more complex functions. For this reason, we investigate the results of adding to the original set the three more complex data functions described below. These functions were chosen to focus the search in regions of function space that are likely to be more relevant.

The expanded set of data functions consists of the original set plus *geo*, *mem*, and *arch*5. The function *geo* returns the following weighted average of 10 lags of past data:

$$geo(data)(\alpha) \equiv \sum_{j=0}^{9} \alpha[(1-\alpha)^j]data_{t-j} \tag{13.1}$$

This function can be derived from the prediction of an IGARCH specification with parameter α, where we constrain α to satisfy $0.01 \leq \alpha \leq 0.99$ and lags are truncated at 10. That is, the IGARCH prediction for future volatility ($\sigma_t^2 = \alpha\varepsilon_t^2 + (1-\alpha)\sigma_{t-1}^2$) can be rewritten as the infinite sum of past shocks with geometrically declining weights by recursively substituting for lagged variance terms ($\sigma_t^2 = \alpha\varepsilon_t^2 + \alpha(1-\alpha)\varepsilon_{t-1}^2 + \alpha(1-$

$\alpha)^2\varepsilon_{t-2}^2 + \alpha(1-\alpha)^3\varepsilon_{t-3}^2 ...)$. The geo function mimics this pattern of geometrically declining weights. Similarly, the mem function mimics the linearly declining weights on past volatility that would be used by a long memory specification. The function *mem* returns a weighted sum similar to that which would be obtained from a long memory specification. It takes the form

$$mem(data)(d) \equiv \sum_{j=0}^{9} h_j data_{t-j} \tag{13.2}$$

where $h_0 = 1$, $h_j = c(1/j!)(d+j-1)(d+j-2)\dots(d+1)d$, and $c = \left[1 + \sum_{j=1}^{9}((1/j!)(d+j-1)(d+j-2)\dots(d+1)d)\right]^{-1}$, for $j > 0$. This formulation provides linearly declining weights and constrains the sum of the coefficients h_j to equal one so that the output is of the same magnitude as recent volatility. The parameter d is determined by the genetic program and is numerically constrained to satisfy $-1 < d < 1$, so that the predicted volatility process is mean reverting. Values of d larger than 1 (-1) are replaced with 0.99 (-0.99) in the evaluation of the function. Finally, the function *arch*5 permits a flexible weighting of the five previous observations, where the values for h_j are provided by the genetic program and constrained to lie within {-5, 5} and to sum to one. The function has the form

$$arch5(data)(h) \equiv \sum_{j=0}^{4} h_j data_{t-j}, \quad h = (h_0, h_1, ..., h_4) \tag{13.3}$$

Figure 13.1 illustrates a simple example of a hypothetical tree determining a forecasting function. The function first computes the maximum of the sum of squared intraday returns over the last five days. This number is multiplied by 0.1 and the result entered as the argument x of the function $(8/\pi)\arctan(x) + 4$. This latter function is common to all trees and maps the real line into the interval (0, 8). It ensures that all forecasts are nonnegative and bounded above by the largest observed in-sample volatility.

We now turn to the form of the fitness criterion. The functions generated by the genetic program produce forecasts of volatility. Since true volatility is not directly observed, it is necessary to use an appropriate proxy in order to assess the performance of the genetic program. One possibility is to use the ex post squared daily return. However, as Andersen and Bollerslev (1998) have pointed out, this is an extremely noisy measure of the true underlying volatility, and is largely responsible for

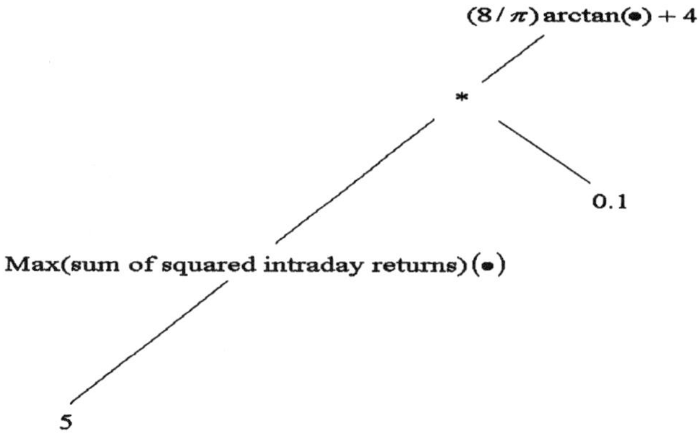

Figure 13.1. An example of a hypothetical forecast function.

the apparently poor forecast performance of GARCH models. A better approach is to sum squared intraday returns to more accurately measure true daily volatility (i.e., integrated volatility). We construct such a measure of integrated volatility using five irregularly spaced intraday observations. If S_{it} is the i-th observation on date t, we define

$$R_{i,t} = 100 \cdot \ln(\frac{S_{i+1,t}}{S_{i,t}}) \tag{13.4}$$

$$\sigma_{i,t}^2 = \sum_{i=1}^{5} R_{i,t}^2 \tag{13.5}$$

Thus $\sigma_{i,t}^2$ is the measure of integrated volatility on date t.[1] Using five intraday observations represents a compromise between the increase in accuracy generated by more frequent observations and the problems of data handling and availability that arise as one moves to progressively higher frequencies of intraday observation.

In constructing the rules, the genetic program maximized the negative mean square forecast error (MSE) as the fitness criterion. We recognize that there are potential difficulties with the use of this criterion in the present context. A heteroskedasticity-corrected fitness measure proved unsatisfactory in experiments. With three to five observations per day, there were instances where the integrated daily volatility was very small; the heteroskedasticity correction caused the measure to be inappropriately sensitive to those observations.

Given the well-documented tendency for the genetic program to overfit the data, we investigated the effect of modifying the fitness criterion by adding a penalty for complexity. This consisted of subtracting an

Table 13.1. Data Type and Source

Time	Source	Type of Price
1000	Swiss National Bank	triangular arbitrage on bid rates
1400	Federal Reserve Bank of New York	mid point of bid and ask
1600	Bank of England	triangular arbitrage, unspecified
1700	Federal Reserve Bank of New York	mid point of bid and ask
2200	Federal Reserve Bank of New York	mid point of bid and ask

amount (0.002 * number of nodes) from the negative MSE. This modification is intended to bias the search toward functions with fewer nodes, which are simpler and therefore less prone to overfit the data.

2. Data and Implementation

As mentioned above, the measure of integrated volatility was constructed as the sum of squared returns from irregularly spaced observations throughout the day. We consider two currencies against the dollar, the German mark (DEM) and Japanese yen (JPY), over the period June 1975 to September 1999. Thus, the final nine months of data for the DEM represents the rate derived from that of the euro, which superseded the DEM in January 1999. The timing of observations was 1000, 1400, 1600, 1700, and 2200 GMT. Days with fewer than three valid observations, or no observation at 1700 were treated as missing. In addition, weekends were excluded. The sources of the data for both exchange rates are summarized in Table 13.1. In addition to the integrated volatility series, we used three further data series as input to the genetic program: daily returns, sum of absolute intraday returns, and the number of days until the next trading day.

The full sample is divided into three subperiods: the training period June 1975 - December 1979; the selection period January 1980 - December 30, 1986; the out-of-sample period December 31, 1986 - September 21, 1999. The role of these subperiods is described below.

In implementing the genetic program we follow the procedures below.

i. Create an initial generation of 500 randomly generated forecast functions.

ii. Measure the MSE of each function over the training period and rank according to performance.

iii. Select the function with the lowest MSE and calculate its MSE over the selection period. Save it as the initial best forecast function.

iv. Select two functions at random, using weights attaching higher probability to more highly-ranked functions. Apply the recombination operator to create a new function, which then replaces an old function, chosen using weights attaching higher probability to less highly-ranked functions. After new functions are created, they are immediately eligible for replacement in the same generation. Repeat this procedure 500 times to create a new generation of functions.

v. Measure the MSE of each function in the new generation over the training period. Take the best function in the training period and evaluate the MSE over the selection period. If it outperforms the previous best forecast, save it as the new best forecast function.

vi. Stop if no new best function appears for 25 generations, or after 50 generations. Otherwise, return to stage 4.

The stages above describe one trial. Each trial produces one forecast function. The results of each trial will generally differ as a result of sampling variation. For this reason it is necessary to run a number of trials and then to aggregate the results. The aggregation methods are described in the following section.

3. Results

The benchmark results are those from a GARCH(1, 1) model, estimated over the period June 1975 to December 30, 1986. We generate forecasts from this model, in and out of sample, at horizons of one, five, and twenty days.[2] The GARCH(1,1) model—which has been shown to provide reasonable volatility forecasts across a variety of economic and financial series (West and Cho 1995; Andersen et al. 2001)—can be written as

$$\sigma_t^2 = \omega + \alpha\varepsilon_t^2 + \beta\sigma_{t-1}^2, \quad \varepsilon_t \sim N(0, \sigma_{t-1}^2) \tag{13.6}$$

where σ_t^2 is the variance of the price process. When $\alpha + \beta < 1$, the variance process displays mean reversion to the unconditional expectation of σ_t^2 , $\omega/(1-\alpha-\beta)$. We apply the GARCH model to the daily integrated volatility measure defined in Equation 13.5.

We also forecast with a genetic program whose training and selection periods coincide with the in-sample estimation period for the GARCH model. For each case of the genetic program we generated ten trials, each

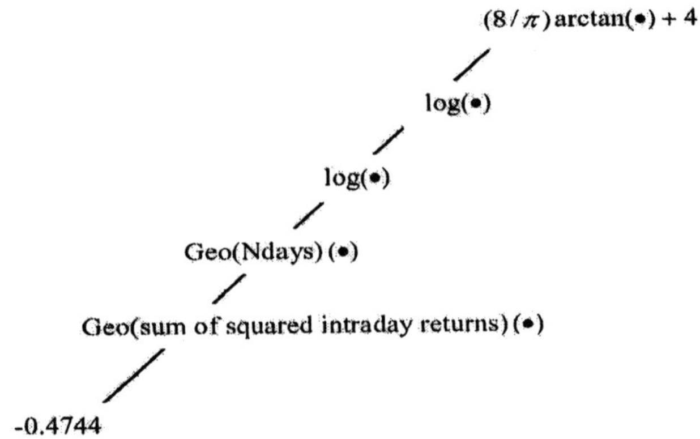

Figure 13.2. An example of a one-day ahead forecasting functions for the DEM found by the genetic program.

of which produced a forecast function. The cases were distinguished by the following factors: (1) forecast horizon — one, five and twenty days; (2) the number of data functions — five or eight; and (3) penalty for complexity — absent or present. The forecasts were aggregated in one of two ways. The equally-weighted forecast is the arithmetic average of the forecasts from each of the ten trials. The median-weighted forecast takes the median forecast from the set of ten forecasts at each date. We report five measures of out-of-sample forecast performance: mean square error, mean absolute error, R^2, mean forecast bias, and kernel estimates of the error densities.

Before discussing the results, we first present a simple example of the forecasting rules produced by the genetic program. Figure 13.2 illustrates a one-day-ahead forecasting function for the DEM. Its out-of-sample MSE was 0.496. The function is interpreted as follows. The number -0.4744 at the terminal node enters as the argument of *geo* (sum of squared intraday returns). Since the argument of *geo* is constrained to lie between 0.01 and 0.99, it is set to 0.01. The number generated by this function then enters as the argument in *geo* (Ndays) where Ndays refers to the data series "number of days to the next trading day". We caution that this example was chosen largely because of its relatively simple form; some trials generated considerably more complex rules with as many as 10 levels and/or 100 nodes.

Table 13.2 reports in-sample results for the baseline case with five data functions and no penalty for complexity. The figures for MSE for the DEM are very similar for the GARCH and equally weighted genetic program forecasts at the 1- and 5-day horizons, but the genetic program is appreciably better at the 20-day horizon. The median weighted

Table 13.2. In-Sample Comparison of Genetic Program and GARCH: The Baseline Case

Exchange Rate	Horizon (days)	MSE EW GP	MSE MW GP	MSE GARCH	R^2 EW GP	R^2 GARCH
DEM	1	0.502	0.533	0.499	0.177	0.165
DEM	5	0.559	0.593	0.560	0.124	0.103
DEM	20	0.609	0.630	0.666	0.076	0.041
JPY	1	0.561	0.578	0.602	0.224	0.139
JPY	5	0.655	0.655	0.729	0.059	0.025
JPY	20	0.661	0.667	0.709	0.045	0.014

The in-sample mean square error (MSE) and R^2 from a GARCH(1,1) and genetic program (GP) forecast on DEM/USD and JPY/USD data at 3 forecast horizons: 1-day, 5-days, and 20-days. The GP forecast was generated using five data functions and without a penalty for complexity. In columns 3 and 6 we report the equally weighted (EW) values of MSE and R^2. In column 4 we report the median weighted (MW) value of MSE. The in-sample period was June 1975 to December 30, 1986.

forecast is generally somewhat inferior to the equally weighted forecast, but follows the same pattern over the forecast horizons relative to the GARCH model. That is, its best relative performance is at the twenty-day horizon. For the JPY, the genetic program produces equally weighted MSE figures which are in all cases lower than for the GARCH model. Similarly, the equally weighted genetic programming rules have higher R^2s over each horizon than the GARCH models. This is not especially surprising given the flexibility of the non-parametric procedure and its known tendency to overfit in sample.

Table 13.3 presents a more interesting comparison, out-of-sample performance over the period December 31, 1986 to September 21, 1999. The MSE for the DEM cases are a bit higher for the equally weighted genetic program forecasts than they are for the GARCH forecasts at the one- and five-day horizons, but the genetic program performs rather better at the twenty-day horizon. For the JPY the GARCH forecast has lower MSE at the one-and twenty-day horizon, but this ranking is reversed at the five-day horizon. This is not, however, reflected in the R^2, for which the GARCH models are better in each case. But the genetic program has lower MAE than does the GARCH model at all horizons and for both currencies. To summarize: With MSE as the performance criterion, neither the genetic program nor the GARCH model is clearly

Table 13.3. Out-of-Sample Comparison of Genetic Program and GARCH: The Baseline Case

Exchange Rate	Horizon (days)	MSE EW GP	MSE MW GP	MSE GARCH	MAE EW GP	MAE GARCH	R^2 EW GP	R^2 GARCH
DEM	1	0.347	0.377	0.334	0.297	0.326	0.094	0.118
DEM	5	0.377	0.416	0.360	0.308	0.353	0.059	0.081
DEM	20	0.411	0.417	0.444	0.311	0.431	0.012	0.017
JPY	1	1.347	1.352	1.294	0.422	0.472	0.139	0.160
JPY	5	1.483	1.480	1.558	0.431	0.518	0.033	0.039
JPY	20	1.476	1.477	1.429	0.449	0.551	0.021	0.047

The out-of-sample MSE and R^2 from a GARCH(1,1) and genetic program (GP) forecast on DEM/USD and JPY/USD data at 3 forecast horizons: 1-day, 5-days, and 20-days. The GP forecast was generated using five data functions and without a penalty for complexity. In columns 3, 6, and 8 we report the equally weighted (EW) values of MSE, MAE, and R^2. In column 4 we report the median weighted (MW) value of MSE. The out-of-sample period was December 31, 1986 to September 21, 1999.

superior. The GARCH model achieves higher R^2s in each case. But the MAE criterion clearly prefers the genetic programming forecasts.

To determine whether the out-of-sample performance of the genetic program could be improved by a simple correction designed to reduce overfitting, we examined the effect of introducing an ad hoc penalty for complexity into the fitness criterion. With the penalty, the genetic program sought to maximize the negative MSE less the product of 0.002 times the number of nodes in the rule. This had very little effect and if anything led to a slight deterioration in out-of-sample performance. For this reason we do not report these results.

Table 13.4 reports the out-of-sample performance of the genetic program forecasts using the augmented set of data functions, which include *geo*, *mem*, and *arch*5. For ease of comparison Table 13.4 repeats the out-of-sample figures for the GARCH model. In the case of the JPY the equally weighted forecasts produced by the genetic program at one- and five-day horizons are now superior to those of the GARCH model. This ranking is matched by that from the R^2 criterion. Things are less clear-cut in the case of the DEM, with the genetic program producing lower MSE at the twenty-day horizon, but otherwise equalling or slightly underperforming the GARCH model. As in the baseline case reported in Table 13.3, the median weighted forecast consistently performs less well

Table 13.4. Out-of-Sample Results Using the Data Functions *geo*, *mem*, and *arch5*

Exchange Rate	Horizon (days)	MSE EW GP	MSE MW GP	MSE GARCH	MAE EW GP	MAE GARCH	R^2 EW GP	R^2 GARCH
DEM	1	0.370	0.440	0.334	0.293	0.326	0.116	0.118
DEM	5	0.365	0.371	0.360	0.301	0.353	0.047	0.081
DEM	20	0.379	0.376	0.444	0.305	0.431	0.008	0.017
JPY	1	1.275	1.308	1.294	0.429	0.472	0.176	0.160
JPY	5	1.447	1.462	1.558	0.465	0.518	0.041	0.039
JPY	20	1.489	1.623	1.429	0.438	0.551	0.044	0.047

The out-of-sample MSE and R^2 from a GARCH(1,1) and genetic program (GP) forecast on DEM/USD and JPY/USD data at 3 forecast horizons: 1-day, 5-days and 20-days. The GP forecast was generated using eight data functions including *geo*, *mem*, and *arch5* (for descriptions see equations 13.1 - 13.3 in the text) and without a penalty for complexity. In columns 3 and 8 we report the equally weighted (EW) values of MSE and R^2. The out-of-sample period was December 31, 1986 to September 21, 1999.

than the equally weighted. The figures for the MAE of the genetic program are very little changed from Table 13.3, and are still substantially better than those of the GARCH model.

Given that we have established that neither imposing a penalty for complexity nor expanding the set of data functions leads to any appreciable improvement in the performance of the genetic program—at least in the ways we have done it—we concentrate in what follows on out-of-sample forecasts in the baseline case presented in Table 13.3, where no penalty for complexity was imposed and only five data functions were used. We present kernel estimates of the densities of out-of-sample forecast errors at the various horizons in Figures 13.3 to 13.5.

The most striking feature to emerge from these figures is the apparent bias in the GARCH forecasts when compared to their genetic program counterparts. At all forecast horizons and for both currencies there is a positive shift in the error distributions of the GARCH forecasts that lead the modes of the forecast densities to be farther from zero. The appearance of greater bias in the GARCH forecasts is illusory, however. Table 13.5 shows that though both forecasts are biased in the mean, the magnitude of the bias is considerably greater for the genetic program. Tests for the bias-carried out with a Newey-West correction for serial correlation-show that all the forecasts are biased in a statistically significant way (Newey and West 1987 and 1994). Given the evidence from

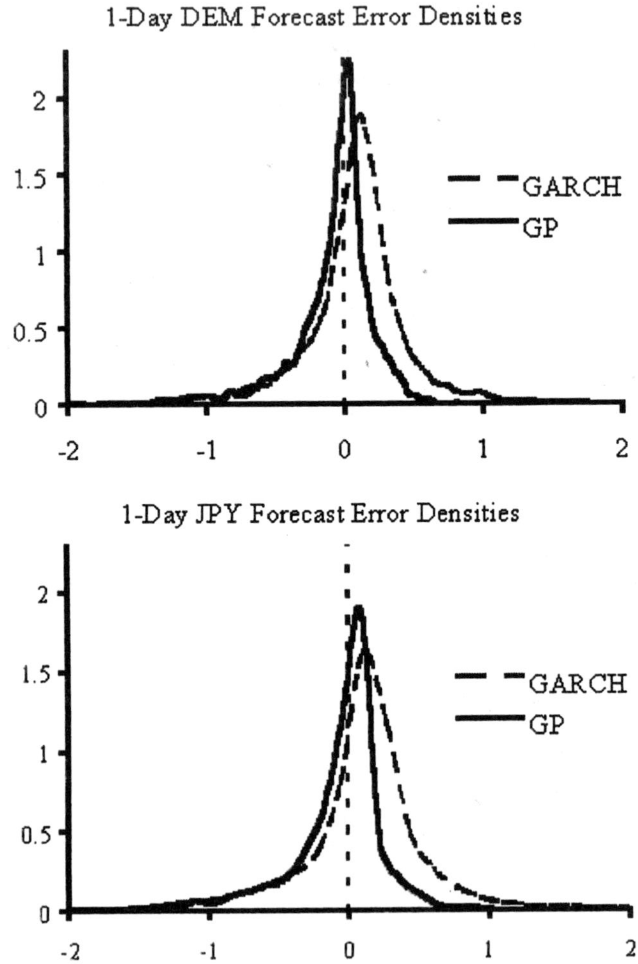

Figure 13.3. The kernel estimates of the densities of the one-day forecast errors (forecast minus realized volatility) for the DEM and JPY for genetic program and GARCH(1, 1) model over the out-of-sample period, December 31, 1986 to September 21, 1999. The dotted vertical line denotes zero.

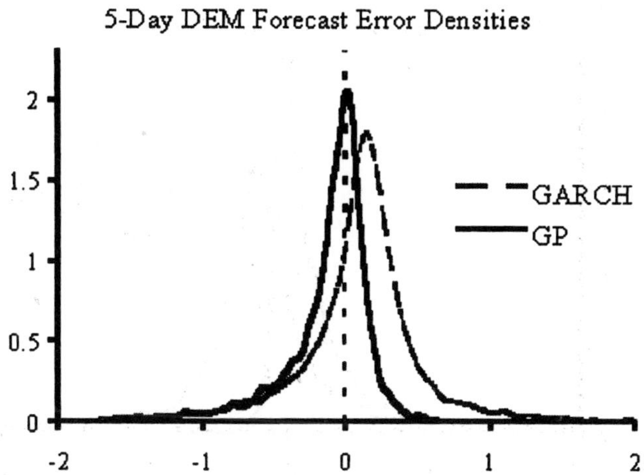

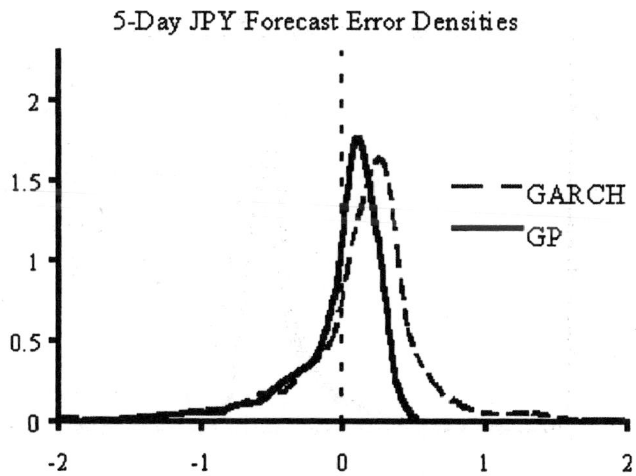

Figure 13.4. The kernel estimates of the densities of the five-day forecast errors (forecast minus realized volatility) for the DEM and JPY for genetic program and GARCH(1, 1) model over the out-of-sample period, December 31, 1986 to September 21, 1999. The dotted vertical line denotes zero.

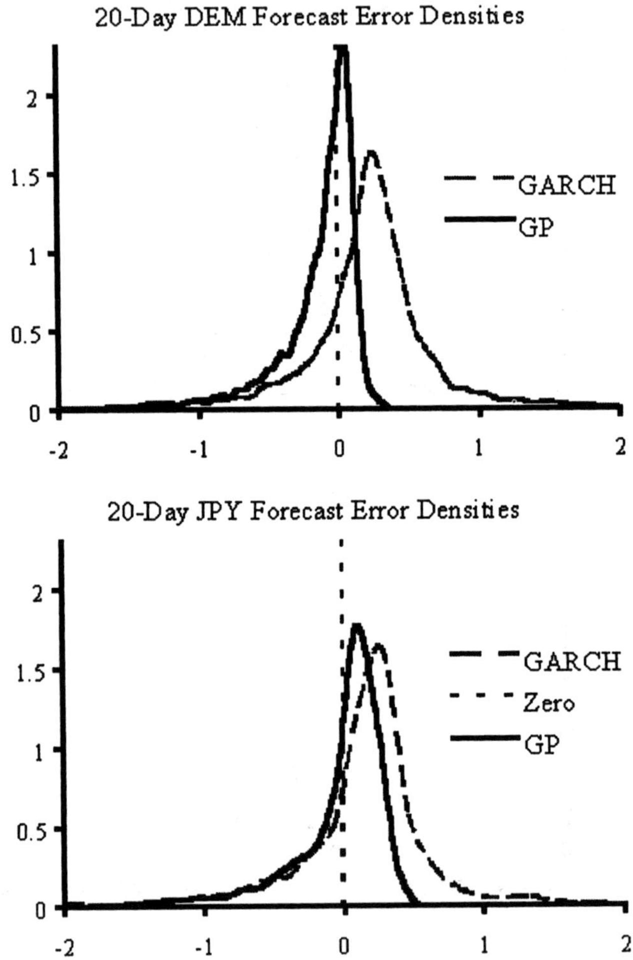

Figure 13.5. The kernel estimates of the densities of the twenty-day forecast errors (forecast minus realized volatility) for the DEM and JPY for genetic program and GARCH(1, 1) model over the out-of-sample period, December 31, 1986 to September 21, 1999. The dotted vertical line denotes zero.

Table 13.5. Tests for Mean Forecast Bias

Exchange Rate	Horizon (days)	Mean Volatility	Predicted Volatility EW GP	Bias EW GP	Bias p-value EW GP	Predicted Volatility GARCH	Bias GARCH	Bias p-value GARCH
DEM	1	0.434	0.287	-0.147	0.000	0.459	0.026	0.000
DEM	5	0.434	0.232	-0.201	0.000	0.487	0.054	0.000
DEM	20	0.434	0.195	-0.239	0.000	0.592	0.159	0.000
JPY	1	0.557	0.333	-0.224	0.000	0.574	0.017	0.064
JPY	5	0.557	0.367	-0.190	0.000	0.594	0.037	0.068
JPY	20	0.557	0.420	-0.137	0.003	0.650	0.093	0.017

In column 3 mean volatility is the mean daily integrated volatility over the out-of-sample period December 31, 1986 to September 21, 1999. Columns 4 and 7 report the mean forecast of integrated volatility over the same period for genetic program (GP) and GARCH(1, 1) model. Columns 5 and 8 report the mean forecast error (predicted volatility minus realized volatility). The genetic program forecast is the equally-weighted forecast for the baseline case described in Table 13.3. Columns 5 and 9 report the p-value for a heteroskedasticity-corrected test of unbiasedness with Newey-West correction for serial correlation.

Figures 13.3 to 13.5—that the modes of the genetic programming error distribution are closer to zero than those of the GARCH model—this indicates that the bias in the genetic programming forecasts is being substantially influenced by a small number of negative outliers.

4. Discussion and Conclusion

We choose to use the problem of forecasting conditional volatility in the foreign exchange market to illustrate the strengths and weaknesses of genetic programming because it is a challenging problem with a well accepted benchmark solution, the GARCH(1,1) model. The genetic program did reasonably well in forecasting out-of-sample volatility. While the genetic programming rules did not usually match the GARCH(1,1) model's MSE or R^2 at 1- and 5-day horizons, its performance on those measures was generally close. But the genetic program did consistently outperform the GARCH model on mean absolute error (MAE) and modal error bias at all horizons. The genetic programming solutions appeared to suffer from some in-sample overfitting, which was not mitigated, in this case, by an ad hoc penalty for rule complexity.

Our results suggest some interesting issues for further investigation. The superiority of the genetic program according to the MAE criterion is perhaps surprising given that we used MSE as the fitness criterion. This raises the possibility that further improvement in the forecasting performance of the genetic program relative to the GARCH model could be achieved by using MAE as the fitness criterion. Also, given that increasing the frequency of intraday observations has been shown to improve the accuracy of forecasts based on the GARCH model (Andersen et al., 2001), it is important to know whether the results of this investigation survive in that context.

Acknowledgments

The views expressed are those of the authors and do not necessarily reflect official positions of the Federal Reserve Bank of St. Louis, or the Federal Reserve System. The authors thank Janis Zvingelis for programming assistance with the kernel density plots.

Notes

1. More precisely, daily volatility is calculated from 1700 GMT to 1700 GMT.

2. Note that the forecasted variable at the five-day (twenty-day) horizon is the integrated volatility five (twenty) days in the future. It is not the sum of the next five (twenty) days of integrated volatility.

References

Andersen, T. and T. Bollerslev (1998). "Answering the Skeptics: Yes, Standard Volatility Models Do Provide Accurate Forecasts," *International Economic Review*, 39, 885–905.

Andersen, T., T. Bollerslev, F. Diebold, and P. Labys. (2001). Modeling and Forecasting Realized Volatility. NBER Working Paper 8160. *What Moves Stock Prices? Journal of Portfolio Management*, 4–12.

Baillie, R. and T. Bollerslev. (1989). *Artificial Economic Life: A Simple Model of a Stockmarket.* Physics D, 75, 264–274.

Baillie, R. and T. Bollerslev. (1991). "Intra-Day and Inter-Market Volatility in Foreign Exchange Rates," *Review of Economic Studies*, 58, 565–585.

Bollerslev, T. (1986). "Generalized Autoregressive Conditional Heteroskedasticity," *Journal of Econometrics*, 31, 307–327.

Engle, Robert F. (1982). "Autoregressive Conditional Heteroskedasticity with Estimates of the Variance of U.K. Inflation," *Econometrica*, 50, 987–1008.

Friedman, D. and J. Rust. (eds. 1991). "The Message in Daily Exchange Rates: A Conditional-Variance Tale," *Journal of Business and Economic Statistics*, 7, 297–305.

Holland, J. (1975). *Adaptation in Natural and Artificial Systems.* Ann Arbor, MI: University of Michigan Press.

Koza, J. (1992). *Genetic Programming: On the Programming of Computers by Means of Natural Selection.* Cambridge, MA: MIT Press.

Neely, C. J. (1999). "Risk-Adjusted, Ex Ante, Optimal, Technical Trading Rules in Equity Markets," Federal Reserve Bank of St. Louis Working Paper WP 99-015C.

Neely, C. J. and P. A. Weller (1999). "Technical Trading Rules in the European Monetary System," *Journal of International Money and Finance*, 18, 429–458.

Neely, C. J. and P. A. Weller (2001). "Technical Analysis and Central Bank Intervention," *Journal of International Money and Finance* (forthcoming).

Neely, C. J., P. A. Weller, and R. Dittmar (1997). "Is Technical Analysis in the Foreign Exchange Market Profitable? A Genetic Programming Approach," *Journal of Financial and Quantitative Analysis*, 32, 405–426.

Newey, W. K. and K. D. West (1987). "A Simple Positive Semi-Definite, Heteroskedasticity and Autocorrelation Consistent Covariance Matrix," *Econometrica*, 55, 703–708.

Newey, W. K. and K. D. West (1994). "Automatic Lag Selection in Covariance Matrix Estimation," *Review of Economic Studies*, 61, 631–653.

West, K. D. and D. Cho (1995). "The Predictive Ability of Several Models of Exchange Rate Volatility," *Journal of Econometrics*, 69(2), 367–391.

Chapter 14

EVOLUTIONARY DECISION TREES FOR STOCK INDEX OPTIONS AND FUTURES ARBITRAGE

How Not To Miss Opportunities

Sheri Markose
Economics Department and Institute of Studies in Finance
University of Essex
scher@essex.ac.uk

Edward Tsang
Computer Science Department
University of Essex
edward@essex.ac.uk

Hakan Er
Economics Department
University of Essex
her@essex.ac.uk

Abstract **EDDIE-ARB** (**EDDIE** stands for Evolutionary Dynamic Data Investment Evaluator) is a genetic program (**GP**) that implements a cross market arbitrage strategy in a manner that is suitable for online trading. Our benchmark for **EDDIE-ARB** is the Tucker (1991) put-call-futures (**P-C-F**) parity condition for detecting arbitrage profits in the index options and futures markets. The latter presents two main problems. (*i*) The windows for profitable arbitrage opportunities exist for short periods of one to ten minutes. (*ii*) From a large domain of search, annually, fewer than 3% of these were found to be in the lucrative range of £500-£800 profits per arbitrage. Standard *ex ante* analysis of arbitrage suffers from the drawback that the trader awaits a contemporaneous signal for a profitable price misalignment to implement an arbitrage in the same direction. Execution delays imply that this naive strategy may fail. A methodology of random sampling is used to train **EDDIE-ARB** to pick up the fundamental arbitrage patterns. The further novel aspect of **EDDIE-ARB** is a constraint satisfaction feature supplementing the fitness function that enables the user to train the **GP** *how not to miss opportunities* by learning to satisfy a minimum and maximum set on the number of arbitrage opportunities being sought. Good **GP** rules generated by **EDDIE-ARB** are found to make a 3-fold improvement in profitability over the naive *ex ante rule.*

Keywords: Genetic Programming, Machine Learning, Genetic Decision Trees, Arbitrage, Options, Futures, Constraint Satisfaction

Introduction

This paper presents an extension of a Genetic Programming (**GP**) tool called **EDDIE** (Evolutionary Dynamic Data Investment Evaluator) developed by Tsang et al. (1998) at the University of Essex for purposes of financial forecasting and engineering (see, Tsang et al. 2000, for a survey). The application we study here is of **EDDIE** within the context of cross market arbitrage, specifically in the FTSE-100 index options and futures, in a manner that is suitable for online trading. Hence, the **GP** merits the acronym of **EDDIE-ARB**. There are two main problems in conducting the FTSE-100 index market arbitrage in real time. (i) The windows for profitable arbitrage opportunities exist for short periods of one to ten minutes, and (ii) from a large domain of search annually fewer than 3% of these were found to be in the lucrative range of £400 - £800 profits per arbitrage. The static representation of an arbitrage position, as is well known, is based on contemporaneous price signals and entails a riskless position where no explicit role for forecasting appears to exist. However, this standard ex ante analysis of arbitrage suffers from the drawback that the trader awaits a contemporaneous signal for a profitable price misalignment to implement an arbitrage in the same direction. Execution delays imply that this so called naive strategy may fail. Hence, a forecasting tool is necessary to evaluate market conditions and make trading recommendations in anticipation of price misalignments in a way that is robust to execution delays. The recognition or detection of arbitrage positions can be aided by known formulaic conditions derived from theory. But this is not sufficient if temporary market price anomalies exist. **GP**s that perform better than the naive strategy can be trained to detect arbitrage positions from a range of economic variables determining prices in the two markets.

However, the problem we encountered is that arbitrage opportunities with sizeable profits though they exist were infrequent. They are as few as only two on average in any trading day. Thus, for the arbitrageur to net a sizeable annual income, the crucial aspect for **EDDIE-ARB** to be successful is that not only must it give correct predictions whilst making recommendations to trade; **EDDIE** must not miss a requisite number of arbitrage opportunities!

EDDIE is built on Genetic Programming (Koza 1992, 1994; Koza et al. 1996) which is a powerful variant of genetic algorithms (Holland 1975; Goldberg 1989; Mitchell 1996). The strengths of **GP** especially for financial applications arise from its use of decision tree representations instead of strings of chromosomes. The decision tree representations referred to as Genetic Decision Trees (**GDT**s) will be able to handle rule sets of variable size and these rules are easy to understand and evaluate by human users.[1] This makes this approach more attractive than neural networks, most of which are black boxes (Goonatilake and Treleaven 1995).

The use of Genetic algorithms (**GA**s) and **GP**s is now widespread in financial markets for purposes of forecasting and financial engineering. The efficacy of these adaptive computational methods in financial forecasting and financial engineering is seen to lie in their flexibility to account for potentially complex non-linear relationships which are not captured well by traditional linear and parametric methods. As excellent surveys already exist on application of **GA**s and **GP**s in financial applications, we will only briefly list some of our precursors here. Bauer (1994) reported his **GA** intelligent systems which aimed at finding tactical market timing strategies; Allen and Karjalainen (1995) applied the **GP** technique to find profitable technical trading rules for trading on the S&P 500 index; Chen and Yeh (1997) attempted to formalize the notion of unpredictability in the efficient market hypothesis in terms of search intensity and chance of success in the search conducted by genetic programming; Mahfoud and Mani (1996) presented a new genetic-algorithm-based system and applied it to the task of predicting the future performances of individual stocks; Neely et al. (1997) and Oussaidene et al. (1997) applied genetic programming to foreign exchange forecasting and reported some success.

In earlier work (Tsang et al. 1998; Li & Tsang 1999a; Li & Tsang 1999b) **EDDIE** was trained primarily in the art of financial forecasting with different objectives to be satisfied. Initially, **EDDIE**'s focus was on predicting whether a price series will increase by r% or more within the next n periods. **EDDIE**'s performance was found to compare favourably with random rules, commonly used individual technical rules and C4.5 rule sets with respect to prediction accuracy and average annualised rate of return.

The objective in this paper is to develop and implement **EDDIE-ARB** on intra daily tick data for cross market stock index arbitrage in a manner that is suitable for online trading when windows of profitable arbitrage opportunities exist for short periods from one minute to ten minutes. Recent work by Markose and Er (2000) on the FTSE-100 stock index options and futures clearly indicates that these contracts with even fewer than 40 days to maturity can present arbitrage opportunities. A methodology of random sampling is used to train **EDDIE-ARB** to pick up the fundamental arbitrage patterns. As the arbitrageur maximizes return by exploiting as many profitable arbitrage opportunities and avoiding as many positions that are loss making, we will show how the fitness function of **EDDIE-ARB** will enable the arbitrageur to tune it to obtain the most favourable trade off. In other words, in arbitrage activity avoiding positions that are loss making or minimizing rate of failure **(RF)** alone which was the focus of earlier work by Tsang et al. (1998) is not sufficient. Thus, a novel aspect of **EDDIE-ARB** is a constraint satisfaction feature supplementing the fitness function that enables the user to train the **GP** how not to miss opportunities by learning to satisfy a minimum and maximum set on the number of arbitrage opportunities being sought. Historical sample data on arbitrage opportunities enables us to set these minimum and maximum bounds. Good **GP/GDT** rules generated by **EDDIE-ARB** are found to make a 3-fold improvement in profitability over the naive ex ante rule.

The rest of the paper is organized as follows. In Section 2, the methodology of Put-Call-Futures index options and futures arbitrage is given This determines the preprocessing requirements of intra daily tick data for purposes of training and testing the **EDDIE-ARB** program to learn from contemporaneous conditions for arbitrage opportunities and to predict these from up to ten minutes in advance. In Section 3, we give some details of the **EDDIE-ARB** program, in particular, how the fitness function has to be supplemented and fine tuned to obtain the favourable trade off between the search intensity for arbitrage opportunities and loss making recommendations. Section 4 reports the empirical results of the application of **EDDIE-ARB** to stock index options and futures arbitrage. Some robustness tests are done to see how well the previously trained **GDT** rules performed in a period when with LIFFEConnect and electronic trading greatly increased the turnover in FTSE-100 index futures and options markets.

1. Methodology of Put-Call-Futures Stock Index Arbitrage

1.1. Cross Market Arbitrage on the Stock Index

As the FTSE 100 spot index has a stock index futures and a European style index option traded on it, we will briefly outline why the most cost effective arbitrage is one that bypasses the cash/spot leg and involves only the two stock index derivatives with the same maturity date. This is illustrated in Figure 14.1 along the arrow C between the two derivatives on the underlying spot index. Figure 14.1 also indicates the three important equilibrium pricing and arbitrage criteria that relate the index options and futures on the same underlying stock index.

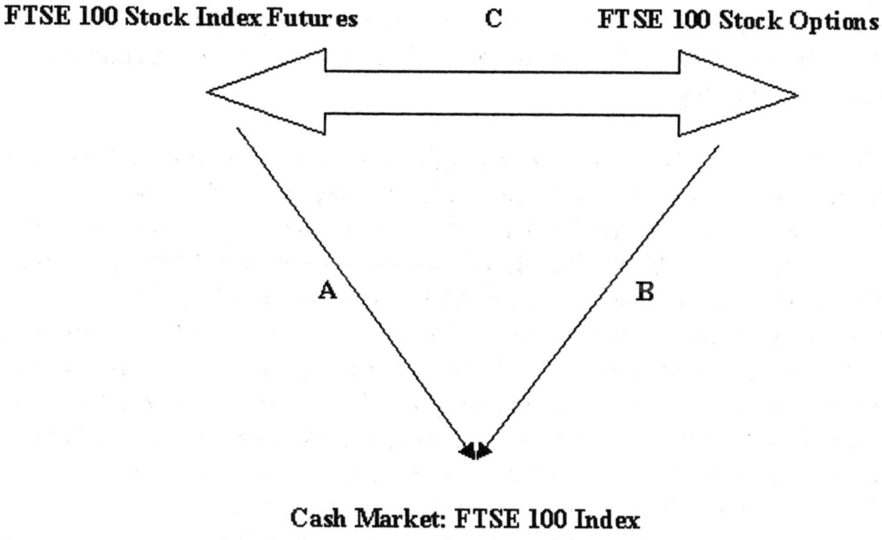

Figure 14.1. Cross market arbitrage.
A: Index Arbitrage based on the cost of carry fair futures formula. B: Option-Cash Index Arbitrage based on Stoll (1969) Put-Call Parity condition. C: Index Options-Futures Arbitrage based on Tucker (1991) Put-Call- Futures Parity condition.

Along arrow A in Figure 14.1, which relates the FTSE-index futures to the spot index, we have the so called index arbitrage based on a theoretical cost of carry formula which is used to detect arbitrage profit opportunities (eg. Modest and Sunderasan 1983; Cornell and French 1983; Yadav and Pope 1990). However, there are high costs of trading in the stock index portfolio and in particular problems of liquidity can arise when shorting stock in the conduct of a short index arbitrage strategy when the futures price trades at a discount to the spot index. The equilibrium pricing relationship between the index option and the stock index, defined by arrow B in Figure 14.1, is based on the Stoll (1969) Put-Call (P-C) parity relationship. Here, typically the option price is derived from the underlying stock index. However, it is well known from studies on the efficiency of index options (see Evnine and Rudd 1985;[2] Gemmill 1993; Gwilm and Buckle 1999), that the high costs of hedging index option positions with a portfolio based on the constituent shares of the spot index may lead to a number of problems for the tests of market efficiency of index options. In addition, Fung et al. (1994) have shown that the distribution of stock dividends which affect arbitrage strategies along the arrows A and B do not affect the **P-C-F** arbitrage along arrow C of Figure 14.1. For the above reason of the high cost of using the spot index in an arbitrage, it is widespread that arbitrage in stock index options follows arrow C in Figure 14.1 and bypasses the cash/spot index leg.

The Tucker (1991) result that combinations of put and call options with the same exercise price can replicate payoffs of futures positions with the same expiration and on the same cash asset underpins the put-call-futures (**P-C-F**, for short) parity parity relationship. Thus, arbitrage strategies based on the **P-C-F** parity (arrow C in Figure 14.1), involves time synchronized triplets of prices on a put, a call and a futures contract as specified above. A short futures position can be replicated by a combination of a short position in a call index option and a long one in a put index option. Conversely, a long futures position is equivalent to a combination of a long call and a short put position. The existence of these synthetic futures contracts involving combinations of options generates conditions for risk free arbitrage profits for the futures and options markets on the same underlying cash asset. For these reasons the put-call-futures parity relationship has also become the basis of recent tests for efficiency in the index options prices (Lee and Nayar 1993; Fung et al. 1997; Bae et al. 1998). We will now turn to the construction of a risk free **P-C-F** arbitrage position.

1.2. P-C-F Short Hedge Arbitrage

A risk free arbitrage portfolio can be constructed by combining a short futures contract and a long synthetic futures position by buying a call, shorting a put and by borrowing the present discounted value of the futures price and lending the same for the exercise price. Panel A in Table 14.1 shows how a zero net return is accomplished by this short hedge arbitrage strategy should $S_T > X$ or $S_T < X$. The upper bound of the FTSE-100 index futures bid price denoted by F_{bt} is given by

$$F_{bt}e^{-r_a\tau} \leq C_{at} - P_{bt} + Xe^{-r_b\tau} + TC \tag{14.1}$$

Here, $\tau = (T - t)$ is the remaining time to maturity; C_{at} is the call premium at the ask, P_{bt} is the put premium at the bid and TC denotes the transactions costs (that will be specified). Note, the interest rate on the futures price is the offer/ask rate as this amount has to be borrowed by the arbitrageur and that on the exercise price is the bid interest rate as it has to lent. If the condition in (14.1) is violated then the arbitrageur by definition will make a risk free profit equal to

$$[F_{bt}e^{-r_a\tau} - (C_{at} - P_{bt} + Xe^{-r_b\tau} + TC)] > 0 \tag{14.2}$$

by shorting the futures and by creating the synthetic long futures given on the R.H.S of (14.1).

1.3. P-C-F Long Hedge Arbitrage

Likewise we have a risk free arbitrage portfolio by combining a long futures contract and a short synthetic futures position obtained by shorting a call at the bid, being long in the put at the ask and by lending the present discounted value of the futures price and borrowing the same for the exercise price. Panel B shows how a zero net return is obtained by this long hedge arbitrage should $S_T > X$ or $S_T < X$. The lower bound of the ask futures price is given by

$$F_{at}e^{-r_b\tau} \geq C_{bt} - P_{at} + Xe^{-r_a\tau} - TC \tag{14.3}$$

If the present value of the ask futures price is less than the R.H.S of (14.3), then the arbitrageur buys the futures and sells the portfolio on the R.H.S of (14.3). This nets a risk free profit equal to

$$[(C_{bt} - P_{at} + Xe^{r_a\tau} - TC) - F_{at}e^{-r_b\tau}] \tag{14.4}$$

Table 14.1. A Contingency Table for Two-Class Classification Prediction Problem

A:			
P-C-F Short Arbitrage TRANSACTIONS UNDERTAKEN	**Cash Flows Today(at t)**	**Cash Flows at Expiration (at T)**	
		$S_T > X$	$S_T < X$
Short a futures at F_t	0	$F_t - S_T$	$F_t - S_T$
Borrow $F_t e^{r(t-T)}$	$+F_t e^{r(t-T)}$	$-F_t$	$-F_t$
Long call	$-C_t$	$S_T - X_t$	0
Short put	$+P_t$	0	$-(X - S_T)$
Lend $Xe^{r(t-T)}$	$-Xe^{r(t-T)}$	$+X$	$+X$
	$F_t e^{r(t-T)}$ $-(C_t - P_t + Xe^{r(t-T)})$	0	0

B:			
P-C-F Short Arbitrage TRANSACTIONS UNDERTAKEN	**Cash Flows Today(at t)**	**Cash Flows at Expiration (at T)**	
		$S_T > X$	$S_T < X$
Long a futures at F_t	0	$S_T - F_t$	$S_T - F_t$
Lend $F_t e^{r(t-T)}$	$-F_t e^{r(t-T)}$	$+F_t$	$+F_t$
Short call	C_t	$-(S_T - X)$	0
Long put	$-P_t$	0	$X - S_T$
Borrow $Xe^{r(t-T)}$	$+Xe^{r(t-T)}$	$-X$	$-X$
	$(C_t - P_t + Xe^{r(t-T)})$ $-F_t e^{r(t-T))}$	0	0

Note, here that the bid interest rate applies to the amount for the futures price that is lent in the arbitrage and the ask/offer interest rate applies to the value of exercise price as that has to be borrowed. As trade prices are not specified in terms of whether a buy or sell is involved the condition for a profitable short arbitrage is given as

$$F_t e^{-r\tau} - [C_t - P_t + Xe^{-r\tau}] - TC > 0 \tag{14.5}$$

Here, r is the mid interest rate. In trade data with long **P-C-F** arbitrage, the condition for its profitability is given as

$$[C_t - P_t + Xe^{-r\tau}] - Fe^{-r\tau} - TC > 0 \qquad (14.6)$$

In the following, analysis is confined to the short **P-C-F** arbitrage.

2. Data, Transactions Costs and Ex Post Evidence of Efficiency Violations

2.1. LIFFE Intradaily Tick Data for FTSE-100 Index Options and Futures

We first use the intraday historical tick data on the FTSE-100 European style index option traded on the LIFFE (London International Financial Futures and Options Exchange) from March 1, 1991 to June 30, 1998. Intraday tick data for a further period from June 30, 1998 to March 30, 1999 was earmarked to do additional robustness tests to see if the introduction of the LIFFEConnect electronic trading had changed the fundamental conditions for the **GP**.

The tick data records contain both two way bid/ask quote data and the transactions/trade price data. Each record is time stamped with the following information: (i) option identity, i.e. whether a call or put; (ii) the maturity date; (iii) the exercise price; (iv) the price of contract (bid and ask for quoted prices and transaction/trade price if indicated as a trade record); (v) the time synchronized underlying FTSE-100 stock index price.

The risk free interest rate used for borrowing funds in the hedge portfolio is the London Interbank Offer Rate (LIBOR) maturing on the day closest to the expiration date of the option. The tick data on the FTSE 100 is obtained from Data Stream. The bid or lending interest rate is conventionally taken to be 1/8 of the LIBOR rate. The mid interest rate is the mean of the above two rates.

To process the tick data for arbitrage relating to the put-call-futures parity, we follow a three way matching criteria for puts, calls and futures with the same nearby maturity. For trade data, all calls and puts with the same exercise price and traded within the same minute are matched. This pair is matched further with a futures contract traded within a minute of the time stamp of the call-put pair. In the LIFFE tick trade data from January 1991 to June 1998, 8073 short arbitrage trade price triplets satisfying condition (14.5) with zero transactions costs were found.

2.2. Transactions Costs

It is conventional for transactions costs for arbitrage to be estimated for at least three categories of traders — the private investor, institutional investors and for broker/market makers. The first two incur commission/broker fees in addition to the exchange fees. Further, there are the bid-ask spreads, these have been estimated to be approximately £57.[3] This study focuses on the arbitrage profits that can accrue to the broker/market maker who is estimated to have transactions costs of less than .1 % value of the value FTSE100 index futures contract. The latter works out to about £60 per **P-C-F** arbitrage. Note that analysis of marking to market and the management of margin requirements which is important in the practical implementation of the **P-C-F** arbitrage is omitted here. The assumption is that the arbitrageur can post treasury bills for the margin requirements and hence does not incur interest rate costs on this.

2.3. Ex Post Efficiency Violations

Table 14.2 gives the ex post analysis of efficiency violations for short arbitrage positions using the 8073 **P-C-F** price triplets for the sample period (1991-1998) at costs appropriate to the market maker/broker.[4] Some 42% of the sample yields profitable short **P-C-F** arbitrage opportunities. The profits reported are those that accrue if the arbitrageur could have obtained as quotes the trade prices recorded at these points in time.[5]

Table 14.2. Ex Post Analysis of Efficiency Violations For Short Arbitrage at Transactions Costs of 0.1% of FTSE-100 Index

	Number of Mispricings Short Arbitrage	Ex PostProfits Mean (Eq. 5) Short Arb. (£s)
All	3357 42%*	368.89 (29.79)
50-80 Days Before	1145 14%*	496.87 (67.79)
20-50 Days Before	1206 15%*	316.55 (45.65)
Spot Month	1006 12%*	285.96 (29.02)

Note: (·) gives the 5% confidence interval; * gives % of total sample of 8073

In Table 14.2, we see that on average for all periods to maturity the profits from the arbitrage is substantial and statistically significant. The spot month nets on average £285 and the 20-50 day period to maturity and further yields over £316. We will now turn to the question of whether arbitrageurs, in a real time setting, can correctly identify and exploit profitable short arbitrage opportunities shown in Table 14.2.

3. Ex Ante Analysis of Arbitrage Profits and Application of EDDIE-ARB

3.1. The Objective of EDDIE-ARB

The standard ex ante analysis of arbitrage (see, Bae et al. 1998) is based on the following scenario. The naive premise is that the arbitrageur waits for a contemporaneous profit signal in the category of either short or long arbitrage (given by Equations 14.1-14.6) above and then continues with arbitrage trades in the same direction in a given time interval. It will be assumed that after a time execution delay of about a minute, the arbitrageur proceeds to execute his arbitrage position within a time interval of ten minutes. He, thus, faces execution price risk, should the trio of the **P-C-F** prices diverge. The question is can a **GP** be trained to improve on the naive strategy as not every contemporaneous **P-C-F** profit signal will continue to be profitable after the execution delay and more importantly can a profit signal be predicted in advance of the required period.

The case for the mechanization of the complex **P-C-F** arbitrage can easily be made. The domain of search involves four asset prices (put price, call price, futures price, and the interest rate), three maturity dates for the derivatives and over thirty exercise prices for the index options. With data arriving continuously, the detection of contemporaneous profitable arbitrage opportunities with the three way **P-C-F** price matching is a considerable task. With the assumed time delay in execution of an arbitrage from an observed contemporaneous profit signal an effective forecasting tool is needed to assess the success rate of such a strategy. Thus, the objective we set **EDDIE-ARB** in the generation of decision rules is as follows: **at any point in time corresponding with the occurrence of a matched P-C-F price triplet,[6] predict which of these have adjacent 10 minute intervals in which profitable arbitrage is possible in a given direction after a one minute execution delay**. Note, the recommended arbitrage positions are judged profitable or not by the criteria given in Equations (14.1-14.6). However, as indicated in the introduction, **EDDIE-ARB** implements a trade off between missed profit opportunities and loss making recommendations.

3.2. Background of EDDIE

Like other standard **GPs**, **EDDIE** maintains a population (set) of candidate solutions, each of which is a decision tree for financial forecasting. Candidate solutions are selected randomly, biased by their fitness, for involvement in generating members of the next generation. General mechanisms (referred to as genetic operators, e.g. reproduction, crossover, mutation) are used to combine or change the selected candidate solutions to generate offspring, which will form the population in the next generation. For details of **GAs** and **GPs**, readers are referred to Holland (1975), Goldberg (1989) and Koza (1992).

In the **GP EDDIE**, a candidate solution is represented by a genetic decision tree (**GDT**). The basic elements of **GDTs** are rules and forecast values. A single rule consists of one useful indicator for prediction, one relational operator such as "reater than" or "less than", etc, and a threshold (real value). Such a single rule interacts with other rules in one **GDT** through logic operators such as "Or", "And", "Not", and "If-Then-Else". Forecast values in this example are either a positive position (i.e. positive return within specified time interval can be achievable) or negative position (i.e. negative return within a specified time interval will be achievable).

3.3. The FGP Fitness Criterion

Since **GDTs** are used to predict whether a profitable arbitrage can exist from any point in time and within the next prespecified minutes, the prediction actually can be categorised as a two-class classification problem. Each time point can be classified into either a positive position or a negative position. For each **GDT**, we define **RC** (Rate of Correctness), **RMC** (Rate of Missed Chances), and **RF** (Rate of Failure)as its prediction performance criteria. Formula for each criterion is given through a contingency table (Table 14.3) as follows:

3.4. Linear Fitness Function

In earlier work by Tsang et al. (1998), the fitness function mainly used in **EDDIE** was the rate of correctness **RC**. By using the following fitness function, the user may satisfy individual objectives by adjusting the weights w_rc, w_rmc and w_rf:

Table 14.3. A Contingency Table for Two-Class Classification Prediction Problem

Predicted Negative Positions (N_-)	**Predicted Positive Positions (N_+)**	
# of True Negative (**TN**)	# of False Positive (**FP**)	**Actual Negative Positions (O_-)**
# of False Negative (**FN**)	# of True Positive (**TP**)	**Actual Positive Positions (O_+)**

RC$=\frac{TP+TN}{O_+ + O_-}=\frac{TP+TN}{N_+ + N_-}$; **RMC**$=\frac{FN}{O_+}$; **RF**$=\frac{FP}{N_+}$;
Where $O_+=FN+TP$; $O_-=TN+FP$; $N_-=TN+FN$; $N_+=FP+TP$

$$f_1 = w\_rc * \mathbf{RC} - w\_rmc * \mathbf{RMC} - w\_rf * \mathbf{RF} \tag{14.7}$$

It involves three performance values, i.e. **RC**, **RMC** and **RF**, each of which is assigned a different weight. Obviously, the performance of a **GDT** is no longer assessed by **RC** only, but by a synthetic value, which is the weighted sum of its three performance rates. By proper adjustment of the three weights, the experimenter is able to put his emphasis on one aspect of the performance of the **GDT** than on the others. In order to achieve a low **RF**, one may assign a higher value to w_rc and w_rf and a smaller or zero value to w_rmc. To a certain extent, the fitness function $f_{(1)}$ does allow us to reduce **RF**. However, the fitness function in (14.7) appears to have two drawbacks: 1) **EDDIE**'s performance is very sensitive to the three weights; 2) results are unstable. For example, in one of our series of preliminary experiments in which we used the following three weights:

$w\_rc = 1$; $w\_rmc = 0$ and $w\_rf = \alpha$ where $0 < \alpha \leq 1$

We found that merely altering α in the linear fitness function did not generate reliable performance of the generated **GDT**s. In some cases even a small positive α resulted in **GDT**s that did achieve a lower **RF** (even zero over training period) by making no positive recommendations over the test period. This was probably due to over-fitting. In contrast, a reduction of α from an α as high as 0.62 resulted in generating **GDT**s that it did not show any improvement on **RF**. We refer to this as the no-effect problem. For example, among 10 runs, only two runs generated a **GDT** that predicted a few correct positive positions in the test period. The remaining 8 runs showed either the over-fitting or no-effect problem.

3.5. Putting constraints into GP

We can further improve the linear fitness function $f_{(1)}$ by introducing constraints into it. We introduce a new parameter to **EDDIE**, $\Re = [P_{min}, P_{max}]$, which defines the minimum and maximum percentage of recommendations that we instruct **EDDIE** to make in the training data (with the assumption that the test data exhibits similar characteristics, an assumption that is made in most machine learning methods). We denote the new fitness function by $f_{(2)}$.

Choosing appropriate values for $\Re$ and the weights for $f_{(2)}$ remains a non-trivial task, which we approached by trial and error. When appropriate parameters were chosen, **EDDIE** managed to reduce **RF** and avoid the over-fitting and no-effect problems. In particular, the intensity of search by **EDDIE-ARB** so as not to miss arbitrage opportunities was set at the minimum and maximum of the number of profitable opportunities found on average in the sample data set. With this in place, the α in (7) was typically left at a low level so as not to increase the **GDT**s' rate of failure, **RF**, as the range of $\Re = [P_{min}, P_{max}]$ was increased. The results pertaining to this are discussed in the next section.

4. Experimentation Results

4.1. Preprocessing of the P-C-F Arbitrage Data

The objective of these tests is to compare the effectiveness of **GDT**s generated by **EDDIE** against a naive arbitrage rule which executes an arbitrage trade whenever there is a contemporaneous profit signal. For the purpose of training **GDT**s, the following data pertaining to the short **P-C-F** arbitrage was fed into **EDDIE**: the strike price, the call premium, the put premium, the underlying index value, futures price, time to maturity of the contract and profit after transactions costs, **TC**. The total number of observations is 8073. **EDDIE** did not give any recommendations on this set. The data was then altered to be more informative in terms of economic theory. The strike price, was converted into the "moneyness" variable, viz. as the ratio of the underlying to the strike price. This variable is introduced as in, at and out of the money for call and put options as this is known to have an impact on the arbitrage profits. The second variable added was basis, the formula for basis is as follows:

$$\text{basis}=(\text{futures-spot})/\text{futures}.$$

This variable controls for the mispricing in the futures-spot leg of the arbitrage. The difference between the call and the put price, (C-P) is also added to the model. Finally, the following variables, i.e. the underlying, (C-P) and profit after transaction costs, were input as a percentage of the futures price. The results of a preliminary run on this data set are summarized in Table 14.4.

Table 14.4. Trial Run Results

Training 4178 rows from 3 to 4180							
29 Jan 91 To 30 Dec 96			Contingency Table for Selected Data				
RC	**RMC**	**RF**	0	1			
94.18%	64.66%	80.38%	3894	168	0	4062	
			75	41	1	116	
			3969	209	5.00%	4178	2.78%
Testing 3895 rows from 4181 to 8075							
1 Jan 97 To 18 Jun 98			Contingency Table for Selected Data				
RC	**RMC**	**RF**	0	1			
27.55%	27.72%	93.01%	867	2743	0	3610	
			79	206	1	285	
			946	2949	75.71%	3895	7.32%

RC: Rate of Correctness; **RMC**: Rate of Missed Chances; **RF**: Rate of Failure

Table 14.4 (continued)

	EDDIE	**Naive**
Total Profit	122727	147257.2
Number of Recommendations	2949	1848
Average Profit	41.62	79.68

As the rate of failure, **RF**, was so high at 93% and the prediction accuracy of the recommendations made by **EDDIE** was so low, we decided to further process the data. Indeed, we recommend the following methodology for training good **EDDIE-ARB** rules due to some typical problems associated with arbitrage opportunities and the time execution delay.

Table 14.5. Profit Distribution for Short P-C-F Arbitrage (Jan., 1991 - Jun., 1998)

Date	Total Profits/ Losses	No	Mean (£s)	Total Profits (£s)	No	Mean (£s)	Total Losses (£s)	No	Mean (£s)
3−91	−887.7	30	−29.59	60.2	4	15.04	−947.9	26	−36.46
6−91	525.0	10	52.50	594.8	8	74.35	−69.8	2	−34.90
9−91	−2357.5	71	−33.20	163.5	5	32.70	−2521.0	66	−38.20
12−91	−3267.7	108	−30.26	487.4	13	37.49	−3755.1	95	−39.53
3−92	−1689.7	76	−22.23	576.9	14	41.21	−2266.6	62	−36.56
6−92	−1938.4	68	−28.51	0.0	0	0.00	−1938.4	68	−28.51
9−92	331.9	144	2.30	3414.9	45	75.89	−3083.0	99	−31.14
12−92	−346.5	40	−8.66	1118.9	9	124.32	−1465.4	31	−47.27
3−93	−2212.6	77	−28.74	393.1	14	28.08	−2605.7	63	−41.36
6−93	−3346.3	106	−31.57	205.1	5	41.03	−3551.4	101	−35.16
9−93	−6098.0	138	−44.19	0.0	0	0.00	−6098.0	138	−44.19
12−93	−3558.7	79	−45.05	0.0	0	0.00	−3558.7	79	−45.05
3−94	−2849.5	71	−40.13	19.8	1	19.85	−2869.4	70	−40.99
6−94	−1826.2	149	−12.26	1913.4	46	41.60	−3739.6	103	−36.31
9−94	−10955.1	305	−35.92	240.6	20	12.03	−11195.7	285	−39.28
12−94	−6611.3	236	−28.01	877.0	17	51.59	−7488.3	219	−34.19
3−95	−8672.8	268	−32.36	941.4	38	24.77	−9614.1	230	−41.80
6−95	9639.8	209	46.12	16376.3	35	467.89	−6736.4	174	−38.72
9−95	−4449.8	111	−40.09	0.0	0	0.00	−4449.8	111	−40.09
12−95	2573.7	135	19.06	6722.3	27	248.97	−4148.6	108	−38.41
3−96	1629.5	228	7.15	10461.2	8	1307.64	−8831.7	220	−40.14
6−96	−1012.1	159	−6.37	4328.3	6	721.38	−5340.3	153	−34.90
9−96	−3383.2	527	−6.42	15530.5	74	209.87	−18913.7	453	−41.75
12−96	11038.1	833	13.25	40913.9	108	378.83	−29875.8	725	−41.21
3−97	−21460.2	772	−27.80	4006.3	109	36.75	−25466.5	663	−38.41
6−97	74697.3	855	87.37	95803.6	241	397.53	−21106.3	614	−34.38
9−97	7688.2	435	17.67	19901.9	101	197.05	−12213.7	334	−36.57
12−97	36055.5	494	72.99	42875.3	282	152.04	−6819.9	212	−32.17
3−98	534927.6	483	1107.51	536097.6	409	1310.75	−1170.1	73	−16.03
6−98	394354.8	856	460.69	398458.8	706	564.39	−4103.8	148	−27.73

5. Methodology for Training EDDIE-ARB With Historical Tick Arbitrage Data

I. The distribution of contemporaneous **P-C-F** arbitrage signals is given in Table 14.5 in terms of contracts and profitability evaluated using Equation (14.5) with about .1% FTSE-100 as transactions costs. In the LIFFE FTSE-100 **P-C-F** arbitrage triplets, it is clear that arbitrage opportunities were sparse in the early years from 1991-1994 with December contracts being the most voluminous. As trading volume in the index options increased, we have a three fold increase of **P-C-F** arbitrage opportunities by 1994 and by the end of 1998, there is over a tenfold increase since the inception of electronic index options trading in LIFFE. On average the numbers of profitable **P-C-F** arbitrage opportunities are far outnumbered by loss making **P-C-F** opportunities in all years. However, in the years after 1995, the average total profitability of **P-C-F** positions become positive with the loss making arbitrages generating smaller losses than the gainful ones. In other words, the returns to **P-C-F** arbitrage as already seen from Table 14.2 are significant if the arbitrageur can successfully 'pick' the cherry.

II. The naive rule recommends that any profitable **P-C-F** signal is followed by an arbitrage in the same direction in a 9 minute window after a one minute execution delay. In the early years, the sparseness of **P-C-F** arbitrage price triplets meant that the few contemporaneous profitable signals that existed do not have follow ups in the given 10 minute window. Many of those that did have loss making follow ups. In fact, only 20% of the total sample of 8073 **P-C-F** triplets have any followups at all. It is only after 1996 that there are profitable follow ups from any **P-C-F** time stamp.

III. The above problem indicates that **EDDIE** has to be trained in the 20% of the total sample of short arbitrage **P-C-F** price triplets that have follow ups in the given window of opportunity. Thus, the sample size is reduced to 1641. Unlike the naive arbitrage strategy, **EDDIE** has to predict profitable follow up trades in the prescribed time window from any given time stamp of a **P-C-F** price triplet.

IV. The follow up based sample of 1641 lines is divided into test and training areas using a randomized procedure rather than one that is time based on a time series. Each **P-C-F** triplet is assigned one random number between 0 to 1. If the random number for the triplet is less than 0.63 then that triplet with all its adjacent followups in the next ten minutes is included in the training part of the sample. All of the remaining triplets, i.e. the triplets with a random number greater than 0.63 along with their adjacent ten minute followups, were included in the test part. This sampling procedure ensured that we have enough observations in the training part for **EDDIE** to function properly (the recommended

Table 14.6. Naive Strategy Trade Recommendations: Performance (January, 1991 – June 30, 1998)

	Total Trade Signals	**Overall Profit/ Loss**	**Profitable Signals**	**Total Profits**	**Average Profits**
Training	290	274.76	159	82858.97	521.13
Testing	196	338.10	108	68092.68	630.49

Table 14.6 (continued)

	Loss Making Signals	**Total Losses**	**Average Losses**
Training	131	-3177.41	-24.26
Testing	88	-1824.52	-20.73

*All profit/losses calculated on the basis of 0.1% of the FTSE-100 transactions costs.

minimum for training this class of **GPs** is 1000 observations). This sampling has also resulted in an equal distribution of arbitrage opportunities in training and test areas to counter the problems raised in point I.

For the 1641 sample points in which any **P-C-F** price triplet had a followup in the next 10 minutes, the naive strategy produced the results given in Table 14.6.

From Table 14.6 , we see that in the test sample, the naive rule made 196 recommendations to trade of which 88 were wrong. Hence the rate of failure is roughly 45%. The average profit on an arbitrage is £338 and the total profits net losses were £66268.16. The problem we were confronted with was to get the **GDT**s generated by **EDDIE-ARB** not only to yield higher average net profit per arbitrage than the naive rule, but to find enough profitable arbitrage positions that netted total profits of over £66268.16. As we will see, **EDDIE-ARB** had to be fitted with an additional constraint, $\Re = [P_{min}, P_{max}]$, where these parameters reflected the intensity of search that was desired without much deterioration of its performance in terms of rate of failure, **RF**.

6. EDDIE-ARB Evaluated

The **EDDIE-ARB** runs trained on this randomized data set provided good results. The results are summarized in Table 14.7.

As we move from a conservative setting to a more ambitious one in the search intensity parameters $\Re = [P_{min}, P_{max}]$, **RF** increases and **RMC** decreases. As expected in the low setting of $\Re = [5\%, 10\%]$, the **GDT**s produced zero **RF**, but having missed over 84% of the profitable arbitrage opportunities, these rules failed to produce total profits greater than the naive rule. In every setting after that, most decision rules generated by **EDDIE-ARB** beat the naive strategy which provided 196 arbitrage signals, £66,268.16 total profit, and £338.10 average profit in the test area. Average profit generated by **GDT**s are highest when the most conservative setting is applied for the tests. However, the total profit increases as the ambitious level increases. The parameters $\Re = [20\%, 25\%]$, roughly reflect the percentage of profitable arbitrage opportunities in the test area of the data with follow ups. The best **GDT**(4) in this setting produces 154 arbitrage signals with a 25% rate of failure (ie. 27 wrong predictions) with very large net total profits of £112683.86 and average profits per arbitrage of £731.71. For the full breakdown of the performance of **GDT**(4), see Table 14.8.

Our final analysis is to put the best performing **GDT**(4) to a robustness test to see how it performs in a period in which electronic trading was introduced into the LIFFE index markets. The robustness test period was from March 30, 1998 – March 30, 1999, when three additional contracts, viz. the September 1998,the December 1998 and the March 1999 contracts were studied. There were 2140 **P-C-F** time matched price triplets in the trade tick data, reflecting the increased turnover in the index markets from 1998. Of these almost as many had followups in the next ten minutes, again a reflection of the increased turnover. But, only 328 were profitable (at .1% of FTSE-100 as transactions costs) in next ten minutes from any **P-C-F** time matched price triplets. The naive strategy makes a total of 273 recommendations in the robustness period of which 27 were wrong, ie. a 10% error rate. The interesting point is that average profit of £952 per arbitrage position from the naive strategy, see Table 14.8, is substantially higher in the period after mid 1998. Hence, clearly, though the markets offer fewer arbitrage opportunities, those that exist are highly profitable. The remarkable result from Table 14.8 is that the best **GDT**(4) rule trained in the period when market turnover was low produced a very robust result for the period after the introduction of electronic trading. The **GDT**(4) rule made 249 recommendations to trade of which only 13 were wrong, ie. an error rate

Table 14.7. EDDIE-ARB **GDT**s: Training and Testing Results with Different $\Re = [Pmin, Pmax]$

$\Re$=[5%–10%]

	RF	RMC	RC	No. of Signals	MEAN PROFIT	TOTAL PROFIT
RULES	**TRAIN TEST**	**TRAIN TEST**	**TRAIN TEST**	**TRAIN TEST**	**TEST**	**TEST**
GDT1	0 0	0.6240 0.6226	0.8499 0.8441	91 60	1046.49	62789.57
GDT2	0 0	0.5868 0.5660	0.8588 0.8583	100 69	929.54	64138.22
GDT3	0 0	0.5992 0.5849	0.8559 0.8535	97 66	968.46	63918.51
GDT4	0 0	0.5992 0.5849	0.8559 0.8535	97 66	968.46	63918.51
GDT5	0 0	0.6157 0.6226	0.8519 0.8441	93 60	1009.24	60554.29
GDT6	0 0	0.5868 0.5849	0.8588 0.8535	100 66	968.46	63918.51
GDT7	0 0	0.5992 0.5472	0.8559 0.8630	97 72	904.25	65105.89
GDT8	0 0	0.5868 0.5535	0.8588 0.8614	100 71	904.39	64211.48
GDT9	0 0	0.5909 0.5472	0.8579 0.8630	99 72	903.54	65055.18
GDT10	0 0	0.5868 0.5849	0.8588 0.8535	100 66	968.46	63918.51
MEAN	0 0	0.5975 0.5799	0.8563 0.8548	97.4 66.8	957.10	63752.90
SD	0 0	0.0131 0.0275	0.0032 0.0069	3.1693 4.3665	47.64	1297.61

Table 14.7 (continued)

$\Re$=[10%–15%]

	RF	RMC	RC	No. of Signals	MEAN PROFIT	TOTAL PROFIT
RULES	**TRAIN** **TEST**	**TRAIN** **TEST**	**TRAIN** **TEST**	**TRAIN** **TEST**	**TEST**	**TEST**
GDT1	0 0	0.4752 0.4591	0.8857 0.8850	127 86	771.68	66364.11
GDT2	0.0338 0.0100	0.4091 0.3774	0.8966 0.9039	148 100	668.46	66846.35
GDT3	0 0	0.4380 0.4465	0.8946 0.8882	136 88	753.78	66332.20
GDT4	0.0726 0.0500	0.5248 0.5220	0.8648 0.8630	124 80	842.08	67366.07
GDT5	0 0	0.4959 0.478	0.8807 0.8803	122 83	796.08	66074.30
GDT6	0 0	0.5165 0.4843	0.8757 0.8787	117 82	804.37	65958.05
GDT7	0.0152 0.0112	0.4628 0.4465	0.8867 0.8866	132 89	744.31	66243.73
GDT8	0.0877 0.0286	0.5702 0.5723	0.8529 0.8535	114 70	892.51	62475.56
GDT9	0 0	0.5207 0.4969	0.8748 0.8756	116 80	826.75	66140.26
GDT10	0 0	0.4669 0.4591	0.8877 0.8850	129 86	770.43	66256.81
MEAN	0.0209 0.0100	0.4880 0.4742	0.8800 0.8800	126.5 84.4	787.00	66005.70
SD	0.0333 0.0168	0.0472 0.0515	0.0135 0.0139	10.393 7.725	61.12	1308.05

Table 14.7 (continued)

$\Re$=[15%–20%]

	RF	RMC	RC	No. of Signals	MEAN PROFIT	TOTAL PROFIT
RULES	**TRAIN TEST**	**TRAIN TEST**	**TRAIN TEST**	**TRAIN TEST**	**TEST**	**TEST**
GDT1	0.3415 0.3077	0.4421 0.4340	0.8241 0.8283	205 130	420.91	54718.71
GDT2	0.2199 0.2031	0.3843 0.3585	0.8658 0.8693	191 128	521.28	66723.61
GDT3	0.2135 0.2214	0.3760 0.3585	0.8688 0.8646	192 131	508.36	66595.46
GDT4	0.2414 0.2241	0.4545 0.4340	0.8489 0.8504	174 116	388.43	45057.48
GDT5	0.2513 0.3023	0.3967 0.4340	0.8559 0.8299	195 129	515.24	66465.95
GDT6	0.1623 0.1500	0.3388 0.3585	0.8877 0.8819	191 120	560.20	67223.75
GDT7	0.2246 0.2593	0.4008 0.3711	0.8618 0.8520	187 135	491.91	66408.02
GDT8	0.2328 0.2188	0.4008 0.3711	0.8598 0.8630	189 128	519.97	66555.53
GDT9	0.2083 0.2406	0.3719 0.3648	0.8708 0.8583	192 133	499.77	66469.69
GDT10	0.1961 0.2391	0.3223 0.3396	0.8827 0.8630	204 138	488.41	67401.27
MEAN	0.2292 0.2366	0.3888 0.3824	0.8626 0.8561	192 128.8	491.40	63361.90
SD	0.0467 0.0462	0.0407 0.0366	0.0179 0.0167	8.705 6.5794	50.50	7464.85

Table 14.7 (continued)

$\Re$=[20%–25%]

	RF	RMC	RC	No. of Signals	MEAN PROFIT	TOTAL PROFIT
RULES	**TRAIN TEST**	**TRAIN TEST**	**TRAIN TEST**	**TRAIN TEST**	**TEST**	**TEST**
GDT1	0.4492	0.3058	0.7903	305	415.20	79303.87
	0.4346	0.3208	0.7890	191		
GDT2	0.4023	0.3554	0.8101	261	495.89	78351.04
	0.3734	0.3774	0.8126	158		
GDT3	0.4580	0.3595	0.7833	286	442.11	79137.55
	0.4302	0.3585	0.7890	179		
GDT4	0.2520	0.2397	0.8807	246	731.71	112683.86
	0.1753	0.2013	0.9071	154		
GDT5	0.3730	0.3471	0.8231	252	397.27	65947.63
	0.3554	0.3270	0.8252	166		
GDT6	0.3373	0.3182	0.8400	249	439.22	66760.98
	0.3092	0.3396	0.8409	152		
GDT7	0.4633	0.3058	0.7823	313	406.01	79171.06
	0.4462	0.3208	0.7827	195		
GDT8	0.4498	0.3884	0.7863	269	459.08	78043.79
	0.4353	0.3962	0.7843	170		
GDT9	0.3485	0.2893	0.8390	264	430.58	78796.89
	0.3770	0.2830	0.8205	183		
GDT10	0.4327	0.3554	0.7962	275	436.09	78497.09
	0.4167	0.3396	0.7969	180		
MEAN	0.3966	0.3264	0.8131	272	465.30	79669.40
	0.3753	0.3264	0.8148	172.8		
SD	0.0687	0.0433	0.0324	23.0314	97.68	12701.92
	0.0827	0.0542	0.0380	15.1936		

Table 14.8. Robustness Test: **GDT**(4) Performance vs. Naive Strategy

	Total Trade Signals	Overall Profit/ Loss	Profitable Signals	Total Profits	Average Profits
Training	246	617.82	184	154768.16	841.13
Testing	154	731.71	127	113847.68	896.44
Robustness Period* GDT(4)	249	1043.35	236	260507.03	1103.84
Robustness Period* Naive Rule	273	952.88	246	261480.83	1062.93

Table 14.8 (continued)

	Loss Making Signals	Total Losses	Average Losses
Training	62	-2784.61	-44.91
Testing	27	-1163.82	-43.10
Robustness Period* GDT(4)	13	-712.509	-54.81
Robustness Period* Naive Rule	27	-1345.46	-49.83

*The robostness test period was March 30, 1998 – March 30, 1999.

of about 5% which is less than that for the naive rule. Thus, though the total net profit remained on par with that of the naive rule, **EDDIE-ARB** can clearly "pick" the cherry quite competently with the average return on an arbitrage exceeding that for the naive rule.

7. Conclusion

The main conclusion is that the conduct of a complex arbitrage in a fast moving market requires a mechanized strategy that cannot rely only on arbitrage recommendations that arise from contemporaneous profit signals. **EDDIE-ARB** that uses genetic programming was developed to predict arbitrage opportunities in the FTSE-100 index futures and options market for a period of up to ten minutes. This gives an arbitrageur a time advantage that cannot be obtained on the basis of contemporaneous profit signals. Technically, the success of **EDDIE-ARB** lies in the unique fitness function that was developed that enabled the genetic program to search more intensively without much deterioration in the rate of failure of its recommendations.

Acknowledgments

We acknowledge the receipt of a RPF grant from the University of Essex for this research.

Notes

1. Typically, in contrast genetic algorithms that operate with strings use strings of fixed lengths.
2. Evnine and Rudd (1985) note that unlike the option on a stock, an option on an index involves more complex procedures for arbitrage, 'it is conceivable that arbitrage forces are not as powerful with index options and hence index options may display greater pricing errors than stock options'.
3. See, Markose and Er (2000) for a full discussion on the derivation of the bid ask spreads in the index option and futures markets.
4. Markose and Er (2000) find that the P-C-F short arbitrage shows more efficiency violations than the long arbitrage.
5. Markose and Er (2000) discuss issues relating to whether these reported profits in trade prices are robust to the well known direction of trade problem.
6. Note that here as the prediction of profitable arbitrage in the next ten minutes is not constrained to be one that follows from a profitable contemporary price signal as in the naive rule, the **GP** is in principle trained to anticipate arbitrage opportunities within a given time interval from any point in time. With electronic trading, any observed **P-C-F** arbitrage opportunity can be executed immediately. As quotes do not 'live' longer than 1-3 minutes even with electronic trading, the methodology given here for generating robust **GDTs** that anticipate profitable arbitrage in a given direction with a ten minute lead is advantageous.

References

Allen, F. and R. Karjalainen (1999). "Using Genetic Algorithms to Find Technical Trading Rules," *Journal of Financial Economics*, 51(2), 245–271.

Alexander, S. S. (1964). "Price Movement in Speculative Markets: Trend or Random Walks," in Cootner, P. (ed.), *The Random Character of Stock Market Prices*, No. 2, 338–372. Cambridge, MA: MIT Press.

Backus, J. W. (1959). "The Syntax and Semantics of the Proposed International Algebraic Language of Zurich," ACM-GAMM Conference, ICIP.

Bae, K. H., Chan, K., and Cheung, Y. L. (1998). "The Profitability of Index Futures Arbitrage: Evidence from Bid-Ask Quotes," *Journal of Futures Markets*, 18, 743–763.

Bauer, R. J. Jr. (1994). *Genetic Algorithms and Investment Strategies*, New York: John Wiley & Sons, Inc.

Brock, W., J. Lakonishok, and B. LeBaron (1992). "Simple Technical Trading Rules and the Stochastic Properties of Stock Returns," *Journal of Finance*, 47, 1731–1764.

Chen, S.-H. and C.-H. Yeh (1997). "Toward a Computable Approach to the Efficient Market Hypothesis: An Application of genetic programming," *Journal of Economic Dynamics and Control*, 21, 1043–1063.

Cornell, B. and K. French (1988). "Taxes and the Pricing of Stock Index Futures," *Journal of Finance*, 38, 675–694.

Evnine, J. and A. Rudd (1985). "Index Options: The Early Evidence." *Journal of Finance*, 40, 743–756.

Fama, E. F. and M. E. Blume (1966). "Filter Rules and Stock-Market Trading," *Journal of Business*, 39(1), 226–241.

Fung, J., W. Chan, and C. Kam (1994). "On the Arbitrage Free Relationship Between Index Futures and Index Options: A Note," *Journal of Futures Markets*, 14, 957–962.

Fung, J., L. Cheng, T. W. Chan, and C. Kam (1997). "The Intraday Pricing Efficiency of Hong Kong Hang Seng Index Option and Futures Markets," *Journal of Futures Markets*, 17, 797–815.

Gemmill, G. (1993). *Options Pricing*, Maidenhead, UK: Mc Graw-Hill.

Gwilm, O. P. and Buckle M. (1999). "Volatility Forecasting in the Framework of the Option Expiry Cycle," *The European Journal of Finance*, 5, 73–94.

Goldberg, D. E. (1989). *Genetic Algorithms in Search, Optimization and Machine Learning*. Addison-Wesley.

Holland, J. H. (1975). *Adaptation in Natural and Artificial System*. University of Michigan Press.

Koza, J. R. (1992). *Genetic Programming: On the Programming of Computers by Means of Natural Selection*. MIT Press.

Koza, J. R. (1994). *Genetic Programming II: Automatic Discovery of Reusable Programs*. MIT Press.

Koza, J., D. Goldberg, D. Fogel, and R. Riolo (1996). *Proceedings of the First Annual Conference on Genetic programming*. MIT Press.

Lee, J. H., and N. Nayar (1993). "A Transactions Data Analysis of Arbitrage between Index Options and Index Futures," *Journal of Futures Markets*, 13, 889–902.

Li, J. and E. P. K. Tsang (1999a). "Improving Technical Analysis Predictions: An Application of Genetic Programming," *Proceedings of The 12th International Florida AI Research Society Conference*, 108–112.

Li, J. and E. P. K. Tsang (1999b). "Investment Decision Making Using FGP: A Case Study," *Proceedings of Congress on Evolutionary Computation (CEC'99)*, 1253–1259.

Mahfoud, S. and Mani, G. (1997). "Financial Forecasting Using Genetic Algorithms," *Journal of Applied Artificial Intelligence*, 10(6), 543–565.

Markose, S. and H. Er (2000). "The Black (1976) Effect and Cross Market Arbitrage in FTSE-100 Index Futures and Options," Working Paper, No. 522, Economics Department, University of Essex.

Mitchell, M. (1996). *An Introduction to Genetic Algorithms.* MIT Press.

Modest, D. and M. Sunderesan (1983). "The Relationship between Spot and Futures Prices in Stock Index Futures Markets: Some Preliminary Evidence," *Journal of Futures Markets*, 3, 15–41.

Stoll, H. R. (1969). "The Relationship between Put and Call Option Prices," *Journal of Finance*, 25, 801–824.

Neely, C., P. Weller, and R. Ditmar (1997). "Is Technical Analysis in the Foreign Exchange Market Profitable? A Genetic Programming Approach," *Journal of Financial and Quantitative Analysis*, 32, 405–426.

Oussaidene, M., B. Chopard, O. Pictet, and M. Tomassini (1997). "Practical Aspects and Experiences - Parallel Genetic Programming and Its Application to Trading Model Induction," *Journal of Parallel Computing*, 23(8), 1183–1198.

Saad, E., D. Prokhorov, and D. Wunsch (1998). "Comparative Study of Stock Trend Prediction Using Time Delay, Recurrent and Probabilistic Neural Networks," *IEEE Transactions on Neural Networks*, 9, 1456–1470.

Sweeney, R. J. (1988). "Some New Filter Rule Tests: Methods and Results," *Journal of Financial and Quantitative Analysis*, 23, 285–300.

Tucker, A. L. (1991). *Financial Futures, Options and Swaps*, St. Paul, MN: West Publishing Company.

Tsang, E. P. K. , J. Li, and J. M. Butler (1998). "EDDIE Beats the Bookies," *International Journal of Software, Practice and Experience*, 28(10), 1033–1043.

Tsang, E. P. K. , J. Li, S. Markose, E. Hakan, A. Salhi, and G. Iori (2000). "EDDIE in Financial Decision Making," *Journal of Management Economics,* 4(4).
<http://www.econ.uba.ar/www/servicos/publicaciones/journal3/>.

Yadav, P. K. and P. Pope (1990). "Stock Index Futures Arbitrage: International Evidence," *Journal of Futures Markets*, 10, 573–603.

V

AGENT-BASED COMPUTATIONAL FINANCE

Chapter 15

A MODEL OF BOUNDEDLY RATIONAL CONSUMER CHOICE

Thomas Riechmann

University of Hannover, Department of Economics, Königsworther Platz 1

30167 Hannover, Germany

riechmann@vwl.uni-hannover.de

Abstract The paper presents an extended version of the standard textbook problem of consumer choice. As usual, agents have to decide about their desired quantities of various consumption goods, at the same time taking into account their limited budget. Prices of the goods are not fixed but arise from a Walrasian interaction of total demand and a stylized supply function for each of the goods. After showing that this type of model cannot be solved analytically, three different types of evolutionary algorithms are set up to answer the question whether agents' behavior according to the rules of these algorithms can solve the problem of extended consumer choice. There are two important answers to this question: a) The quality of the results learned crucially depends on the elasticity of supply, which in turn is shown to be a measure of the degree of state dependency of the economic problem. b) Statistical tests suggest that for the agents in the model it is relatively easy to adhere to the budget constraint, but that it is relatively difficult to reach an optimum with marginal utility per Dollar being equal for each good.

Keywords: Consumer Choice, Evolutionary Algorithms, State Dependency

Introduction

In introductory courses to microeconomics, when it comes to the problem of consumer choice, a question often heard is "Do people *really* behave this way?" The standard answer to this question is "Not *really*, but they can learn to achieve the optimal outcome, anyway." But, is this really true? Can boundedly rational people learn how to choose their optimal consumption bundle?

This paper tries to answer this question. In order to do this, the standard text book problem of consumer choice is extended: The assumption of fixed prices is dropped, which makes the individual problem even harder to solve. This model is simulated, applying three different learning techniques in form of three different evolutionary algorithms. These kinds of algorithms have often been applied to similar economic problems. Interpretations of evolutionary algorithms as metaphors for various types of learning schemes have been given by various authors (Dawid 1999 and Riechmann 1999).

It can be shown, that the learnability of an optimal solution of the extended consumer choice problem not only depends on the learning technique chosen, but also on the degree of state dependency of the extended consumer choice problem, which, in turn, can be measured by the slope of the supply functions of the model.

More than this, it can be learned, that it is easier to adhere to the budget constraint than it is to find the optimal consumption bundle.

Apart from answering a core economic problem, this paper shows some new features of evolutionary programming, which have rarely been used in economic modelling before. The most important new feature is the simulation of simultaneous constrained optimization over more than just one variable, and the use of a penalty function.

1. The Economic Model

1.1. The General Structure of the Problem

Let there be a number of n agents (households) facing the standard textbook problem of consumer choice, i.e. selecting a bundle of consumption goods which, under the restriction of a limited budget, maximizes utility.

Agents are assumed to have identical utility functions, which do not change over time t. Utility is derived from consuming a bundle of m different goods indexed k. For convenience, the utility function is assumed to be of Cobb-Douglas type.

$$u_{i,t} = A \prod_{k=1}^{m} q_{i,k,t}^{\alpha_k}; \quad A > 0;\ \alpha_k > 0\,\forall\, k = 1, \ldots, m\,. \tag{15.1}$$

$q_{i,k,t}$ gives the quantity of good k agent i consumes in period t. A and α_k, $k = 1, \ldots, m$ are parameters of the model.

The budget M is assumed to be the same for every agent i in every period t, so the budget constraint is given by

$$M \geq \sum_{k=1}^{m} p_{k,t}\, q_{i,k,t}\,. \tag{15.2}$$

$p_{k,t}$ represents the market price for good k in t.

Thus, every agent i aims to solve the constrained maximization problem

$$\max_{q_{i,k,t}} u_{i,t} \qquad \forall\, k = 1, \ldots, m \tag{15.3}$$

$$\text{s.t.} \qquad M \geq \sum_{k=1}^{m} p_{k,t}\, q_{i,k,t} \tag{15.4}$$

$$q_{i,k,t} \geq 0 \quad \forall\, k = 1, \ldots, m \tag{15.5}$$

Equation (15.5) gives the usual non-negativity constraints.

Different from the standard textbook model, prices will not generally be held fixed, but will be subject to a stylized Walrasian mechanism. For each good k, in every period t, the price will be determined as the equilibrium price resulting from the interaction of aggregate demand for good k in t and aggregate supply of the good.

Aggregate demand for good k in t, $Q_{k,t}$ is simply the sum of individual demand for k in t:

$$Q_{k,t} = \sum_{i=1}^{n} q_{i,k,t}\,. \tag{15.6}$$

Aggregate supply will be modelled by a time invariant standard supply function for each good k, so that the equilibrium price $p^{\star}_{k,t}$ results as

$$p^{\star}_{k,t} = B_k + m_k \sum_{i=1}^{n} q_{i,k,t}\,. \tag{15.7}$$

1.2. The Basic Model

The basic model is a model of fixed prices. This means that in Equation (15.7) $m_k = 0$, yielding

$$p^{\star}_{k,t} = B_k\,. \tag{15.8}$$

Note, that this case does not represent an "economic" problem in the sense of agents' fitness being state dependent. Each agent's utility only depends on her own actions, but in no way on the actions of any other agent.

For fixed prices the solution to the problem of consumer choice can easily be derived.

By standard calculus, the solution can be determined as

$$q^{\star}_{i,k,t} = \frac{\alpha_k}{\sum_{k=1}^{n} \alpha_k} \frac{M}{B_k}; \quad \forall\, k;\ \forall\, i;\ \forall\, t. \tag{15.9}$$

This result yields two crucial characteristics common to every optimal bundle of consumption goods. The first characteristic is efficiency: The whole budget is being spent. In an optimal situation, (15.2) becomes binding:

$$M = \sum_{k=1}^{m} p_{k,t}\, q_{i,k,t}\,. \tag{15.10}$$

The second crucial characteristic of the optimal consumer choice is the fact, that marginal utility per Dollar is the same for every pair of goods k, l,

$$\frac{\partial u_{i,t}\,(\cdot)/\partial q_{k,i,t}}{p_{k,t}} = \frac{\partial u_{i,t}\,(\cdot)/\partial q_{l,i,t}}{p_{l,t}} \quad \forall k,\ l \in \{1,\ldots,m\}\,. \tag{15.11}$$

Both of these well known standard results will become of greater importance in the second part of the paper.

1.3. The Enhanced Model

In the enhanced case, prices are no longer fixed, i.e. $m_k \neq 0$ in (15.7). As a consequence, the problem of consumer choice becomes a problem of state dependent fitness: Aggregate demand now has an influence on the market price and the market price has an influence on every single agent's economic success.

With the introduction of flexible prices the problem turns into a kind of n–person Cournot game of the demand side and thus becomes analytically intractable. This means that for flexible prices, an explicit numerical solution analogous to (15.9) for the fixed price case cannot be found. Nevertheless, at least for uniformly behaving populations, the characteristics (15.10) for efficiency and (15.11) for optimality still apply.

2. The Evolutionary Algorithm

2.1. The Basics

An evolutionary algorithm aiming to model the above setting in an agent based manner must above all be capable of coping with two problems: a) The maximization problem in focus involves the optimization of more than just one independent variable, and b) the maximization problem in focus has to be solved subject to a constraint.

Both of the problems have been solved in natural sciences before,[1] but up to now there is no economic model making use of these results.

In this paper, a hybrid evolutionary algorithm will be employed, making use of principles from two worlds, from the world of genetic algorithms (GAs) and from the world of evolution strategies (ESs). From GAs, the well known operators of crossover and selection / reproduction are used. In economics, they have been broadly interpreted as forms of social learning by communication and imitation as well as the functioning of the market (Riechmann 1999). From evolution strategies, the operator of mutation is used, being interpreted as a form of isolated individual learning by experiment. Moreover, the variables in focus will not be coded as bit strings, but as real valued numbers, which is also a feature of evolution strategies.

To make things clearer, an agent i will be fully described by her economic plans, i.e. the vector of demanded quantities for each good k in period t, $q_{i,k,t}$. In the simulations, there will only be three different goods available to the agents. Thus, an agent is characterized by a vector $q_{i,t}$,

$$q_{i,t} \in \mathbf{R}^3. \tag{15.12}$$

An example would be

$$q_{10,5}\,(11.2;\ 3.7;\ 17.0) \tag{15.13}$$

meaning that agent number 10 plans to consume 11.2 units of the first, 3.7 units of the second and 17 units of the third good in period 5.

2.2. Standard Operators

Due to the change in the representation of the agents, the operators (often called "genetic operators") have to be changed, too. As these operators are used as metaphors for learning techniques, the changed operators still have to support a sensible economic interpretation.

The "double operator" of selection and reproduction, often being interpreted as learning by imitation connected with the role of the market, needs no changes at all. Agents are selected from their population and reproduced into the next one with a probability equal to their relative fitness. Fitness, in this case, equals individual utility. This means, that in this paper, the standard roulette wheel selection operator (Goldberg 1989) will be used, not one of the selection operators stemming from the tradition of evolution strategies.

Recombination in the form of crossover is usually seen as a metaphor for learning by communication. The change of this operator is quite straightforward. In a real valued rather than a binary representation, agents' economic strategies can be separated into clearly defined economic substrategies. A substrategy in the current model is the consumption quantity of a single good. This means that in the above example (15.13) the agent has three substrategies: 11.2 for good one, 3.7 for good two, and 17 for good three. Crossover now works as usual, recombinating two agents' substrategies. The agents involved in crossover are chosen from their population, a crossover point is selected, the vectors of substrategies are cut at the crossover point, the resulting parts are interchanged and put together again, yielding two new strategies. Figure 15.1 shows an example. Thus, in the world of real valued coding of

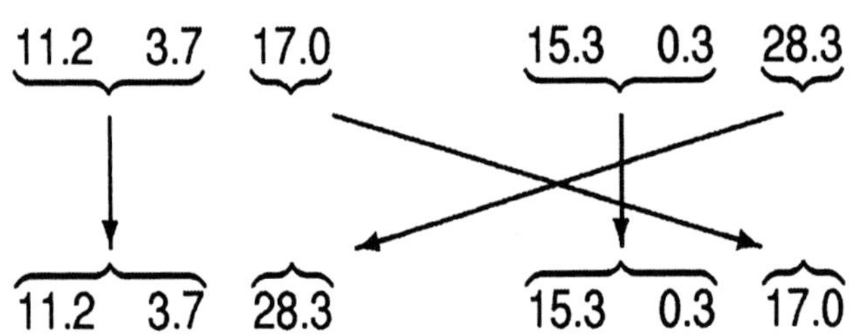

Figure 15.1. Crossover (example).

agents' strategies, the interpretation of crossover as a form of learning by exchanging substrategies becomes even clearer than in the world of binary coding.

The third standard operator, mutation, has to be changed, too. Mutation, often being interpreted as learning by experiment, has to undergo the most severe changes. Real valued coding certainly does not allow for simple bit flipping, as used in GAs with binary coding. Instead, the mutation operator from the tradition of evolution strategies can be used. In this tradition, a substrategy $q_{i,k,t}$ is mutated by adding a term v_k to it, where v_k is normally distributed with mean 0 and finite variance σ^2, resulting in the new substrategy $\tilde{q}_{i,k,t}$.[2]

$$\tilde{q}_{i,k,t} = q_{i,k,t} + v_k \quad \forall\, k \in \{1, \ldots, m\}; \quad v_k \sim N\left(0,\, \sigma^2\right) \qquad (15.14)$$

By endogenizing the mutation variance σ^2, this operator could easily be extended in order to represent some kind of meta learning (Riechmann 2001b), but for the clarity of the economic argument this will not be done within this paper.

2.3. Enhanced Operators

In addition to the standard operators, two more 'enhanced' operators shall be employed. The first one is the well known election operator (Arifovic 1994). Election requires two (or more) agents to meet, jointly work out new strategies, evaluate these strategies and than finally decide which strategy to use in reality. Though election, especially in its basic form, is not undisputed in its economic meaning,[3] it is known to result in stable states of the learning process, which most of the time even represent optimal solutions to the underlying economic problem.[4] Thus—economically meaningful or not—election represents a good benchmark for the test of the performance of other learning operators.

The second enhanced operator is something more than just an operator, as it requires a slight change in the construction and implementation of the agents. For the operator of *preselection*, agents will be equipped with a memory. And although this memory is very limited, it will be shown to help improving agents' learning performance. Precisely, agents will be given the ability to remember one certain strategy together with the level of utility they gained from employing this strategy. The specific strategy an agent remembers is her all time best strategy, i.e. the strategy that has brought her the highest utility during the whole course of the learning process. During preselection an agent choses from two possible strategies: the strategy she used in the last period (the fitness of which she can remember) and her all time best strategy. She decides by the fitness. The strategy with the higher fitness is chosen as the strategy the following learning processes of the current period (i.e. communication and experiment) are based on.

Note, that for non state dependent problems, like the fixed price problem of consumer choice, the preselection algorithm essentially works like an intra agent operator of elitist selection for the basic strategy in each period. In the simple fixed price model, an agent's fitness or utility is independent of what the others do, i.e.

$$u_i\left(q_{i,t} | S_t\right) = u_i\left(q_{i,t}\right) \quad \forall\, S_t \in S'\,. \qquad (15.15)$$

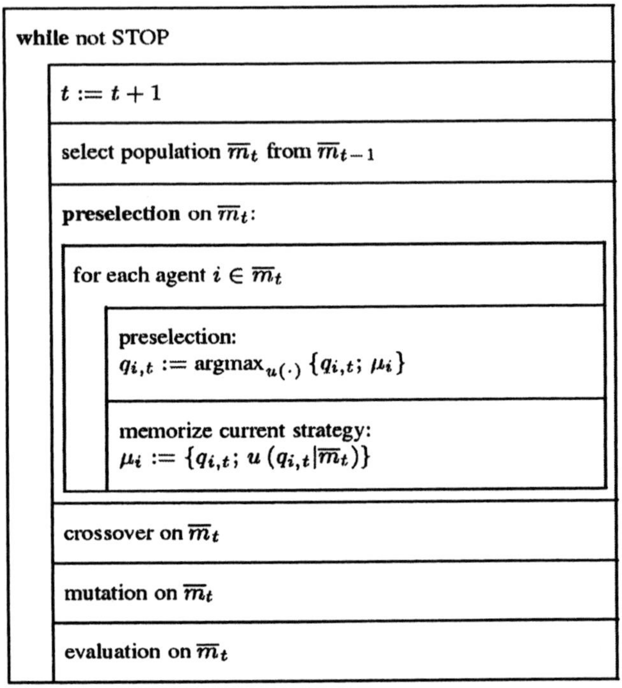

Figure 15.2. Main loop of preselection algorithm.

S_t is the population in period t, S' is the set of all different populations.

In problems of state dependent fitness, on the contrary, all actions of all the other agents in the population have an influence on an agent's fitness. This means, that for $m_k \neq 0$ in (15.7), the strategy an agent thinks of as her all time best one, might—due to the current population—not be as good as she thinks. A strategy that once performed brilliantly may perform very poorly in the context of a different population. This is a fact that the agents in this model are assumed to ignore. Agents use the rule of preselection, because they simply do not know that they may be mistaken.

2.4. Coping with the Constraints

The problem in focus is a problem of constrained optimization. Agents do not only have to maximize utility but also have to be careful not to exceed their budget. In the simulations, the budget constraint will not be *directly* accounted for. This means that there is no good the consumed quantity of which serves as some kind of residual. Agents *do not* do something like determine the quantity of the last good as all they can afford to get by the rest of their budget like

$$q_{i,m,t} = \frac{1}{p_{m,t}} \left(M - \sum_{k=1}^{m-1} p_{k,t} q_{i,k,t} \right) . \tag{15.16}$$

Instead, agents freely decide on the quantity of all three goods, at first hand independently of the budget constraint. This also means, that in the fixed price cases, at the time of making their consumer plans, agents do not even know the prices of the goods, which makes their decision problem even harder.

The compliance to the budget constraint is secured by using a *penalty function*, which is a standard tool in evolutionary optimization (see Michalewicz 1996, p. 321). Using a penalty function means decreasing an agent's fitness if she breaks the budget constraint. In economic terms, this looks like an extension to the utility function, which may be rather unusual to most theoretical economists.

In the simulations the following fitness function was used in order to transform utility $u_{i,t}$ into fitness $R_{i,t}$:[5]

$$R_{q_{i,t}} = \begin{cases} U\left(q_{i,t}\right) + M - \sum_{k=1}^{3} p_{k,t} q_{i,k,t} & \text{for} \quad \sum_{k=1}^{3} p_{k,t} q_{i,k,t} > M \\ U\left(q_{i,t}\right) & \text{for} \quad \sum_{k=1}^{3} p_{k,t} q_{i,k,t} \leq M \end{cases} . \tag{15.17}$$

Using fitness function (15.17) means, that an agent *can* break the constraints, but that she should *learn* not to do so.

3. Simulations and Results

For the model described above, simulations were run using three different algorithms, a) the canonical algorithm, using selection/reproduction, crossover and mutation, b) the election algorithm, using selection/reproduction, crossover, mutation and election, and c) the preselection algorithm, using selection/ reproduction, preselection, crossover and mutation. The simulations were run for three different parameter sets which only differ with respect to the elasticity of supply.

All three sets thus share the values for $A = 1$ and for the budget M, $M = 100$, for B_k, $B_1 = B_2 = B_3 = 1$, and for α_k, $\alpha_1 = 0.2$, $\alpha_2 = 0.3$, and $\alpha_3 = 0.5$. The only difference is that for the first set ("FIXPRICE"), $m_1 = m_2 = m_3 = 0$, yielding $p_1 = p_2 = p_3 = B_k = 1$. Thus, FIXPRICE represents the fixed price case or, put in different words, a case of infinitely high elasticity of supply.[6] The second set ("HIGHELASTICITY") uses $m_1 = m_2 = m_3 = 0.0001$, thus representing a state of high elasticity of supply. The third set ("LOWELASTICITY") uses $m_1 = m_2 = m_3 = 0.1$ which leads to a state of low elasticity of supply.

Simulations were run for populations of $n = 500$ agents and for $t_{\max} = 500$ periods.

3.1. Fixed Prices

For the fixed price case, at least for the case of uniform behavior, optimal outcome can be determined analytically. If all agents within a population behave the same, the best strategy according to (15.9), given parameter set FIXPRICE is

$$q^{\star}_{1,k,t} = 20, \quad q^{\star}_{2,k,t} = 30, \quad q^{\star}_{3,k,t} = 50 \quad \forall \quad k,\, t\,. \tag{15.18}$$

These results represent both, efficiency and optimality. This means that for the results given in (15.18) the budget is fully spent (i.e. the budget residual is zero) and every agent's utility is as high as possible due to the given budget.

The quality of the learning algorithms can be judged by comparing the learned results to the theoretical results given in (15.18). More than this, it seems appropriate to check if, and if so, how good, the budget constraint is met. In order to do this, the average budget residual ρ_t will be used, which is defined as the average amount of money not spent on consumption throughout the population, i.e.

$$\rho_t = \frac{1}{n}\sum_{i=1}^{n}\left(M - \sum_{k=1}^{3} p_{i,k,t}q_{i,k,t}\right)\,. \tag{15.19}$$

Representative simulation results for the three respective algorithms are given in the following figures. Each figure consists of two subfigures which give a plot of the population average of the quantities of good one to three (on the left) and a plot of the average budget residual (on the right).

It is easy to see that for all three algorithms, the quantities tend to converge toward the theoretically optimal quantities. The election algorithm seems to perform best whereas the canonical algorithm performs worst. This impression is supported by the results concerning the budget residual. Election converges to a residual equal to zero, preselection tends to oscillate around zero, whereas the canonical algorithm results in relatively large positive residuals. This means that each of the three learning methods is able not to violate the budget constraint, but only learning with election and learning with preselection enables the agents not to waste a part of their income.

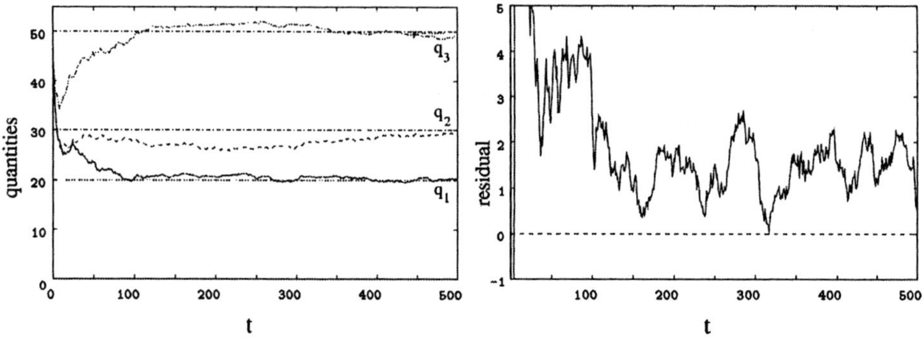

(a) Quantities (b) Budget Residual

Figure 15.3. Canonical Algorithm.

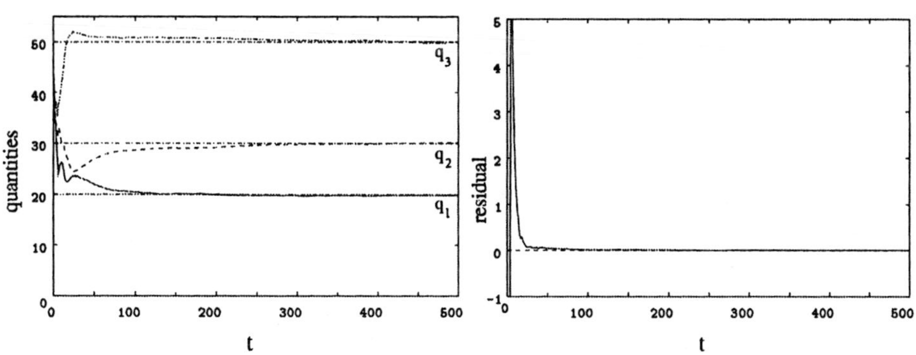

(a) Quantities (b) Budget residual

Figure 15.4. Election Algorithm.

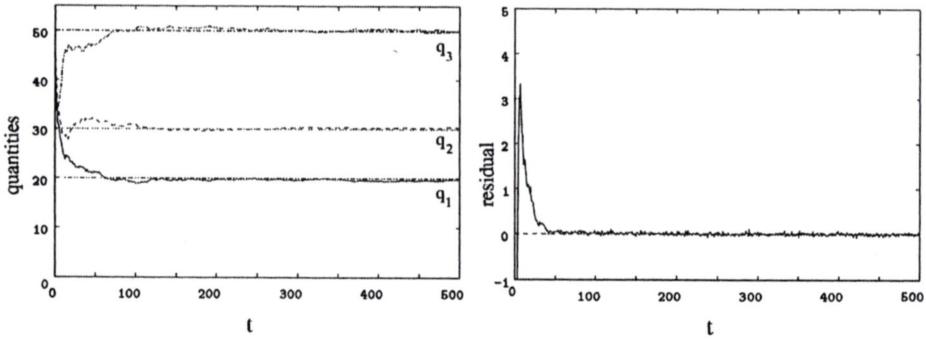

(a) Quantities (b) Budget Residual

Figure 15.5. Preselection Algorithm.

3.2. Flexible Prices, High Elasticity

For the case of flexible prices, it becomes impossible to determine an explicit solution analytically even for the case of homogeneous behavior. In order to judge the quality of the simulation results, two measures are introduced. The first one is the average budget residual as defined in Equation (15.19). The second one is the difference between the marginal utility per Dollar for each pair of the three goods. Although it is impossible to solve the problem explicitly, the fact remains that in an optimal solution of the consumer choice problem marginal utility per Dollar has to be equal for each good. Thus, a perfect solution should have two characteristics: a) Marginal utility per Dollar is the same for each of the three goods, and b) the average budget residual is zero.

The plots of representative simulations for the three algorithms and the HIGHELASTICITY case (Figures 15.6 to 15.8) consist of two subfigures each. The first subfigure gives the average budget residual ρ_t. The second subfigure is a plot of the marginal utility per Dollar for each of the three goods.

A closer look at the figures confirms the results of the FIXPRICE case. The election algorithm performs best. Marginal utilities per Dollar seem to converge completely while the budget residual vanishes. The preselection algorithm generates slightly worse results. While the budget residual fluctuates around zero, marginal utilities per Dollar become similar but not equal. The canonical algorithm performs worst. Marginal utilities per Dollar become similar, but the budget is never fully spent.

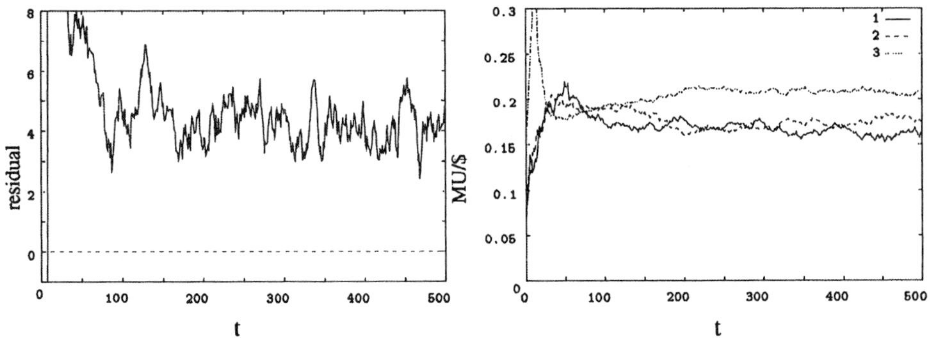

(a) Budget Residual

(b) Marginal Utility per Dollar

Figure 15.6. Canonical Algorithm — high elasticity of supply.

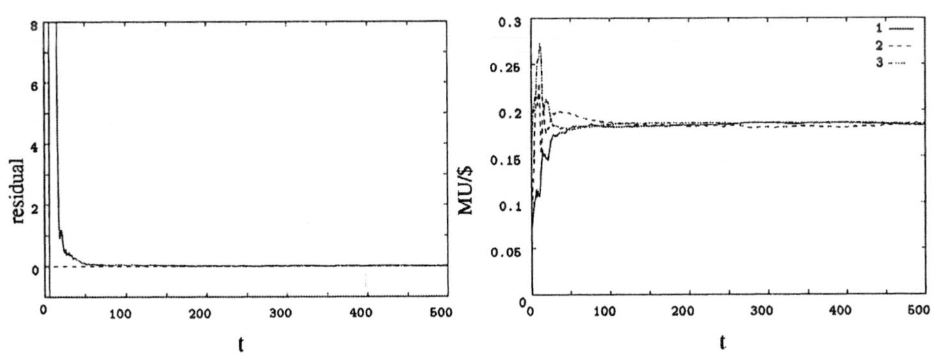

(a) Budget Residual

(b) Marginal Utility per Dollar

Figure 15.7. Election Algorithm — high elasticity of supply.

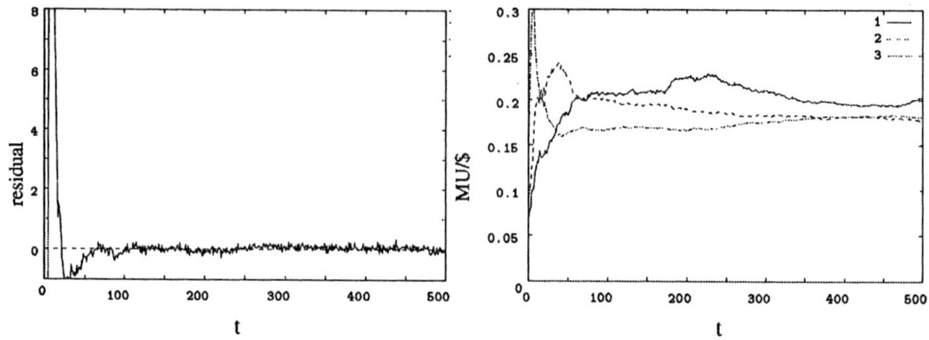

(a) Budget Residual (b) Marginal Utility per Dollar

Figure 15.8. Preselection Algorithm — high elasticity of supply.

3.3. Flexible Prices, Low Elasticity

The third case in focus, flexible prices with a low elasticity of supply, leads to quite different results. It can be seen that the canonical algorithm does not reach a sensible outcome. The budget constrained is never met while marginal utilities per Dollar never get close to each other. In the LOWELASTICITY case, a sensible consumer choice cannot be learned by use of the simple canonical algorithm learning rules.

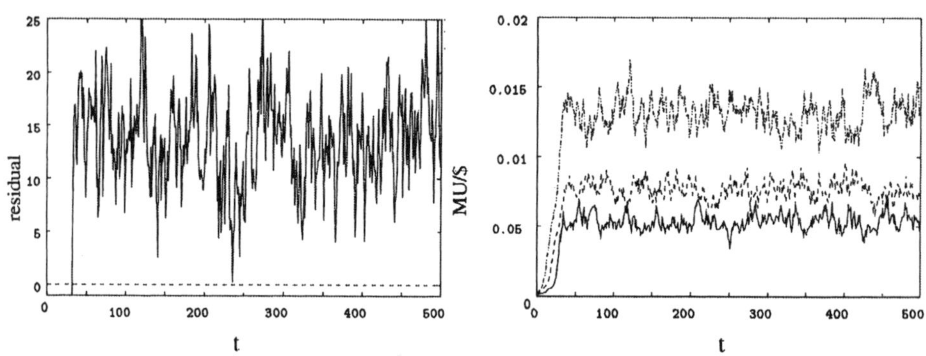

(a) Budget Residual (b) Marginal Utility per Dollar

Figure 15.9. Canonical Algorithm — low elasticity of supply.

Even worse is the result for the election algorithm. In each of the simulations for this case, with only one explicitly shown in this paper, the model collapsed. Election leads to totally unrealistic cases of violation of the budget constraint while marginal utilities per Dollar do not converge at all. For the case of flexible prices and a low elasticity of supply, election is far from leading to any kind of sensible consumer choice.

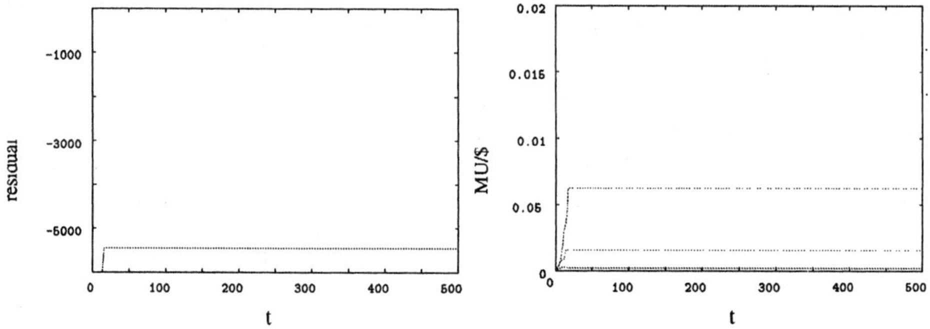

(a) Budget Residual

(b) Marginal Utility per Dollar

Figure 15.10. Election Algorithm — low elasticity of supply.

The only sensible results are achieved by the preselection algorithm, which causes the marginal utilities per Dollar to become at least slightly similar to each other while the budget constraint is not violated too severely.

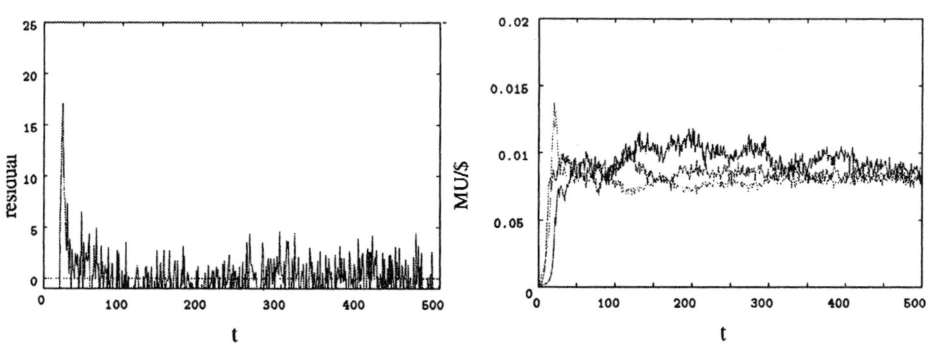

(a) Budget Residual

(b) Marginal Utility per Dollar

Figure 15.11. Preselection Algorithm — low elasticity of supply.

3.4. Summary of Results

A summary of the simulation results can be given as follows.

Concerning the budget constraint it can be found, that it seems to be relatively easy to learn not to violate it. For the FIXPRICE case as well as for HIGHELASTICITY, the canonical algorithm results in states which are significantly positive, while election is significantly very close to zero and preselection is even significantly equal to zero.

For the choice of the correct quantity or the best marginal utility per Dollar, for FIXPRICE and HIGHELASTICITY, the results from the plots do not look too bad, but from the statistics (Appendix 15.A) it can be found, that significances are quite poor. From the three different algorithms, the canonical one shows the highest degree of fluctuations (i.e. the highest variance) and consequently the highest (though still very poor) significance of reaching the optimal result, while election produces nearly no fluctuations at all, resulting in the least significance of an optimal outcome.

For LOWELASTICITY, only the preselection algorithm leads to sensible results at all.

All in all, it seems as if the preselection algorithm performs best over all types of economic situations. This means, that a little memory to the past notably improves learning abilities.

4. Conclusions

4.1. The Influence of State Dependency

As a conclusion to be drawn from the simulation results it can be found that it seems to be much harder to learn a sensible consumer choice in a situation with low elasticity of supply than in situation with high (or even infinitely high) elasticity.

The reason for this is the following. In each round of the algorithm (in each market period) the resulting market prices and by that, the utility gained, reveal some information to each agent. This information is information about the quality of her last period consumption plan. Ceteris paribus, the plan was a good plan if utility was high and so the plan should not be changed too much for the next period. If, on the contrary, utility was not very high, the plan obviously was a bad one and thus should be changed. Unfortunately, this argument is only a ceteris–paribus argument. The information revealed actually consists of two parts which cannot be distinguished. Surely the information is information about the quality of the agent's plan. But it is only information about the quality of the agent's plan in the context of the plans of all other agents in the population. This means, that from period to period the quality of an agent's plan can be changed either by a change in the plan itself or by a change in the plans of the rest of the population. While there certainly is a *direct* impact of the agent on her economic success, there is also an indirect impact caused by the rest of the population. This indirect impact has often been called *state dependency* of agents' fitness.

State dependency can be found to cause noise in the part of the information valuable for the agent, i.e. the information about the quality of her plan. The stronger state dependency, the stronger is the noise, the less valuable is the information to the agent, and thus the more complicated is the consumption decision. Thus, in order to give a reason why less elasticity seems to complicate the problem, it should be shown that less elasticity of supply means more state dependency in the problem of consumer choice.

It can be shown that the gradient m_k of the respective supply function (15.7) is a measure of state dependency. A sketch of the argument runs as follows: The utility function (15.1)[7] can be rewritten as integrating the budget constraint (15.2),[8] using $q_{i,m}$ as a numeraire, yielding

$$V_i = V\left(q_{i,1}, q_{i,2}, \ldots, q_{i,m-1}, M, p_1, p_2, \ldots, p_m\right) . \tag{15.20}$$

A change in utility can be written as

$$\mathrm{d}V_i = \underbrace{\sum_{k=1}^{m-1} \frac{\partial V_i}{\partial q_{i,k}} \,\mathrm{d}q_{i,k}}_{\text{direct effect}} + \frac{\partial V_i}{\partial M}\,\mathrm{d}M + \underbrace{\sum_{k=1}^{m} \frac{\partial V_i}{\partial p_k}\,\mathrm{d}p_k}_{\text{indirect effect}} . \tag{15.21}$$

From this, it can be recognized that a change in an agent's utility can be caused by two effects, by the direct effect of the agent changing her quantities, and by the indirect effect of a change in prices.

The interesting aspect is the change of the prices. From (15.7) it can be deduced that a change in the price of good k is

$$\mathrm{d}p_k = \sum_{j=1}^{n} \frac{\partial p_k}{\partial q_{j,k}}\,\mathrm{d}q_{j,k} = m_k \sum_{j=1}^{n} \mathrm{d}q_{j,k} . \tag{15.22}$$

It can thus be seen that a change in the price of a good is caused by a change in the demand for this good of one or more of the agents.

A change in utility becomes

$$\mathrm{d}V_i = \underbrace{\sum_{k=1}^{m-1} \frac{\partial V_i}{\partial q_{i,k}} \mathrm{d}q_{i,k}}_{\text{direct effect}} + \frac{\partial V_i}{\partial M} \mathrm{d}M + \underbrace{\sum_{k=1}^{m} \left(\frac{\partial V}{\partial p_k} m_k \sum_{j=1}^{n} \mathrm{d}q_{j,k} \right)}_{\text{indirect effect}} \quad (15.23)$$

Abbreviating the direct effect as D and the change of aggregate demand for good k, $\sum_{j=1}^{n} dq_{j,k}$ as dQ_k, (15.23) can be simplified:

$$\mathrm{d}V_i = \underbrace{\sum_{k=1}^{m-1} \frac{\partial V_i}{\partial q_{i,k}} \mathrm{d}q_{i,k}}_{\text{direct effect}} + \frac{\partial V_i}{\partial M} \mathrm{d}M + \underbrace{\sum_{k=1}^{m} \left(m_k \frac{\partial V_i}{\partial p_k} \mathrm{d}Q_k \right)}_{\text{indirect effect}} . \quad (15.24)$$

Focusing on the indirect effect, it is now easy to see that for each market k, it is the parameter m_k that decides the impact of the indirect effect on an agent's utility.

If, for example, $m_k = 0$, there is no indirect effect at all. Only the agent herself has an influence on her utility. In other words: For $m_k = 0$, there is no state dependency. Notice, that this case is the case of fixed prices, which leads to relatively good results for all three types of algorithms.

If, on the contrary, m_k is very high, there is also a large influence of all other agents on each agent's utility. This is the case of high state dependency. This case is equal to the case of flexible prices and low elasticity of supply. In this case, due to the high degree of state dependency, the noise in the information is strong and consequently the learning results are relatively bad.

Summarizing, the higher m (i.e. the lower elasticity of supply), the higher is the impact of state dependency on the change of each agent's utility. This makes it harder for an agent to recognize the impact of her own consumer plan on her economic success which in turn makes it harder to learn a sensible solution to the consumer choice problem.

4.2. The Influence of Different Learning Schemes

4.2.1 Election and the Concept of Potential Fitness. The algorithm using the election operator seems to perform very good in the FIXPRICE and in the HIGHELASTICITY cases, whereas in the LOW-ELASTICITY case performance is extremely poor. In order to find the reasons for this behavior, it is appropriate to recall the central working principle of election. During election, two agents meet and jointly try to find a new strategy. They do this by, among others, calculating a so called potential fitness for the newly created strategies, which should help to find out which strategy is the best one. The calculation of potential fitness requires the knowledge about all the influences on future economic success of the new strategies. As it is impossible to know about all these influences, all these influences are assumed to be unchanged since the last period of time.

This means, that the concept of potential fitness is basically a concept of 'ceteris paribus' fitness: An agent calculates the potential fitness of her strategy assuming all the other agents will not change their behavior. This means that agent i, while finding potential fitness, assumes that $dq_{j,k} = 0 \; \forall \; i = \{1, \dots n\}$, $i \neq j$. In other words, most of the indirect influence on actual fitness is neglected. This, of cause, is absolutely correct for situations without state dependency, like the FIXPRICE case, and this is still quite good for low degrees of state dependency like in the HIGHELASTICITY case. But, the more important the indirect effect, the more severe becomes the difference between actual fitness and potential fitness. In situations with high state dependency, e.g. in the LOWELASTICITY case, this may lead to systematically wrong strategy choices, as can be seen from the simulation results in Figure 15.10.

4.2.2 Preselection. In contrast to election, preselection is not the end of the learning process in a period, but the beginning. Whereas in the process of election, the strategy resulting from the election process is the one to be applied in the market, the strategy resulting from preselection is subject to learning by communication (crossover) and experiment (mutation), before it is used at the market. This means that in situations with only little state dependency, preselection — like election — has the advantage of chosing between two strategies (the all time best and last strategy used last period), but the "pure" result of the preselection process can be slightly changed during the following two learning steps, which may be a disadvantage in an environment with only little noise. But, these following two learning steps become the great advantage of preselection in situations with high degrees of state dependency. This means, that different from election learning, preselection learning does not get stuck in strategies that are successful only due to potential but not to actual fitness, but can still change the preselected strategy in each period.

5. Summary

The paper employs three different learning algorithms in order to find out, if boundedly rational agents can learn to choose the optimal consumption bundle. In the model, the problem is complicated by allowing for flexible prices and allowing for violation of the budget constraint.

It can be found that it is relatively easy to learn not to break the budget constraint, but that it seems to be quite complicated to find the optimal consumption bundle. Simulation results show that the problem becomes even harder if the elasticity of supply decreases. It is shown that a decrease in the elasticity of supply means an increase in the degree of state dependency. For some types of algorithms, state dependency works like an external effect: Agents do not include it into their calculations for their future strategy, thus being badly mistaken in each future period of time.

Can boundedly rational agents learn the optimal consumer choice? — They can, if the problem is not too complicated. And they do even better if they have some memory of the past.

This is all of the message this paper can give, but at least this seems to be a better story to tell the first year students if they ask again ...

Appendix: Data

The following tables show data from one randomly chosen simulation each. All simulations were run for 1 000 periods with a population size of n=500. The data contains information about rounds 501 to 1 000 to eliminate possible startup effects of the algorithms.

FIXPRICE

The t–values are test statistics of an approximate Gauss tests for $q_1 = 20$, $q_2 = 30$, $q_3 = 50$ and $\rho = 0$, respectively.[9]

		Canonical	Election	Preselection
q_1	Mean	20.2484	20.0492	20.2077
	Var.	0.515554	0.000617538	0.0153379
	t	7.73661	44.2923	37.4965
q_2	Mean	30.8866	30.0306	29.9948
	Var.	1.36406	0.00149178	0.00520296
	t	-186.699	-5760.89	-3035.61
q_3	Mean	47.4426	49.9195	49.7968
	Var.	1.33563	0.00358468	0.00447201
	t	-575.641	-11185.8	-9961.81
ρ	Mean	1.4224	0.000674724	0.000770113
	Var.	0.453434	3.88333 E(-7)	0.000613616
	t	47.2334	24.2108	0.69517

HIGHELASTICITY

The t–value is a test statistic of an approximate Gauss test for $\rho = 0$, the values t_{12}, t_{13}, and t_{23} are test statistics of approximate Gauss test for $MU_1/\$ = MU_2/\$$, $MU_1/\$ = MU_3/\$$ and $MU_2/\$ = MU_3/\$$, respectively.

		Canonical	Election	Preselection
$MU_1/\$$	Mean	0.178931	0.185862	0.189568
	Var.	0.0000532438	8.59688 E(-7)	7.43989 E(-6)
$MU_2/\$$	Mean	0.178914	0.183026	0.181765
	Var.	0.0000914054	2.21833 E(-7)	3.54032 E(-6)
$MU_3/\$$	Mean	0.196372	0.184637	0.183881
	Var.	0.0000684217	2.27176 E(-7)	6.94431 E(-7)
t_{12}		0.0316065	60.9781	52.6552
t_{13}		-35.3568	26.2744	44.5869
t_{23}		-30.8784	-53.7592	-22.9925
ρ	Mean	4.1415	0.0108742	-0.0000875281
	Var.	0.63692	0.0000392457	0.00494539
	t	116.038	38.8137	-0.0278312

LOWELASTICITY

		Canonical	Election	Preselection
$MU_1/\$$	Mean	0.00531371	0.000225087	0.00901293
	Var.	3.19915 E(-7)	0	2.0274 E(-7)
$MU_2/\$$	Mean	0.00768015	0.00624466	0.00837344
	Var.	6.2859 E(-7)	0	9.85724 E(-8)
$MU_3/\$$	Mean	0.0131581	0.00156332	0.0080648
	Var.	1.04098 E(-6)	0	4.02153 E(-8)
t_{12}		-54.3326		26.0501
t_{13}		-150.36		43.0121
t_{23}		-94.7983		18.5254
ρ	Mean	13.3108	-5453.53	-0.538288
	Var.	26.2627	0	4.07375
	t	58.0792		-5.96352

Notes

1. A summary of results can be found in Michalewicz (1996).

2. In order to make sure the adherence to the non–negativity constraints (15.7), the pure mutation operator has to be accompanied by some mechanism guaranteeing that $\tilde{q}_{i,k,t} \geq 0$.

3. For a clear formulation of the points of criticism as well as for some suggestions how to fill the election operator with more economic meaning, see Franke (1997) and Birchenhall et al. (1997).

4. Riechmann (2001a) provides an explanation to why election tends to result in stable states.

5. Negative fitness values can arise, so that before entering the selection operator, fitness has to be subject to one of the standard positive transfer mechanisms (Goldberg 1989 and Mitchell 1996).

6. The definition of "elasticity of supply" differs between various textbooks of microeconomic theory. This paper makes use of the definition by Henderson and Quandt (1986), who define the price elasticity of supply as the ratio of relative change in the quantity of supply (nominator) to the relative change of the price (denominator).
For this paper this means that the elasticity of supply for good k, ε_k, is given by $\varepsilon_k = \frac{\partial Q_k / Q_k}{\partial p_k / p_k}$. This definition implies that the change in the quantity is a result of a change in market price. This paper, though, will argue the other way round: A change in price is the reaction on a change in aggregate demand, i.e. a change in quantity.

7. For notational convenience, the period index t is omitted.

8. An important prerequisite for this is the constraint being an equation. As most of the simulations in fact do result in a situation where the budget is fully spent, for (15.2), equality is being assumed.

9. This is a t–test for large populations.

References

Arifovic, J. (1994). "Genetic Algorithm Learning and the Cobweb-Model," *Journal of Economic Dynamics and Control*, 18, 3–28.

Birchenhall, C., N. Kastrinos, and S. Metcalfe (1997). "Genetic Algorithms in Evolutionary Modelling," *Journal of Evolutionary Economics*, 7, 375–393.

Dawid, H. (1999). *Adaptive Learning by Genetic Algorithms*, 2nd ed. Berlin, Heidelberg, New York: Springer.

Franke, R. (1997). "Behavioural Heterogeneity and Genetic Algorithm Learning in the Cobweb Model," Discussion Paper 9, IKSF — Fachbereich 7 — Wirtschaftswissenschaft. Universität Bremen.

Goldberg, D. E. (1989). *Genetic Algorithms in Search, Optimization, and Machine Learning.* Reading, Massachusetts: Addison-Wesley.

Henderson, J. M. and R. E. Quandt (1986). *Microeconomic Theory. A Mathematical Approach*, 3rd ed. New York: McGraw-Hill.

Michalewicz, Z. (1996). *Genetic Algorithms + Data Structures = Evolution Programs*, 3rd ed. Berlin, Heidelberg, New York: Springer.

Mitchell, M. (1996). *An Introduction to Genetic Algorithms.* Cambridge, MA, London: MIT Press.

Riechmann, T. (1998). "Learning How to Learn. Towards an Improved Mutation Operator Within GA Learning Models," In *Computation in Economics, Finance and Engeneering: Economic Systems*, Cambridge, England. Society for Computational Economics.

Riechmann, T. (1999). "Learning and Behavioral Stability — An Economic Interpretation of Genetic Algorithms," *Journal of Evolutionary Economics*, 9, 225–242.

Riechmann, T. (2001a). "Genetic Algorithm Learning and Evolutionary Games," *Journal of Economic Dynamics and Control*, 25(6–7), 1019–1037.

Riechmann, T. (2001b). *Learning in Economics. Analysis and Application of Genetic Algorithms.* Heidelberg, New York: Physica-Verlag.

Chapter 16

PRICE DISCOVERY IN AGENT-BASED COMPUTATIONAL MODELING OF THE ARTIFICIAL STOCK MARKET

Shu-Heng Chen
AI-ECON Research Center, Department of Economics, National Chengchi University
Taipie, Taiwan 116
chchen@nccu.edu.tw

Chung-Chih Liao
Graduate Institute of International Business, National Taiwan University
Taipie, Taiwan 106
liao@aiecon.org

Abstract This paper examines the behavior of price discovery within a context of an *agent-based artificial stock market.* In this model, traders stochastically update their forecasts by searching the *business school* whose evolution is driven by *genetic programming.* We observe how well the market can track the "*true price,*" i.e., the *homogeneous rational expectations equilibrium price* (**HREEP**). It is found that market prices are statistically significantly biased. Furthermore, the pricing error is negatively correlated to *market size.* Excess volatility is also noticeable in these markets.

Keywords: Price Discovery, Homogeneous Rational Expectations Equilibrium, Genetic Programming, Agent-Based Computational Finance, Excess Volatility

Introduction

In their 1999 article "**Frontier of Finance: Evolution and Efficient Markets,**" Farmer and Lo (1999) asserted that *the agent-based computational model of financial markets* "is truly a new frontier whose exploration has just begun." Indeed, the first notable article in this field was not published until 1997, and that is Arthur et al. (1997). Arthur et al. (1997), which is probably the most frequently cited paper in this area, is the fruit of the first economic research project at the Santa-Fe Institute (SFI). This research project was initialized by Brain Arthur and John Holland. The latter is also known as the father of *genetic algorithms*. The project was originally motivated by an attempt to simulate complex adaptive systems, and *the stock market* was later found to be a good starting point.

By all standards, the stock market is qualified to be a complex adaptive system. However, conventional financial models are not capable of demonstrating this feature. On the contrary, the famous *no-trade theorem* shows in the equilibrium how inactive this market can be [Tirole (1982)]. It was therefore invigorating when the SFI artificial stock market infused a colorful life into the model of the stock market. What one can possibly learn from this approach was well summarized in Palmer et al. (1994), which is in fact the first journal publication on the SFI artificial stock market. There are nine observations made in the summary concerning four aspects of the stock market, namely, *price dynamics*, *trading volumes*, *the heterogeneity and complexity of traders' behavior*, and *wealth distribution*. Each of the follow-up studies of Palmer et al. (1994) can be regarded as a contribution to some of these aspects [Taylor (1995), Arthur et al. (1997), Chan et al. (1999), LeBaron et al. (1999), LeBaron (2000), LeBaron (2001)].

Among the four aspects, *price dynamics* is the one under most intensive study. This is not surprising because since the 1960s price dynamics has been the focus of the studies of random walks, the efficient market hypothesis, and market rationality (the rational expectations hypothesis). With the advancement of econometrics, it further became the focus of the study of nonlinear dynamics in the 1980s.[1] Palmer et al. (1994) made three observations on price dynamics. They all concern the relation between the price dynamics generated by the agent-based artificial stock market and the *fundamental value* of stocks. This issue is particularly interesting because in the real market fundamental value is unobservable. Sometimes, it is assumed that the *market price* can track well fundamental value. In the literature, the study of this tracking behavior is known as *price discovery*.[2] Since the true price is never known, empirical studies of price discovery usually start with an estimation of the true price. However, the estimation was mainly based on the market price.[3] Hence, the quality of the implied true price is not assured.

The agent-based artificial stock market is a simulation-based approach. It enables us to simulate a market whose true price is exogenously given and is known. Therefore, measurement is not a problem here. Moreover, the agent-based artificial stock market makes the study of price discovery even more intriguing since the true price can be discovered by a group of interacting agents who are assumed to be "*initially stupid.*"

Despite the fact that *price discovery* is the first three observations made in Palmer et al's summary, few formal results exist in the literature. Attention has been drawn instead to a more advanced question: *can the agent-based artificial stock market replicate the price dynamics observed from the real market?* On price discovery, we have only a hand-waving answer: *an assemblage of initially stupid agents may not be able to track the fundamental value.*

> In sufficiently simple cases — with few agents, or few rules per agent, or a low-variance dividend stream — the agents converge to an equilibrium in which price tracks fundamental value (Eq. 15), volume stays low, and there are no appreciable anomalies such as bubbles or crashes .... On the other hand, in a richer environment, there is no evidence of equilibrium. Although the price frequently stays close to fundamental value, it also displays major upward and downward deviations which may be called bubbles and crashes. (Palmer et al. (1994), p. 272)

Farmer and Lo (1999) had a similar observation.

> Prices do not necessarily reflect "true values;" if we view the market as a machine whose job is to set prices properly, the inefficiency of this machine can be substantial. Patterns in the price tend to disappear as agents evolve profitable strategies to exploit them, but this occurs only over an extended period of time, during which substantial profits may be accumulated and new patterns may appear.

Apart from these very general qualitative descriptions, systematic quantitative treatments are not available in the literature. This chapter is motivated by this need. It reports a series of experiments using a variant of the SFI artificial stock market to investigate whether the agent-based artificial stock market can discover fundamental value in various environments. The rest of the chapter is organized as follows. Section 1 reviews the agent-based artificial stock market employed. Section 2 describes the experimental designs, and Section 3 presents the simulation results, followed by concluding remarks in Section 4.

1. The Analytical and Computational Model

The agent-based artificial stock market employed in this chapter is a variant of the SFI artificial stock market. The SFI artificial stock market is built upon the standard asset pricing model [Grossman (1976), Groassman and Stiglitz (1980)]. We shall first give a brief review of the standard asset pricing model.

1.1. The Standard Asset Pricing Model

Assume that there are N traders in the stock market, and all of them share the same utility function.[4] More specifically, this function is assumed to be a *constant absolute risk aversion* (CARA) utility function,

$$U(W_{i,t}) = -exp(-\lambda W_{i,t}), \tag{16.1}$$

where $W_{i,t}$ is the wealth of trader i at time period t, and λ is the degree of *relative risk aversion*. Traders can accumulate their wealth by making investments. There are two assets available for traders to invest in. One is the riskless interest-bearing asset called *money*, and the other is the risky asset known as *stock*. In other words, at each point in time, each trader has two ways to keep her wealth, i.e.,

$$W_{i,t} = M_{i,t} + P_t h_{i,t}, \tag{16.2}$$

where $M_{i,t}$ and $h_{i,t}$ denote the money and shares in stock held by trader i at time t. Given this portfolio ($M_{i,t}$,$h_{i,t}$), a trader's total wealth $W_{i,t+1}$ is thus

$$W_{i,t+1} = (1+r)M_{i,t} + h_{i,t}(P_{t+1} + D_{t+1}), \tag{16.3}$$

where P_t is the stock price at time period t, D_t the per-share *cash dividends* paid by the companies issuing the stocks, and r the riskless interest rate. D_t follows an exogenous stochastic process as follows:

$$D_t = \bar{d} + \rho D_{t-1} + \xi_t, \tag{16.4}$$

where $\xi_t \sim N(0, \sigma_d^2)$. Given this wealth dynamics, the goal of each trader is to myopically maximize the one-period expected utility function,

$$E_{i,t}(U(W_{i,t+1})) = E(-exp(-\lambda W_{i,t+1}) \mid I_{i,t}), \tag{16.5}$$

subject to

$$W_{i,t+1} = (1+r)M_{i,t} + h_{i,t}(P_{t+1} + D_{t+1}), \tag{16.6}$$

where $E_{i,t}(.)$ is trader i's conditional expectations of W_{t+1} given her information up to t (the information set $I_{i,t}$).

As shown in the appendix, under *CARA* utility and Gaussian distribution for forecasts, trader i's desire demand for holding shares in the risky asset, $h^*_{i,t}$, is linear in the expected *excess returns*:

$$h^*_{i,t} = \frac{E_{i,t}(P_{t+1} + D_{t+1}) - (1+r)P_t}{\lambda \sigma^2_{i,t}}, \tag{16.7}$$

where $\sigma_{i,t}^2$ is the conditional variance of $(P_{t+1} + D_{t+1})$ given $I_{i,t}$. Market clearing price can be found by equating aggregate demand to aggregate supply,

$$\sum_{i=1}^{N} h_{i,t}^*(P_t) = \sum_{i=1}^{N} \frac{E_{i,t}(P_{t+1} + D_{t+1} - (1+r)P_t)}{\lambda \sigma_{i,t}^2} = H. \tag{16.8}$$

Under *full information* and *homogeneous expectations*, it can be shown that the *homogeneous rational expectations equilibrium price* (**HREEP**) is

$$P_t = fD_t + g, \tag{16.9}$$

where $f = \frac{\rho}{1+r-\rho}$ and $g = \frac{1}{r}(1+f)[\bar{d} - \lambda(1+f)\sigma_d^2(\frac{H}{N})]$. Furthermore, in a special case where dividend D_t follows an *iid* process, the ρ in Equation (16.4) is 0, and hence $f = 0$. The **HREEP** in this special case is then simply

$$P_t = \frac{1}{r}[\bar{d} - \lambda \sigma_d^2(\frac{H}{N})] = \frac{1}{r}[\bar{d} - \lambda \sigma_d^2 h], \tag{16.10}$$

where h is shares per capita. Equation 16.10 helps us to examine whether the agent-based artificial stock market can well track fundamental value.

1.2. The Agent-Based Artificial Stock Market

To simulate the agent-based artificial stock market based on the standard asset pricing model, the AI-ECON Research Center at the National Chengchi University developed software known as the AI-ECON artificial stock market (**AIE-ASM**). The AIE artificial stock market differs from the SFI stock market in the computational tool employed. The former applies genetic programming, and the latter genetic algorithms. Generally speaking, genetic programming is more computationally demanding than genetic algorithms but has a greater expression power. Expression power is crucial to agent-based economic modeling because agents' behavior in general can be too complex to be embedded in a finite-dimensional space. In symbolic regression, genetic programming proves to be a useful tool for non-parametric and non-linear modeling.[5] Therefore, it is natural to assume that when agents are adapted to the potentially infinitely complex world, they are not confining themselves to any parametric model but are trying instead to maximize their flexibility with genetic programming.

In **AIE-ASM**, genetic programming is used to model agents' expectations of the price and dividends. Chen and Yeh (2001) gave the details of the implementation. A menu-like introduction to **AIE-ASM** *Ver. 2* can be found in Chen et al. (2002). The simulation conducted in this

Table 16.1. Experimental Designs

CASE	$\bar{d}$	σ_d^2	r	λ	h	HREEP
base-line	10	4	0.1	0.5	1	80
D01	20	4	0.1	0.5	1	180
V01	10	8	0.1	0.5	1	60
V02	10	0	0.1	0.5	1	100
R01	10	4	0.2	0.5	1	40
R02	10	4	0.05	0.5	1	160
R03	10	4	0.025	0.5	1	320
R04	10	4	0.01	0.5	1	800
L01	10	4	0.1	0.75	1	70
L02	10	4	0.1	0.25	1	90
L03	10	4	0.1	0.1	1	96
H01	10	4	0.1	0.5	1.5	70
H02	10	4	0.1	0.5	1.25	75
H03	10	4	0.1	0.5	0.75	85
H04	10	4	0.1	0.5	0.5	90
H05	10	4	0.1	0.5	0.25	95

paper is based on *Ver. 3.* Ver. 3 differs from Ver. 2 mainly in the *terminal set*, i.e., the *variables* on which agents' forecasting is based. Ver. 2 includes in the the terminal set only the time series of price ($\{P_t\}$) and dividends ($\{D_t\}$), whereas Ver. 3 incorporates the time series of price, dividends, and *trading volumes* ($\{V_t\}$).[6]

2. Experimental Design

A series of experiments was conducted to examine whether the agent-based artificial stock market can discover the HREEP in various environments. As shown in Equation 16.10, the HREEP is a function of the following five economic variables: *mean dividends* ($\bar{d}$), *volatility of dividend* (σ_d^2), *interest rate* (r), *degree of risk aversion* (λ), and *shares per capita* (h). Changes in any of these variables can affect the HREEP. The following is a description of the design of our experiments. We first started with a baseline, and then conducted the experiments by changing only one variable at a time.

The baseline is taken from a set of parameter values employed in Chen and Yeh (2001):

$$(\bar{d}, \sigma_d^2, r, \lambda, h) = (10, 4, 0.1, 0.5, 1)$$

Given this baseline, we consider the following changes of each variable.

- *Dividend* ($\bar{d}$): changed from 10 to 20.
- *Volatility* (σ_d^2): changed from 4 to 8 and 0.
- *Interest Rate* (r): changed from 0.1 to 0.2, 0.05, 0.025, and 0.01.
- *Degree of Risk Aversion* (λ): changed from 0.5 to 0.75, 0.25 and 0.1.
- *Shares Per Capita* (h): changed from 1 to 1.5, 1.25, 0.75, 0.5 and 0.25.

This gives us a total of 16 experiments to conduct. Table 16.1 summarizes these 16 settings. For each setting, one can calculate the associated HREEP based on Equation 16.10 (see Table 16.1). The software to conduct these experiments is **AIE-ASM** Version 3. Values of the user-supplied control parameters are listed in Table 16.2.[7]

3. Experimental Results

3.1. General Description

For each setting indicated in Table 16.1, a single run with 5000 trading periods was conducted. The time series plots of the stock price observed in these 16 experiments are given in Figures 16.1 and 16.2. In each diagram, the HREEP is plotted as the horizontal dotted line. A number of observations can be made from these diagrams.

First, *persistent fluctuations.* None of the markets successfully converged to the HREEP (the dotted line). In fact, *persistent fluctuations* are a generic property of the agent-based artificial stock market. To observe persistent fluctuations, one need not introduce any kind of exogenous uncertainty, the endogenous uncertainty induced by agents' speculative behavior would be enough.

Second, *relaxation stability.* While the price series did not converge, they were able to move to a vicinity of the HREEP. For example, in the case of **D01**, we doubled the mean dividends, which changed to 20 dollars from the original 10 dollars. The resultant HREEP is 180. Remember that we initiated all the markets with the price around 100, which means that the asset in this case was just half of its true value,

Table 16.2. Parameters of the Stock Market

The Stock Market	
Share per capita (h)	1.5, 1.25, 1, 0.75, 0.5, 0.25
Initial money supply per capita (m)	100
Interest Rate (r)	0.2, 0.1, 0.05, 0.025, 0.01
Stochastic process (D_t)	IID N(10,4), N(10, 8), N(10,0)
Price adjustment function	tanh
Price adjustment (β_1)	10^{-4}
Price adjustment (β_2)	0.2×10^{-4}
Business School	
Number of faculty members	500
Proportion of trees initiated	
by the full method	0.5
by the grow method	0.5
Function set	$\{+, -, \times, \div, \text{Sin}, \text{Cos}, \text{RExp}, \text{Rlog}, \text{Abs}, \text{Sqrt}\}$
Terminal set	$\{P_t, P_{t-1}, ..., P_{t-10}, P_{t-1} + D_{t-1}, ..., P_{t-10} + D_{t-10}, V_{t-1}, ..., V_{t-10}\}$
Selection scheme	Tournament selection
Tournament size	2
Proportion of offspring trees created	
by reproduction (p_r)	0.1
by crossover (p_c)	0.7
by mutation (p_m)	0.2
Probability of mutation	0.0033
Mutation scheme	Tree mutation
Replacement scheme	Tournament selection
Maximum depth of tree	17
Number of generations	5,000
Maximum in the domain of RExp	1,700
Criterion of fitness (faculty)	MAPE
Evaluation cycle	20
Sample size (MAPE)	10

Table 16.2 (continued)
Parameters of the Stock Market

Traders	
Number of traders	500
Degree of RRA (λ)	0.75, 0.5, 0.25, 0.1
Criterion of fitness (traders)	Increments in wealth
Sample size of σ_t^2	10
Evaluation cycle	1
Sample size	10
Search intensity	5
θ_1	0.5
θ_2	10^{-4}
θ_3	0.0133

but in the subsequent periods, the price moved upward quickly to an area close to the HREEP. In other cases, such as **V01**, **R01**, **R02**, and **R03**, where the initial prices were far away from the HREEP, one can also observe this moving tendency.

There are, however, exceptions, and the most noticeable one is the case of **R04**, in which the interest rate was reduced to only 1%, one-tenth of that in the baseline. The corresponding HREEP is 800. In an earlier experiment, we also initiated this market with prices around 100, and the price in the following periods did move upward as expected, though not high enough. It got stabilized around 300-400 dollars. We could not figure out the reason, so we tried to re-initiate the market from the price around 800, but the price could not get stabilized either. Instead, it dropped down to the same area as we observed before (see Figure 16.1). We are still trying to work out what may cause these *anomalies.*

Third, *persistent bias.* If the price series is persistently lower or higher than the HREEP, then the asset can be over- or under-valued permanently. Normally, this is not what we would expect from an efficient market, but in a few cases, such as **R03**, **R04**, **H04**, and **H05**, we did experience a permanent bias.

3.2. Statistical Analysis

From the price series, we calculated the *mean price* of each series by first using the whole sample $\{P_t\}_{t=1}^{5000}$, and then a subsample consisting

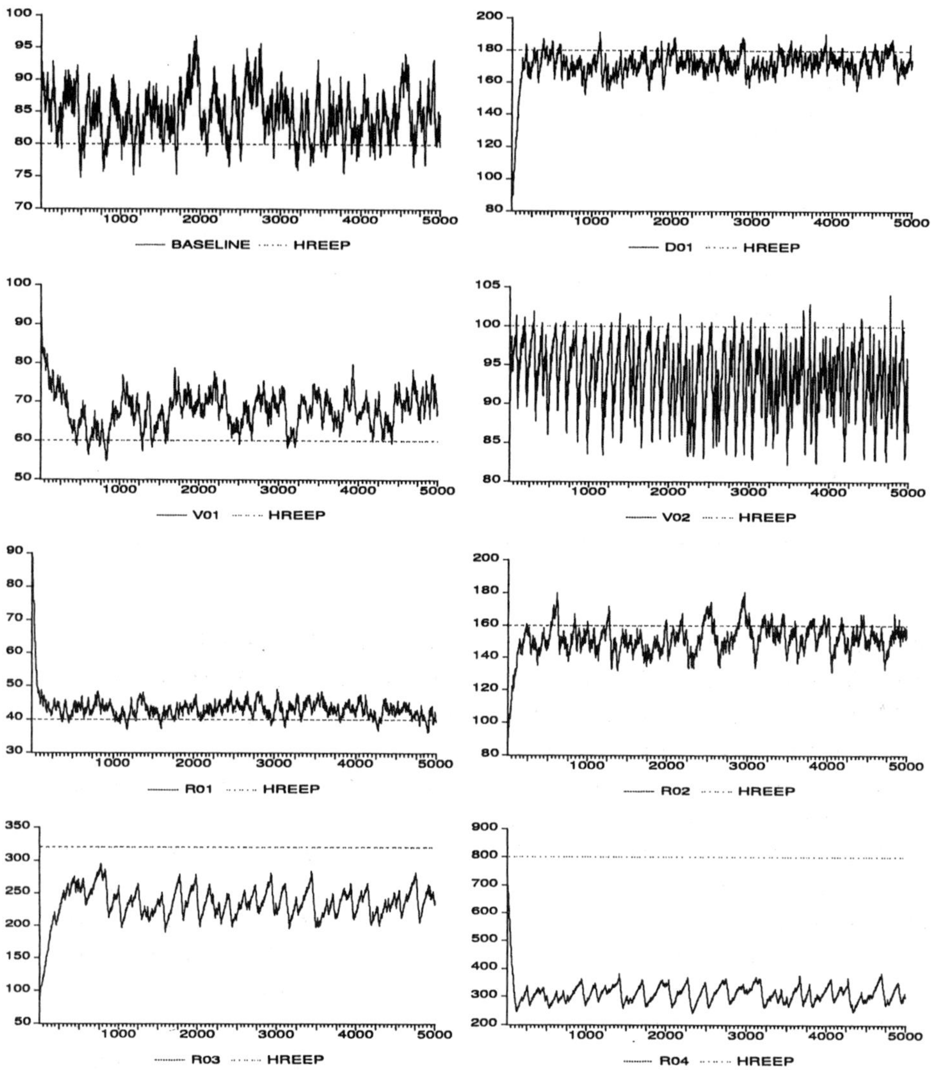

Figure 16.1. Time Series Plots of the Stock Price (I)

of the last 3000 observations, $\{P_t\}_{t=2001}^{5000}$. We denote these two means by $\bar{P}_1$ and $\bar{P}_2$. The reason to report both of these statistics is mainly due to the possible initialization effect. Notice that traders in our market had to learn everything from scratch.

To make these experiments easy to compare, we also calculated the *percentage error* (**PE**) and the *absolute percentage error* (**APE**), defined

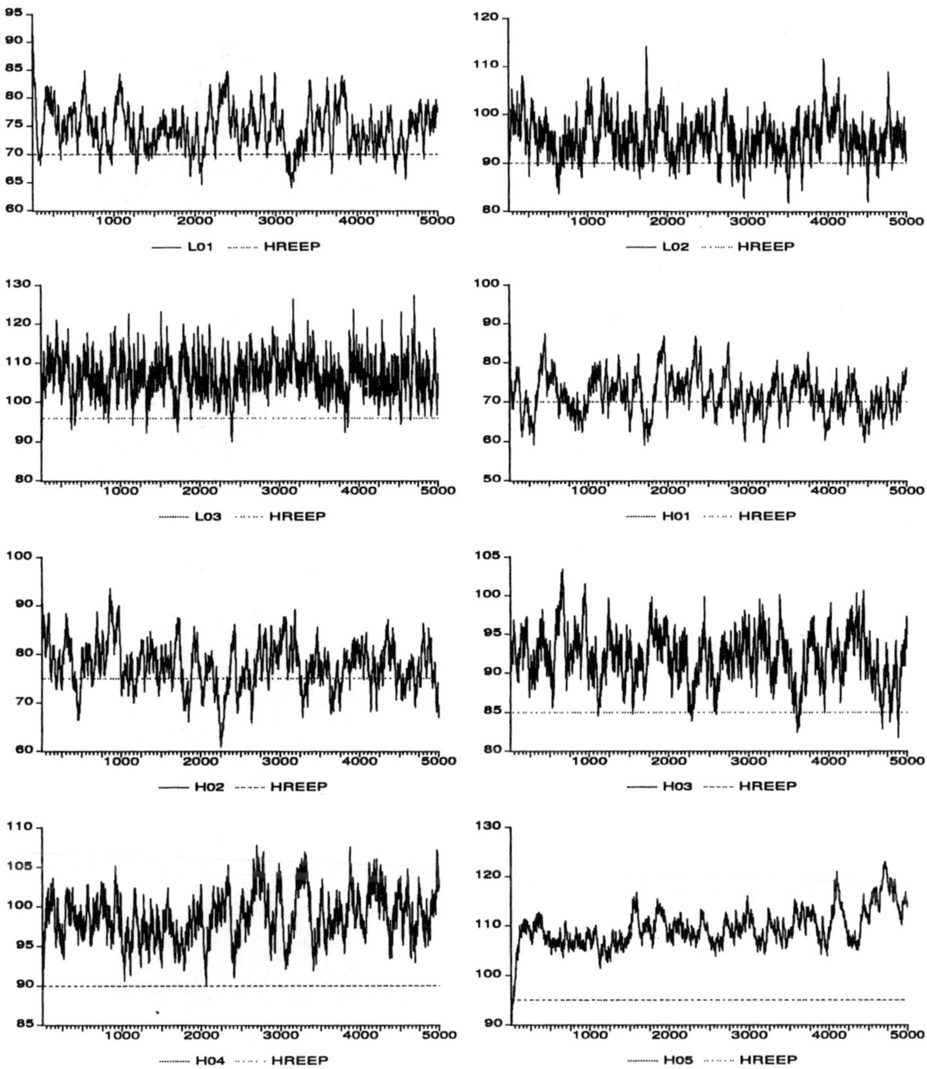

Figure 16.2. Time Series Plots of the Stock Price (II)

as

$$PE_t = \frac{P_t - P^*}{P^*}, \quad APE_t = \mid PE_t \mid, \tag{16.11}$$

where P^* refers to the HREEP. The time series behaviour of the $\{PE_t\}$ is displayed in the left half of Figures 16.3, 16.4, 16.5, and 16.6, and a histogram based on the last 3000 observations of the series is drawn in

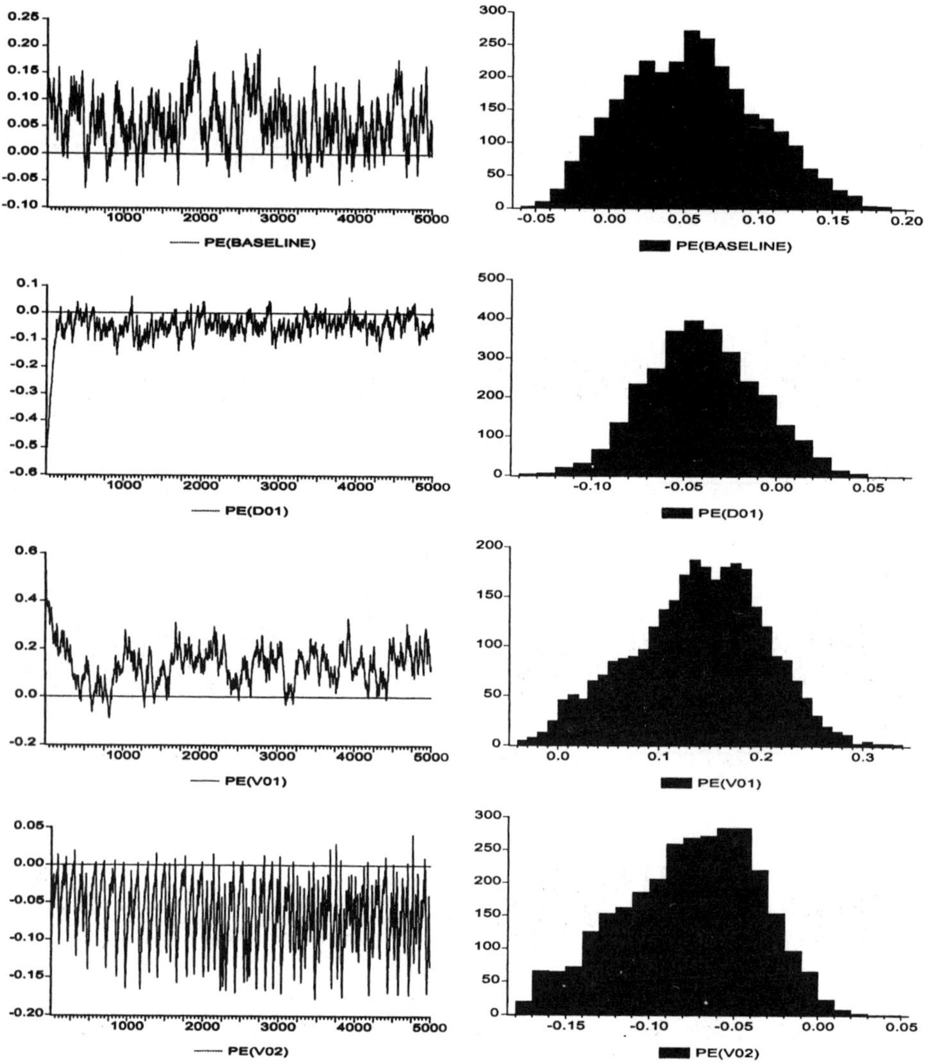

Figure 16.3. Time Series Plots and Histograms of the Percentage Error (I)

the right. Table 16.3 summarizes the mean statistics of the series $\{P_t\}$, $\{APE_t\}$, and $\{PE_t\}$. Here, we report the mean for the whole sample ($\bar{P}_1, MAPE_1$) and for the subsample ($\bar{P}_2, MAPE_2$, MPE_2). The last column gives the *volatility* of the stock price in each market, i.e., the sample standard deviation of $\{P_t\}$, $\sigma(P_t)$.

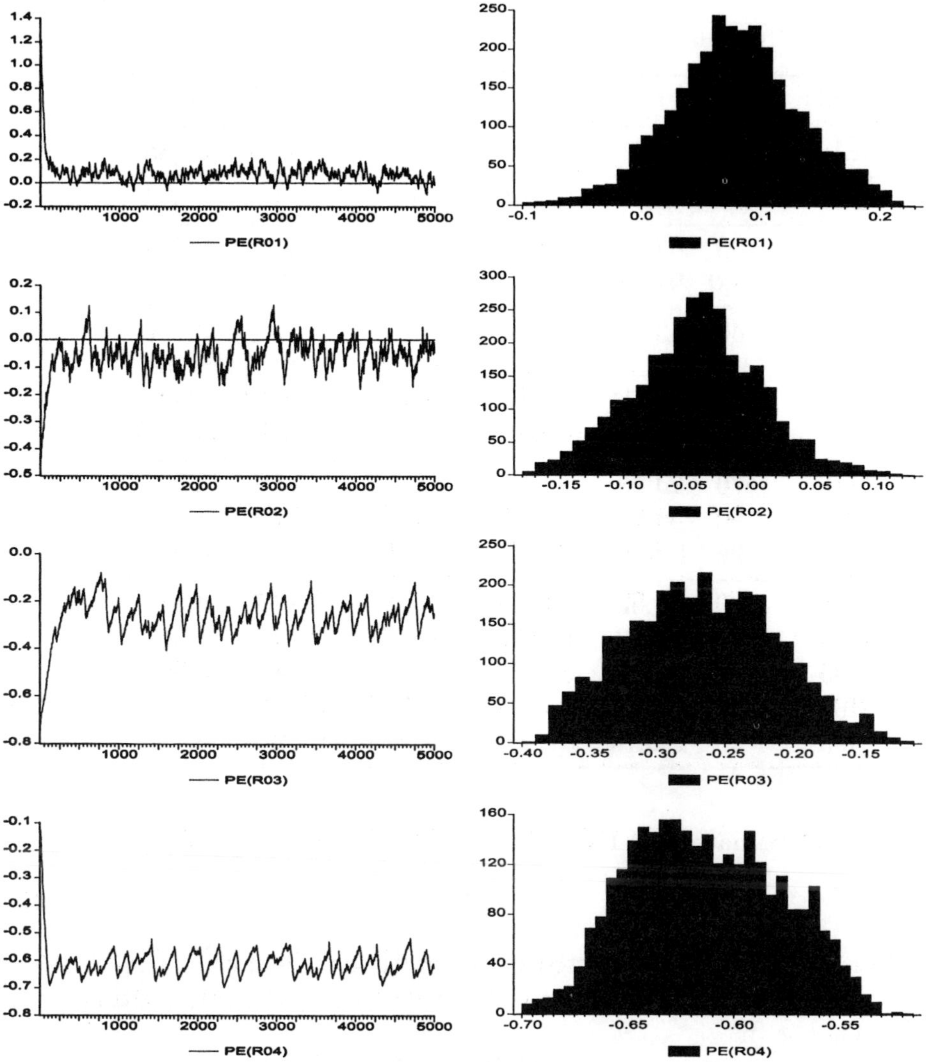

Figure 16.4. Time Series Plots and Histograms of the Percentage Error (II)

These statistics are able to give a picture of the long-term behavior of price discovery of the agent-based artificial market. From the previous subsection, we know that the price series in the limit do not approach the HREEP. Instead, they fluctuated every single day. Therefore, the goal of a long-term analysis is to examine, *on average*, how well the market

Table 16.3. Experimental Results

CASE	P^*	$\overline{P}_2$	$MAPE_1$	$MAPE_2$	MPE_2	σ_P
BASELINE	80	84.41	6.16%	5.88%	5.51%	3.638
D01	180	172.76	5.29%	4.32%	-4.02%	5.512
V01	60	68.29	13.60%	13.87%	13.82%	3.919
V02	100	92.33	6.81%	7.69%	-7.67%	4.062
R01	40	43.11	8.81%	8.19%	7.78%	2.167
R02	160	153.20	6.43%	5.37%	-4.25%	8.058
R03	320	234.75	26.81%	26.64%	-26.64%	17.483
R04	800	309.30	61.08%	61.34%	-61.34%	28.196
L01	70	74.53	7.10%	7.10%	6.47%	3.956
L02	90	94.65	6.16%	5.74%	5.17%	4.103
L03	96	106.60	11.12%	11.09%	11.04%	4.978
H01	70	71.52	6.05%	5.42%	2.17%	4.523
H02	75	76.90	5.78%	5.31%	2.54%	4.530
H03	85	91.55	8.32%	7.78%	7.71%	3.305
H04	90	99.38	9.71%	10.42%	10.42%	3.172
H05	95	110.74	15.36%	16.56%	16.56%	3.722

price approximates the HREEP. Consider Equation 16.12.

$$P_t = P^* + \eta_t \tag{16.12}$$

We are particularly interested in the statistical behavior of η_t. First of all, we want to know whether in the long-run P_t is a good estimator of P^*. That is, if

$$E(P_t) = P^*, \tag{16.13}$$

then we shall expect

$$E(\eta_t) = 0. \tag{16.14}$$

If Equation 16.14 is satisfied, then the mean percentage error (MPE) is also zero. The sixth column of Table 16.3 shows that MPE_2 ranges from -61.34% to 16.56%. The one closest to zero is 2.17% (**H01**). All of them are statistically significantly different from 0. The evidence presented here is strong enough to show that P_t is far from an unbiased estimator of P^*.

Second, we want to examine the distribution of η_t. From Figures 16.3 and 16.4, the pricing error in most of the markets follows a bell-shape

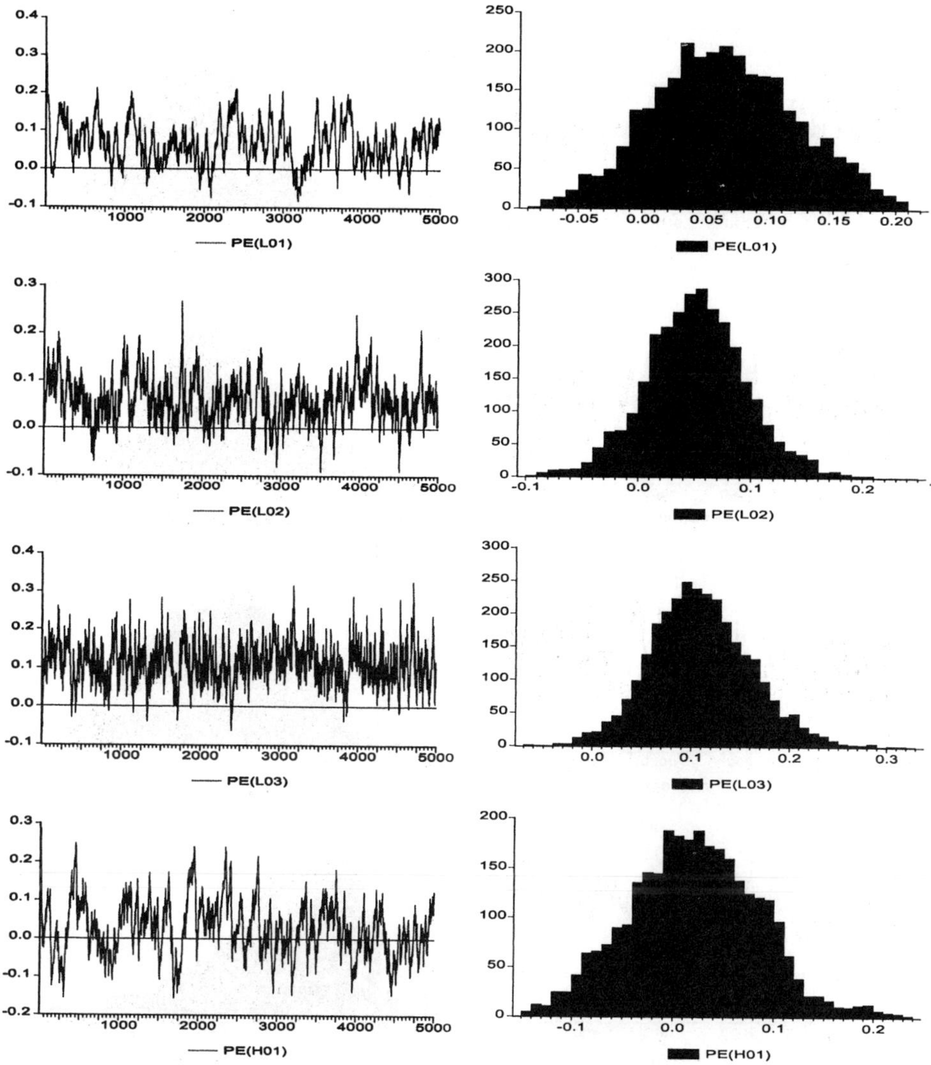

Figure 16.5. Time Series Plots and Histograms of the Percentage Error (III)

distribution. However, few are normally distributed. The *Jarque-Bera* test fails to reject the normality hypothesis for cases **D01**, **R01**, and **R02**. Since P_t is not an unbiased estimator of P^*, and the errors are rarely normally distributed, P_t does not meet the minimum requirements for being a good estimator (predictor) of P^*. In other words, discovering the *true price* is still a daunting task for the agent-based stock market.

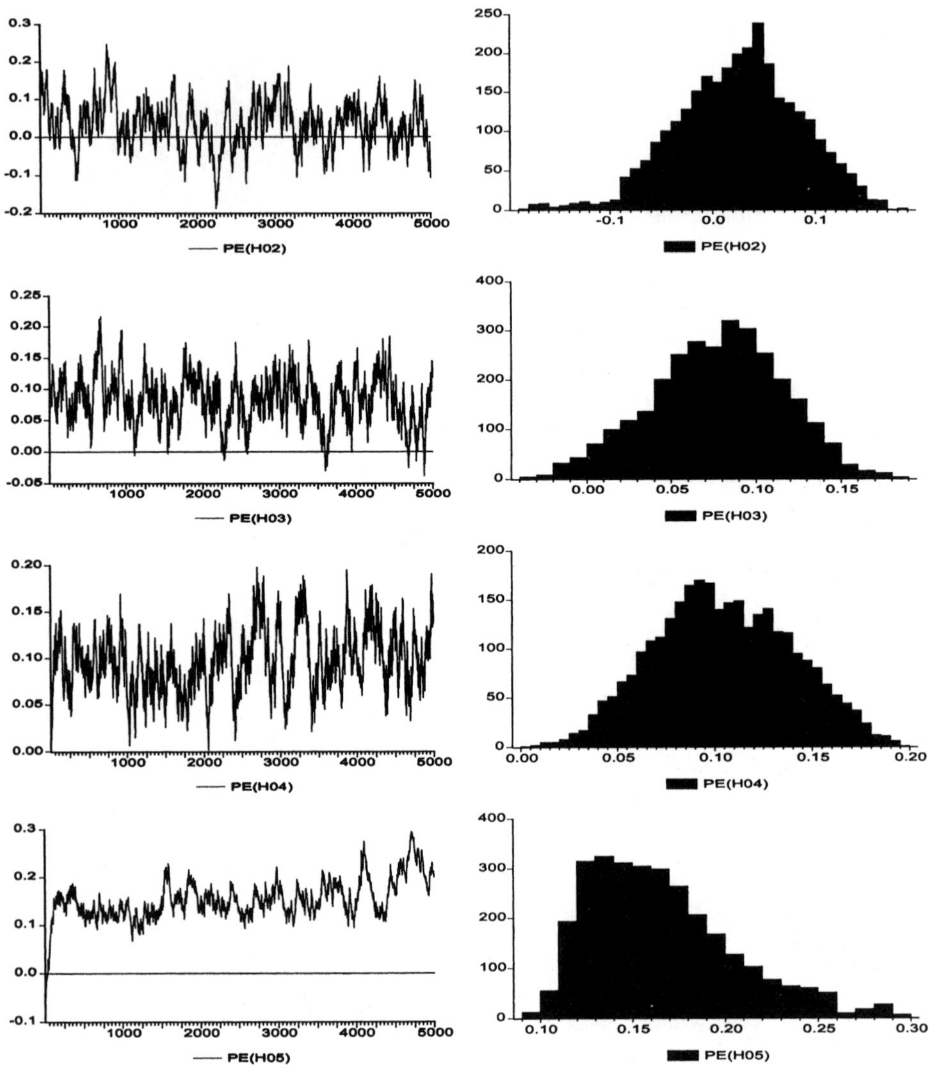

Figure 16.6. Time Series Plots and Histograms of the Percentage Error (IV)

Third, one may wonder if P_t can at least be *asymptotically unbiased* if there is longer evolution of the market, i.e.,

$$\lim_{t \to \infty} E(P_t) = P^*. \tag{16.15}$$

Time matters if it takes time for agents to learn the HREEP. However, whether they are learning anything about the HREEP during the evolution is unclear. If agents are indeed learning the HREEP, then the following inequality is expected.

$$MAPE_2 < MAPE_1. \tag{16.16}$$

Ten out of the sixteen markets satisfy this inequality, but five support the reverse relation. Thus, not all the markets are learning the HREEP. Therefore, longer evolution does not make too much difference, and the property of 16.15 may still fail.

Fourth, what prevents a market from discovering the true price? This is no easy question for it involves the theory of market anomalies, which is beyond the scope of the paper. However, we do like to point out that the agent-based financial market can be a powerful tool to tackle this issue. One can conduct simulations like the ones presented here on an even larger scale and then search for *patterns* for pricing errors. We have just an example on hand. To the bottom of the sixth column of Table 16.3, one can discern a pattern of pricing errors. From cases **H01** to **H05**, shares per capita (h) monotonously decrease from 1.5 to 0.25. If we further take the baseline into account, it is evident that the mean pricing error increases with the decrease in h. The smaller the h, the larger the pricing error. Since shares per capita give the size of the market, the results indicate that *the smaller the market size, the larger the pricing error.* This is nothing surprising for those who study the distinctions between the large companies/thick markets and the small/thin ones. An in-depth analysis in this direction may lend support to the market size hypothesis for anomalies.

Finally, let us look at the last column of Table 16.3, i.e., the standard deviation of the pricing error. The striking feature of these simulations is the overwhelmingly *excessive volatility.* Notice that the **HREEP** determined by Equation 16.10 is a constant and is *deterministic.* Accordingly fluctuations are inconsistent with the rational expectations equilibrium and may be regarded as a kind of *excessive volatility.*[8]

4. Concluding Remarks

Using agent-based artificial stock markets, this chapter examines the possible price deviation from the HREEP when the new-classical assumptions on *homogeneity* and *rationality* are relaxed. Since in the real world these two assumptions can easily fail to hold, our study can be taken as an assessment on how accurately the market price can be approximated (predicted) by the theory (the HREEP) built upon the new-classical simplification. Except for a few extreme cases, the mean percentage error ranges from 4% to 16%. The reader can decide for himself/herself whether it is an acceptable approximation but should be aware of the implications of his/her decision. Notice that the new-classical simplification is not just taken by the standard asset pricing model. It is everywhere in mainstream financial models. If one can accept this approximation and hence other approximations, then one can live with the new-classical simplification, *regardless of how unrealistic it may be* but if one cannot accept this approximation, one may also find it difficult to accept other approximations and may as well start thinking about alternative approaches to finance. We believe that the latter is what we have seen recently and would continue in the years to come.

A financial market is definitely a complex adaptive system,, yet for a long time it has been treated merely as a simple stochastic dynamic system. Many generic properties of this complex adaptive system have been mistaken as market anomalies in mainstream finance. The advent of agent-based computational finance provides an opportunity to change. Nonetheless, agent-based computational finance is still in an embryonic stage. So far, agent-based financial markets have considered only one risky asset. Multi-asset agent-based financial markets have not come into existence. An extension of the research to cover multi-asset markets would be an important step towards an agent-based version of portfolio theory. Of course, to reach maturity, the current artificial stock markets would have to be augmented with futures and options.

A lot of work needs to be done on the *financial agents* too. At the present stage, agents are largely homogeneous. They are assumed to have the same *preferences*, which are exogenously given, and to use the same *language* in their perception of the world. To have a more realistic description of agents, both assumptions should be relaxed. First, agents' preferences should maintain a degree of diversity and can be endogenously determined via an evolutionary process.[9] Second, language varies. It can be crispy or fuzzy.[10] It can be quantitative or qualitative. It can be model-based or data-driven. It can be continuous or discrete. Adding these varieties would introduce us to a novel approach, namely, an agent-based approach to *risk*, which is the heart of finance.

Acknowledgments

The chapter was based on our paper presented at the 7th International Conference of the Society for Computational Economics on Computing in Economics and Finance (**SCE'2001**). We are grateful to Thomas Lux, Chia-Hsuan Yeh, and Chih-Chi Ni for their valuable comments.

Appendix: Homogeneous Rational Expectations Equilibrium Price (HREEP)

In this appendix, we shall provide proof of Equations (16.7) and (16.9), i.e., the determination of the *optimal portfolio* and the *homogeneous rational expectations equilibrium price* (**HREEP**). For Equation (16.9), without loss of generality, we shall only prove the case where $\rho = 0$. When $\rho = 0$, the **HREEP** is simply Equation (16.10).

To do so, we need the *iid normality* assumption, i.e.,

$$D_t \ \ iid \sim N(\bar{d}, \sigma_d^2). \tag{16.A.1}$$

Given the assumption 16.A.1, $W_{i,t}$ is also conditionally normally distributed, conditional on $I_{i,t-1}$. As a result, using the moment generating function of the normal distribution, we have

$$\begin{aligned} E_{i,t}(U(W_{i,t+1})) &= E(-\exp(-\lambda W_{i,t+1})|I_{i,t}) \\ &= -\exp\{-\lambda E_{i,t}(W_{i,t+1}) + \frac{\lambda^2}{2} Var_{i,t}(W_{i,t+1})\} \\ &= -\exp\{-\lambda[E_{i,t}(W_{i,t+1}) - \frac{\lambda}{2} Var_{i,t}(W_{i,t+1})]\} \end{aligned} \tag{16.A.2}$$

Hence maximizing $E_{i,t}[U(W_{i,t+1})]$ is equivalent to maximizing

$$E_{i,t}(W_{i,t+1}) - \frac{\lambda}{2} Var_{i,t}(W_{i,t+1}). \tag{16.A.3}$$

But, since

$$\begin{aligned} W_{i,t+1} &= (1+r)M_{i,t} + h_{i,t}(P_{t+1} + D_{t+1}) \\ &= (1+r)W_{i,t} + h_{i,t}(P_{t+1} + D_{t+1} - (1+r)P_t), \end{aligned}$$

$$\begin{aligned} E_{i,t}(W_{i,t+1}) &= E_{i,t}[(1+r)W_{i,t} + h_{i,t}(P_{t+1} + D_{t+1} - (1+r)P_t)] \\ &= (1+r)W_{i,t} + h_{i,t}E_{i,t}(P_{t+1} + D_{t+1}) - h_{i,t}(1+r)P_t, \end{aligned}$$

and

$$\begin{aligned} Var_{i,t}(W_{i,t+1}) &= Var_{i,t}[(1+r)W_{i,t} + h_{i,t}(P_{t+1} + D_{t+1} - (1+r)P_t)] \\ &= h_{i,t}^2 Var_{i,t}(P_{t+1} + D_{t+1}). \end{aligned} \tag{16.A.4}$$

The necessary condition of maximizing 16.A.3 is

$$E_{i,t}(P_{t+1} + D_{t+1}) - (1+r)P_t - \lambda h_{i,t} Var_{i,t}(P_{t+1} + D_{t+1}) = 0. \tag{16.A.5}$$

Solving Equation (16.A.5), we have

$$h_{i,t}^* = \frac{E_{i,t}(P_{t+1} + D_{t+1}) - (1+r)P_t}{\lambda Var_{i,t}(P_{t+1} + D_{t+1})}. \tag{16.A.6}$$

Furthermore, the market clearing condition implies

$$\sum_{i=1}^{N} h_{i,t}^* = \sum_{i=1}^{N} \frac{E_{i,t}(P_{t+1} + D_{t+1}) - (1+r)P_t}{\lambda Var_{i,t}(P_{t+1} + D_{t+1})} \tag{16.A.7}$$

$$= N \frac{E_t(P_{t+1} + D_{t+1}) - (1+r)P_t}{\lambda Var_t(P_{t+1} + D_{t+1})} = H. \tag{16.A.8}$$

From Equations (16.A.7) and (16.A.8), we impose the second assumption: *homogeneity*. Rearranging Equation (16.A.8), we have

$$\begin{aligned} \frac{H}{N} &= \frac{E_t(P_{t+1}) + E_t(D_{t+1}) - (1+r)P_t}{\lambda Var_t(D_{t+1})} \\ &= \frac{\bar{d} - rP_t}{\lambda \sigma_d^2}. \end{aligned} \tag{16.A.9}$$

Now, it is clear that

$$P_t = \frac{1}{r}(\bar{d} - \lambda \sigma_d^2 \frac{H}{N}). \tag{16.A.10}$$

Notes

1. See Pagan (1996) and Campbell et al. (1997) for a nice review of the field.
2. Price discovery concerns two things: the differential reaction of different markets to new information, and the rate at which new information is incorporated into price.
3. For example, in financial econometrics, the true price is estimated by *state-space models*. See Moody and Wu (1998).
4. While agents in agent-based economic models can be heterogeneous in every dimension, the current literature has not considered the heterogeneity in preference (the utility function).
5. Earlier applications of genetic programming to econometrics can be found in Koza (1992), Schmertmann (1996), Szpiro (1997a), and Szpiro (1997b). Chen (2001) provides a review of this early-stage development.
6. In fact, Ver. 3 also allows users to input other variables which are not endogenously generated from the artificial market.
7. For details of these parameters, see Chen and Yeh (2001) and Chen et al. (2002).
8. The excess volatility of stock prices was first noted in Shiller (1981) and LeRoy and Porter (1981). By excess volatility Robert Shiller means that the variability of price movements is too large to be justified in terms of efficient market models, given the relatively low variability of fundamentals and the correlation of price with fundamentals. Campbell and Shiller (1988) found that the standard deviation of stock returns should be between a quarter and a half of what it is. Chen and Liao (2002) followed the procedure suggested by Shiller to re-calculate the variability of fundamentals, and compared it with market volatility. They found that while excess volatility existed in some of the cases, it was not so excessive as Shiller observed from the real market.
9. This relaxation concerns the debate on the significance of the logarithmic utility function in the determination of survivability of investors. See Blume and Easley (1992), Sandroni (2000), Sciubba (1999), and also the chapter by Caldarelli et al. in this volume.
10. It is only fairly recently that people have started to realize the significance of the fuzzy classifier system in agent-based financial modeling. It enhances agents' adaptation to complexity with *natural language*. See Tay and Linn (2001).

References

Arthur, W. B., J. Holland, B. LeBaron, R. Palmer , and P. Tayler, (1997). "Asset Pricing under Endogenous Expectations in an Artificial Stock Market," In Arthur, W. B., Durlauf, S. and Lane, D. (Eds.), *The Economy as an Evolving Complex System II*, Reading, pp. 15–44. MA: Addison-Wesley.

Blume, E. and E. Easley (1992). "Evolution and Market Behavior," *Journal of Economic Theory*, 58, 9–40.

Campbell, J. and R. Shiller (1988). "Stock Prices, Earnings, and Expected Dividends," *Journal of Finance*, 43, 661–676.

Campbell, J., A. Lo, and A. C. Mackinlay (1997). *The Econometrics of Financial Markets*, NJ: Princeton University Press.

Chan, N. T., B. LeBaron, A. W. Lo, and T. Poggio (1999). "Agent-Based Models of Financial Markets: A Comparison with Experimental Markets," Unpublished Working Paper, MIT Artificial Markets Project, MIT, MA.

Chen S.-H. (2001). "On the Relevance of Genetic Programming to Evolutionary Economics," in Y. Aruka (ed.), *Evolutionary Controversies in Economics: A New Transdisciplinary Approach*, pp. 135–150. Tokyo: Springer-Verlag.

Chen, S.-H. and C.-C. Liao (2002). "Excess Volatility in Agent-Based Artificial Stock Markets," AI-ECON Research Center Working Paper, National Chengchi University.

Chen, S.-H. and C.-H. Yeh (2001). "Evolving Traders and Business School with Genetic Programming: A New Architecture of the Agent-Based Artificial Stock Market," *Journal of Economic Dynamics and Control*, 25, 363–393.

Chen S.-H., C.-H. Yeh, and C.-C. Liao (2002). "On AIE-ASM: Software to Simulate Artificial Stock Markets with Genetic Programming," in S.-H. Chen (ed.), *Evolutionary Computation in Economics and Finance*, pp. 107–122. Heidelberg:Physica-Verlag.

Farmer, J. D. and A. W. Lo (1999). "Frontiers of Finance: Evolution and Efficient Markets," *Proceedings of the National Academy of Sciences*, 96, 9991–9992.

Grossman, S. (1976). "On the Efficiency of Competitive Stock Markets Where Traders Have Diverse Information," *Journal of Finance*, 31, 573–585.

Grossman, S. and J. Stiglitz (1980). "On the Impossibility of Informationally Efficient Markets," *American Economic Review*, 70, 393–408.

Koza J. R. (1992). "A Genetic Approach to Econometric Modeling," in P. Bourgine, and B. Walliser (eds.), *Economics and Cognitive Science*, pp. 57–75. Pergamon Press.

LeBaron, B. (2000). "Agent Based Computational Finance: Suggested Reading and Early Research," *Journal of Economic Dynamics and Control*, 24, 679–702.

LeBaron, B. (2001). "Evolution and Time Horizons in an Agent Based Stock Market," *Macroeconomic Dynamics*, 5, 225–254.

LeBaron, B., W. Arthur, and R. Palmer (1999). "Time Series Properties of an Artificial Market," *Journal of Economic Dynamics and Control*, 23, 1487–1516.

LeRoy, S. and R. Porter (1981). "The Present Value Relation: Tests Based on Variance Bounds," *Econometrica*, 49, 555–577.

Moody, J. and L. Wu (1998). "High Frequency Foreign Exchange Rates: Price Behavior Analysis and 'True Price' Models," In Dunis, C. and Zhou, B. (eds.), *Nonlinear Modelling of High Frequency Financial Time Series*, NY: John Wiley & Sons.

Pagan, A. (1996). "The Econometrics of Financial Markets," *Journal of Empirical Finance*, 3, 15–102.

Palmer, R. G., W. B. Arthur, J. H. Holland, B. LeBaron, and P. Tayler (1994). "Artificial Economic Life: A Simple Model of a Stock Market," *Physica D*, 75, 264–274.

Sandroni, A. (2000). "Do Markets Favor Agents Able to Make Accurate Prediction?" *Econometrica*, 68(6), 1303–1341.

Schmertmann, C. P. (1996). "Functional Search in Economics Using Genetic Programming," *Computational Economics*, 9(4), 275–298.

Sciubba, E. (1999). "The Evolution of Portfolio Rules and the Capital Asset Pricing Model," DAE Working Paper no. 9909, University of Cambridge.

Shiller, R. (1981). "Do Stock Prices Move Too Much to Be Justified by Subsequent Changes in Dividends?" *American Economic Review*, 71, 421–436.

Szpiro, G. G. (1997a). "A Search for Hidden Relationships: Data Mining with Genetic Algorithms," *Computational Economics*, 10(3), 267–277.

Szpiro, G. G. (1997b). "Forecasting Chaotic Time series with Genetic Algorithms," *Physical Review E*, 2557–2568.

Tay, N. and S. Linn (2001). "Fuzzy Inductive Reasoning, Expectation Formation and the Behavior of Security Prices," *Journal of Economic Dynamics and Control*, 25, 321–361.

Taylor, P. (1995). "Modeling Artificial Stock Markets Using Genetic Algorithm," In Goonatilake, S. and Treleaven, P. (Eds.), *Intelligent Systems for Finance and Business*, 365–376. New York, NY: Wiley.

Tirole, J. (1982). "On the Possibility of Speculation under Rational Expectations," *Econometrica*, 50, 1163–1182.

Chapter 17

INDIVIDUAL RATIONALITY AS A PARTIAL IMPEDIMENT TO MARKET EFFICIENCY

Allocative Efficiency of Markets with Smart Traders

Shu-Heng Chen
AI-ECON Research Center, Department of Economics, National Chengchi University
Taipei, Taiwan 11623
chchen@nccu.edu.tw

Chung-Ching Tai
AI-ECON Research Center, Department of Economics, National Chengchi University
Taipei, Taiwan 11623
elliot@aiecon.org

Bin-Tzong Chie
AI-ECON Research Center, Department of Economics, National Chengchi University
Taipei, Taiwan 11623
chie@aiecon.org

Abstract In this chapter we conduct two experiments within an agent-based double auction market. These two experiments allow us to see the effect of learning and smartness on price dynamics and allocative efficiency. Our results are largely consistent with the stylized facts observed in experimental economics with human subjects. From the amelioration of price deviation and allocative efficiency, the effect of learning is vividly seen. However, smartness does not enhance market performance. In fact, the experiment with smarter agents (agents without a quote limit) results in a less stable price dynamics and lower allocative efficiency.

Keywords: Agent-Based Double Auction Markets, Genetic Programming, Quote Limits, Alpha Value, Allocative Efficiency

Introduction

In their seminal paper "**Allocative Efficiency of Market with Zero-Intelligence Trader**," Gode and Sunder posed an interesting question. *How much intelligence is required of an agent to achieve human-level trading performance?* The answer, as it first appeared, is a little surprising: *little.* How little? To make that message clearer, they called their agents *zero-intelligence* agents. These agents, when assigned a bargaining position in a standard double auction market, were simply bidding or asking randomly as if they had no capability to extract any useful information from the market. While these agents *individually* cannot bargain in an intelligent manner, their interactions via the market did *collectively* result in a near-100% allocative efficiency. They attributed this magic to Adam Smith's *invisible hand.*

> "Adam Smith's invisible hand may be more powerful than some have thought; it can generate *aggregate rationality* not only from individual rationality, but also from *individual irrationality.*" (Gode & Sunder, 1993, p.119, Italics added.).

It may be delicate to show that aggregate rationality does not rest upon individual rationality in some economic contexts. However, that does not include Gode and Sunder's case, and as a matter of fact, in their case it is *trivial* to show that allocative efficiency (aggregate rationality) does not rest upon individual rationality. To see this, let us simply assume that all agents are *truth-tellers.* In this regard, buyers would bid with their redemption values and sellers would ask with their unit costs. Obviously, allocative efficiency would then automatically be 100%.

Gode and Sunder did seem to assume that individual rationality would necessarily imply aggregate rationality (see the quotation above). They might also suppose that the allocative efficiency would be near 100% if all traders are *smart enough.* Consequently, their focus was to look for the *minimum* intelligence level which can generate aggregate rationality. What, however, may surprise them is that this minimum level is so low that you only need *dumb agents.*

Our picture is different: we do not assume that individual rationality necessarily implies aggregate rationality. Therefore, instead of the minimum level, we are looking at the other direction. We already have argued that the minimum level is not a problem at all: a group of innocent (honest) traders would result in a 100% allocative efficiency. What may make things uncertain is the case when traders are no longer innocent, but rather sophisticated. Hence, for us, the interesting question to ask is: *would smart or smarter traders reduce allocative efficiency?* As we shall see in this paper, the answer is positive.

In this paper we conduct agent-based simulations of DA markets with different intelligence levels of traders. Traders in one case are endowed with more space to act than traders in the other case. This extra space makes more sophisticated trading strategies possible to emerge, and hence can make traders in one case *potentially smarter* than traders in the other case.[1] Traders who are given this favor are called the *smart trader*, and traders who are not are called the *mediocre trader*. We then examine the allocative efficiency achieved within these two different setups. It is found that markets composed of mediocre traders realized 96% of the potential social surplus, whereas markets composed of smart traders realized only 88% of it.

Our finding for smart traders can be compared to the one Gode and Sunder saw for their zero-intelligence traders. While Gode and Sunder's finding shows that the invisible hand is more powerful than we thought, our finding shows the opposite. Moreover, to attain a higher allocative efficiency, the privilege given to traders should be deprived, and this deprivation can be interpreted as a kind of an intervention used to protect the market. Therefore, one way to summarize our study is as the following: *Smart agents do not necessarily bring goodness to the market. One purpose of regulation is to annihilate the evil-side of smartness.*

The rest of the paper is organized as follows. Section 1 shall briefly review the agent-based double auction markets used in this paper. Section 3 proposes experimental designs and measures of market performance. Section 4 presents and analyzes the simulation results, as Section 5 provides the concluding remarks.

1. Agent-Based Double Auction Markets: AIE-DA

Agent-based double auction (**DA**) markets were first seen in Dawid (1999), who applied the *single-population genetic algorithm* (**SGA**) to evolve traders' bids and asks, but not bargaining strategies per se. Chen (2000) proposed an agent-based DA market which is suitable for studying the evolution of bargaining strategies and which can be implemented with the software **AIE-DA**, developed by the **AI-ECON Research Center**. Chen (2001) and Chen, Chie, and Tai (2001) prepared a documentation accompanying this software, which is written in the language of **Delphi**, and is largely motivated by *object-oriented program-*

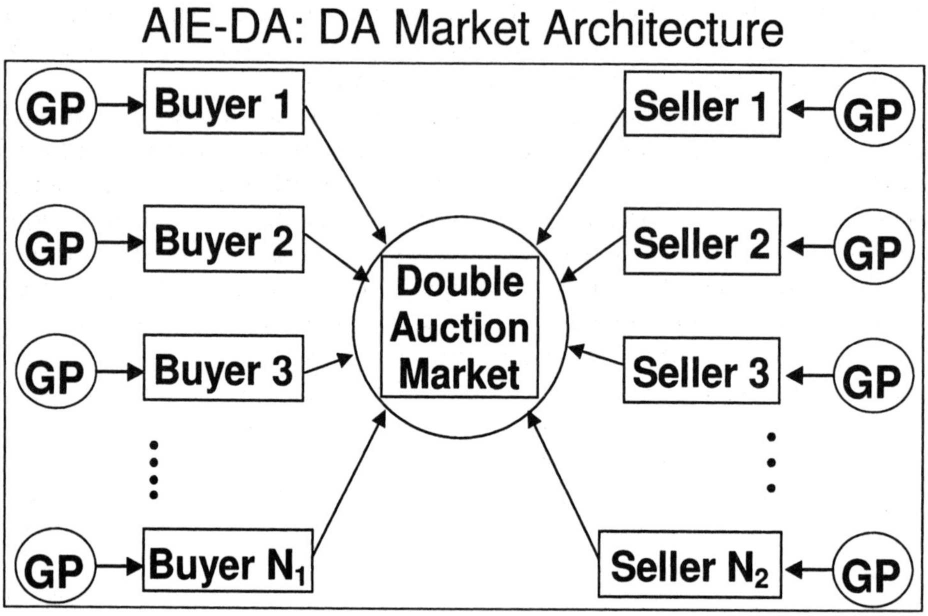

Figure 17.1. The AIE-DA Architecture: Multi-Population Genetic Programming.

ming (**OOP**). The experiments conducted in this paper will be based on this software.

All buyers and sellers in AIE-DA are *artificial adaptive agents* as described in Holland and Miller (1991). Each artificial adaptive agent is built upon *genetic programming.* The architecture of genetic programming used in AIE-DA is what is known as *multi-population genetic programming* (**MGP**). Briefly, we view or model an agent as *a population of bargaining strategies.*[2] Genetic programming is then applied to *evolving* each population of bargaining strategies. In this case, a society of bargaining agents consists of many populations of programs. This architecture is shown in Figure 17.1.

The evolution of bargaining strategies for each agent proceeds as follows. First, consider a counter called *generation.* At generation t, each agent is assigned K bargaining strategies, collectively denoted by Gen_t, whose determination will be explained later. For each trading period h, the agent takes a random strategy I as follows: $I \sim Uniform[1, K]$. At the end of the trading period, the profits of the chosen strategy i ($i \in [1, K]$) will be recorded as $\pi_{i,h}$. If strategy i is not chosen, then its profits will be counted as zero. After every H periods of trading, these k

strategies will be revised and renewed by standard genetic programming. The *fitness* is the mean profits:

$$\pi_i = \frac{\sum_{h=1}^{H} \pi_{i,h}}{\sum_{h=1}^{H} 1_{i,h}}, \tag{17.1}$$

where $1_{i,h}$ is 1 if i is selected at period h, otherwise it is 0.[3] The revision and renew process will generate a new generation of K strategies, and at this point, the counter shall move to generation $t+1$. The new generation of K strategies will be denoted by Gen_{t+1}. This revision and renew procedure is summarized as follows.

$$Gen_{t+1} \Longleftarrow \underbrace{[Reproduction] \bigvee [Crossover] \bigvee [Mutation]}_{Genetic\ \ Operators} \bullet (Gen_t) \tag{17.2}$$

As for the initial generation Gen_0, it will be generated by the *ramped half-and-half* method. This process will continue when t hits a prespecified number T. For the simulations in this paper, K is set to 50, H is 100, and T is 100.[4]

2. Experimental Designs

2.1. Quote Limit

To test the effect of *smartness* on allocative efficiency, a quote limit (upper bound for bid and lower bound for ask) is imposed in one experiment, but not the other. The purpose of a quote limit is to prevent buyers and sellers from bidding and asking an unprofitable price, and so a buyer cannot bid a price higher than his redemption value, and a seller cannot not ask a price lower than his unit cost. One may wonder why traders would be so *foolish* to bid or ask a price outside the quotation limit. Would they definitely make a loss? The answer is negative. Quoting a price outside the limit makes a trader's offer more lucrative, and enhances the chance of making a deal. Once the deal is won, depending on the trading mechanism, the trader may actually fulfill the transaction with a price which is different from his original quote.

To show an example, consider the **AURORA** computerized trading system developed by the Chicago Board of Trade. AURORA rules stipulate that only the holder of current bid (**CB**) or current ask (**CA**) is allowed to trade if CB $\geq$ CA.[5] By the AURORA rule, the actual transaction price (P) to fulfill a transaction will be somewhere between the current ask and the current bid. Let us assume the middle of them,[6]

Table 17.1. Values of Control Parameters for Genetic Programming

Parameter	Value
Number of Generations (T)	100
Population Size (K)	50
Evaluation Cycle (H)	100
Fitness Function	Mean Profits
Elitist Strategy	On
Number of Elites	1
Number of Strategies Chosen in Each Generation	100
Selection Scheme	Tournament
Tournament Size	5
Mutation Rate	0.05
Tree Mutation	0.1
Point Mutation	0.9
MaxDepth	17

i.e.,

$$P = \frac{CA + CB}{2}. \tag{17.3}$$

It is therefore clear that $CA \leq P \leq CB$, and so neither side would have to fulfill the transaction with their original quotes. This explains why the trader may take a risk of quoting a price outside the limit. This aggressive quotation should therefore be considered as a strategic behavior instead of a foolish one. Furthermore, this kind of aggressive bargaining strategy can actually emerges from the evolution of DA markets (Chen and Chie 2001).

2.2. Designs of Markets and Traders

Once the meaning of the quote limit is clear, we run two series of experiments. The first series of experiments are conducted without the quote limit, whereas the other series are conducted with it. This is essentially to say that we do not allow traders in the second series of experiments to evolve and develop aggressive bargaining strategies, while they are free to do so in the first series. By this limit, the "smartness" of the traders in the second series is restricted, whereas traders in the first are not. We therefore call traders in the first series *smart traders*, and traders in the second series *mediocre traders.*[7]

Twenty experiments were conducted for each series. In each experiment a token-value table, which is randomly generated, is applied to both

Table 17.2. List of Primitives, the Terminal Set and Function Set

Terminal Set	Function Set
Highest Token (**HT**) Next Token (**NT**) Lowest Token (**LT**)	Arithmetic Operators (+, −, ×, ÷)
Current Ask (**CASK**) Current Bid (**CBID**)	Absolute Value (**abs**)
Time Left Before the Termination of a Trading Period (**T1**) and Time Elaspe Since the Last Successful Trading (**T2**)	Logarithmic and Exponential Functions (**exp, log**)
The Average, Minimum and the Maximum Price of the Previous Trading Period (**PAvg, PMin, PMax**)	Trigonometric Functions (**sin, cos**)
The Average, Minimum and the Maximum Bid of Previous Trading Period (**PAvgBid, PMinBid, PMaxBid**)	Logical Operators (**If-Then-Else, If-Bigger-Then-Else**)
The Average, Minimum and the Maximum Ask of the Previous Trading Period (**PAvgAsk, PMinAsk, PMaxAsk**)	Extreme Operators (**max, min**)
Quiet (**Pass**)	Comparison Operator (>)
Ephemeral Random Constants (**C**)	

series. This table allows for four buyers and four sellers each with four units of tokens to trade. The 20 token-value tables (markets) generated are depicted in Figure 17.2. In each market, buyers and sellers' trading strategies evolved with genetic programming, whose control parameters are specified in Table 17.1 and 17.2.

2.3. Measures of Market Performance

In experimental economics, two measures are frequently used for market performance. One is the so-called *alpha value*, and the other is the *efficiency ratio*. The alpha value takes the *competitive equilibrium price* (p^*) as a benchmark, and attempts to measure the deviation degree between *market prices* and the *competitive equilibrium price*. More pre-

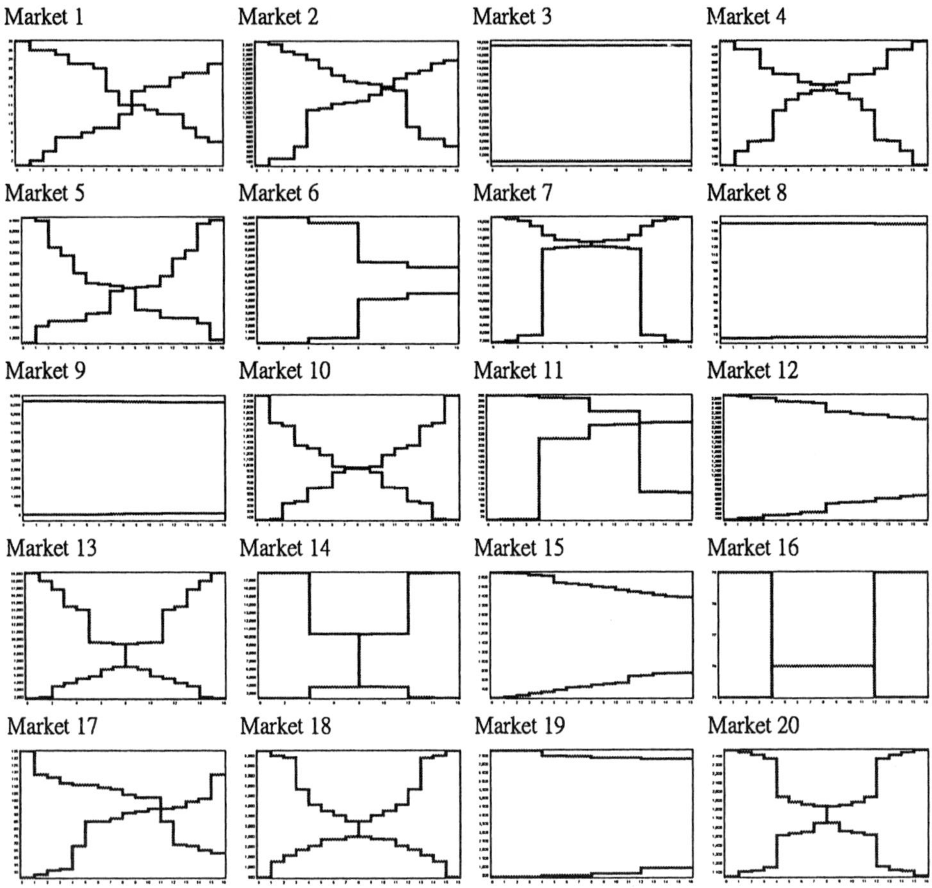

Figure 17.2. 20 Different Markets Used in the Experiments.

cisely, it is defined as follows,

$$\alpha = \frac{\sqrt{\frac{\sum_{i=1}^{n}(P_i - P*)^2}{n}}}{P*}, \tag{17.4}$$

where n is the number of transaction made in each trading period, and P_i is the market price of the ith transaction. One technical issue involved in this definition is the location of P^*. The meaning of competitive equilibrium price is clear when the demand and supply curves intersect at a single point, e.g., Markets 10 and 16 (See Figure 17.2). However, it is less clear when they intersect at an interval (Markets 7 and 20) or have no intersection at all (Markets 3 and 9). For the former case, we

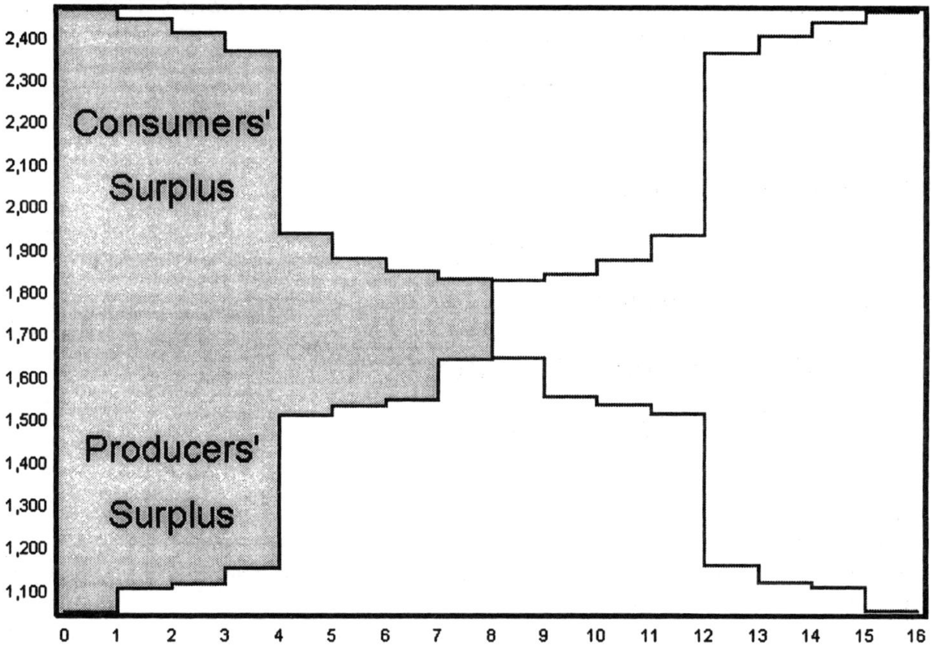

Figure 17.3. Consumers' Surplus & Producers' Surplus.

would define P^* as the midpoint of the intersecting interval, while for the latter case it is the midpoint between the lowest redemption value and the highest unit cost.

As to the second measure, we first take the sum of consumers' surplus (CS) and producers' surplus (PS) as the *potential social surplus* (See Figure 17.3). We then divide the realized surplus by the potential surplus, and post-multiply the quotient by 100%. The result is then a measure for allocative efficiency. It is mathematically described as follows. Let π_j be trader j's profits at a specific trading period,

$$\pi_j = \sum_{q \in bought} (V_{j,q} - P_q), \quad j \in Buyer, \tag{17.5}$$

and

$$\pi_j = \sum_{q \in sold} (P_q - V_{j,q}), \quad j \in Seller, \tag{17.6}$$

where $V_{j,q}$ is the redemption value of the qth unit for the jth buyer, or the unit cost of the qth unit for the jth seller. Term P_q is the transaction price of that unit. The *realized surplus* (RS) is then simply the sum of profits earned by all traders, namely,

$$RS = \sum_j \pi_j, \quad j \in traders. \tag{17.7}$$

The efficiency ratio β is

$$\beta = \frac{RS}{PS}, \tag{17.8}$$

where PS is the potential surplus. By this definition, $0 \leq \beta \leq 1$.

3. Experimental Results

3.1. Price Dynamics

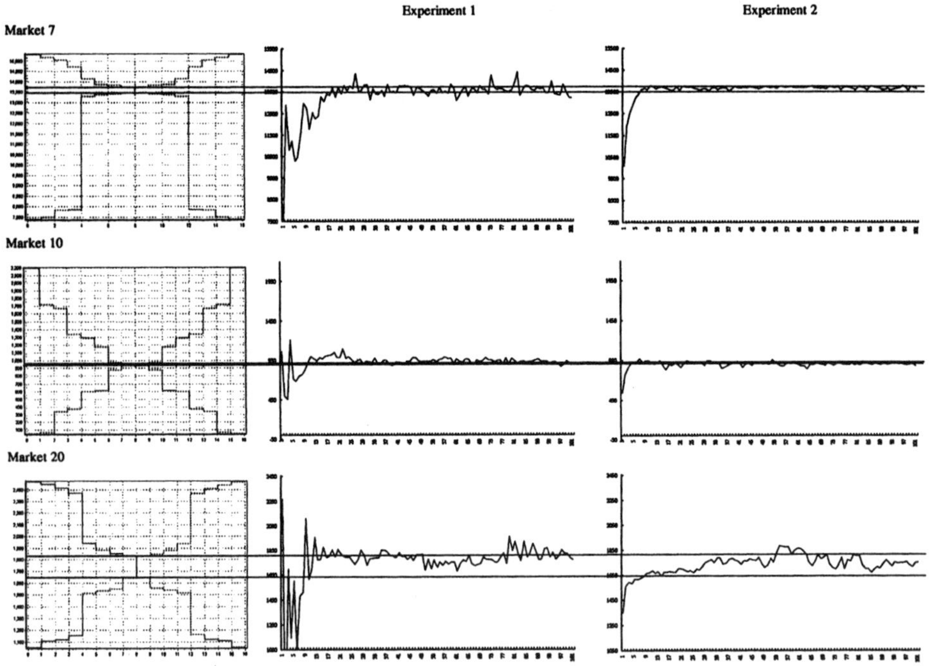

Figure 17.4. CASE 1: Time Series Plots of Price for Market 7, 10, and 20.

Our analysis of the simulation results will be based on the two measures introduced above. However, before proceeding to those measures, it will be useful to first look at the price dynamics of those markets. For this purpose, we demonstrate time series plots of the price for the following six markets: Markets 3, 7, 9, 10, 16, and 20 (Figures 17.4, 17.5, 17.6). For each figure, there are three plots. The leftmost plot is the market with its equilibrium price or equilibrium price interval. The middle plot is the time series of the price of experiment 1, whereas the rightmost plot is that of experiment 2. We shall distinguish these results by three separate cases. The first case refers to Markets 7, 10, and 20 (Figure 17.4). In these markets, we either have a unique equilibrium price (Market 10) or a tight equilibrium interval (Markets 7 and 20). Market prices in this case quickly move toward the equilibrium price (or price interval), and then slightly fluctuate around there. This result is basically consistent with what we learned from experimental economics with human subjects (Smith 1991).

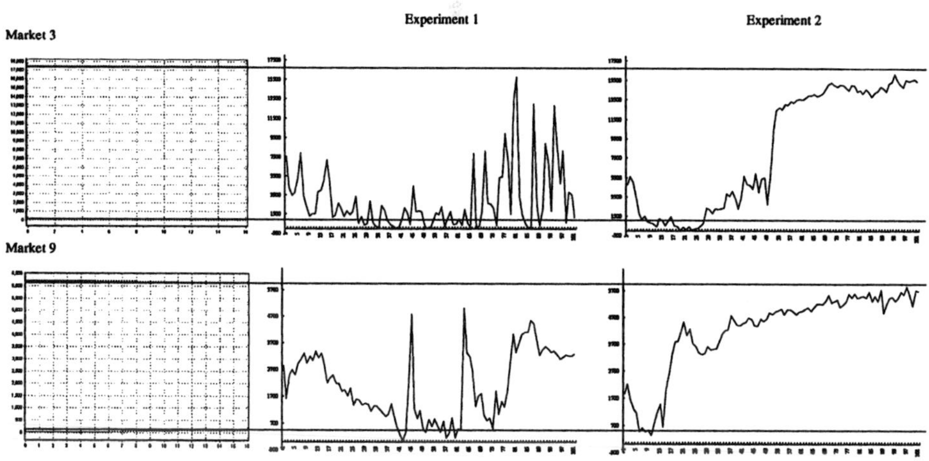

Figure 17.5. CASE 2: Time Series Plots of Price for Market 3 and 9.

The distinguishing feature of our DA markets actually starts from the second case, Markets 3 and 9 (Figure 17.5). The common characteristic of these two markets is that demand and supply curves are completely flat with no intersection. This case, while very intriguing, has almost been neglected in experimental economics literature.[8] How the price shall be determined under this circumstance is still an open question. Our simulation results from both markets seem to indicate that the price can shop around the whole interval, and is difficult to settle down to a narrow niche, but the price does not just randomly fluctuate. In fact, in Experiment 2, we see evidence of a slowly-upward moving trend. It begins with a price favorable to buyers, but then eventually turns to the seller side. For Experiment 1, Market 9 in particular, the price path is even more complicated. It starts with a downward trend, but then ends up with an upward trend. In the middle, the trend breaks occur several times.[9]

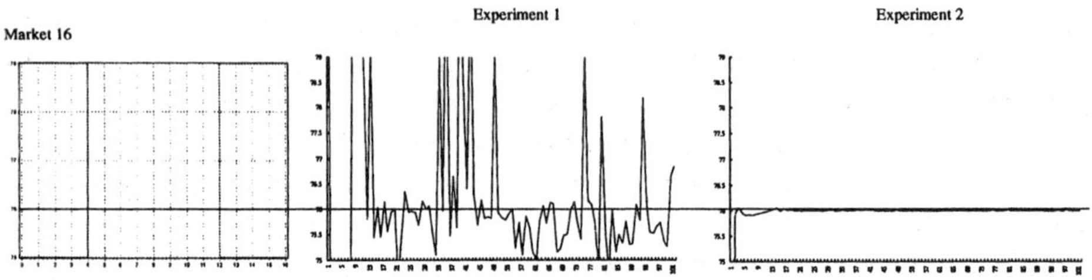

Figure 17.6. CASE 3: Time Series Plots of Price for Market 16.

We have so far not seen any qualitatively differences between Experiments 1 and 2. Their divergence appears to be clear in Case 3, Market 16 (Figure 17.6). In this market, demand and supply do intersect at a unique price. Based on our experience with Case 1, one might predict that the price will converge to this competitive equilibrium price; nevertheless, these two curves are completely flat before the intersection, and, as we have encountered in Case 2, the price may find it difficult to converge. The perplexity of this situation is also seen in our simulation results: Experiment 2 confirms the first conjecture, whereas Experiment 1 supports the second. Here is a good time to ask: *would a market with smart traders (a market without a quote limit) destabilize rather than stabilize the market?* We shall now deal with this question with more delicate statistics.

3.2. Alpha Value

As defined in Equation 17.4, the Alpha value measures the price deviation of a market period. Let $\alpha_{t,h}$ be the α value observed at the hth market period in the tth generation. In an ideal case, where $\lim_{t \to \infty} P_{t,h} \to P^*$,

$$\lim_{t \to \infty} \alpha_{t,h} = 0, \quad h = 1, ..., H. \tag{17.9}$$

The distribution (histogram) of $\{\alpha_{t,h}\}_{h=1}^{H}$ might also be expected to degenerate to the point zero. To see how far we are away from the ideal case, we draw the histogram of $\{\alpha_{t,h}\}_{h=1}^{H}$ for every 10 generations, i.e., t=0, 10,...100. To economize the presentation, the histogram is drawn

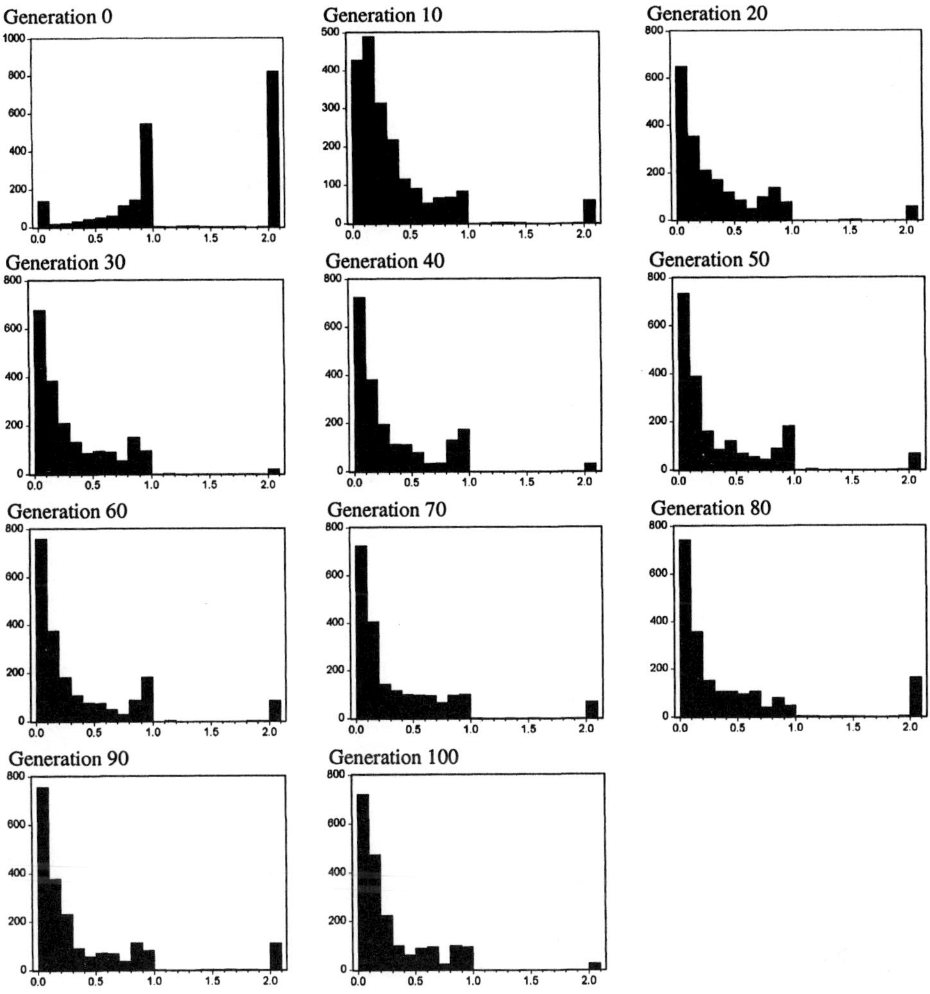

Figure 17.7. The Distribution of α Values: Experiment 1.

by pooling the $\{\alpha_{t,h}\}_{h=1}^{H}$ of all twenty markets. Figures 17.7 and 17.8 give the results of Experiments 1 and 2, respectively.[10]

These two figures roughly do indicate that these distributions exhibit a mass at zero and a skew to the right. Furthermore, Figure 17.9 plots the time series of the medium of $\{\alpha_{t,h}\}_{h=1}^{H}$ for both experiments. Here, we can see the effect of learning on price deviation: alpha values in both experiments tend to decrease in time. Drawing the two lines together also makes the effect of smartness transparent: the experiment with

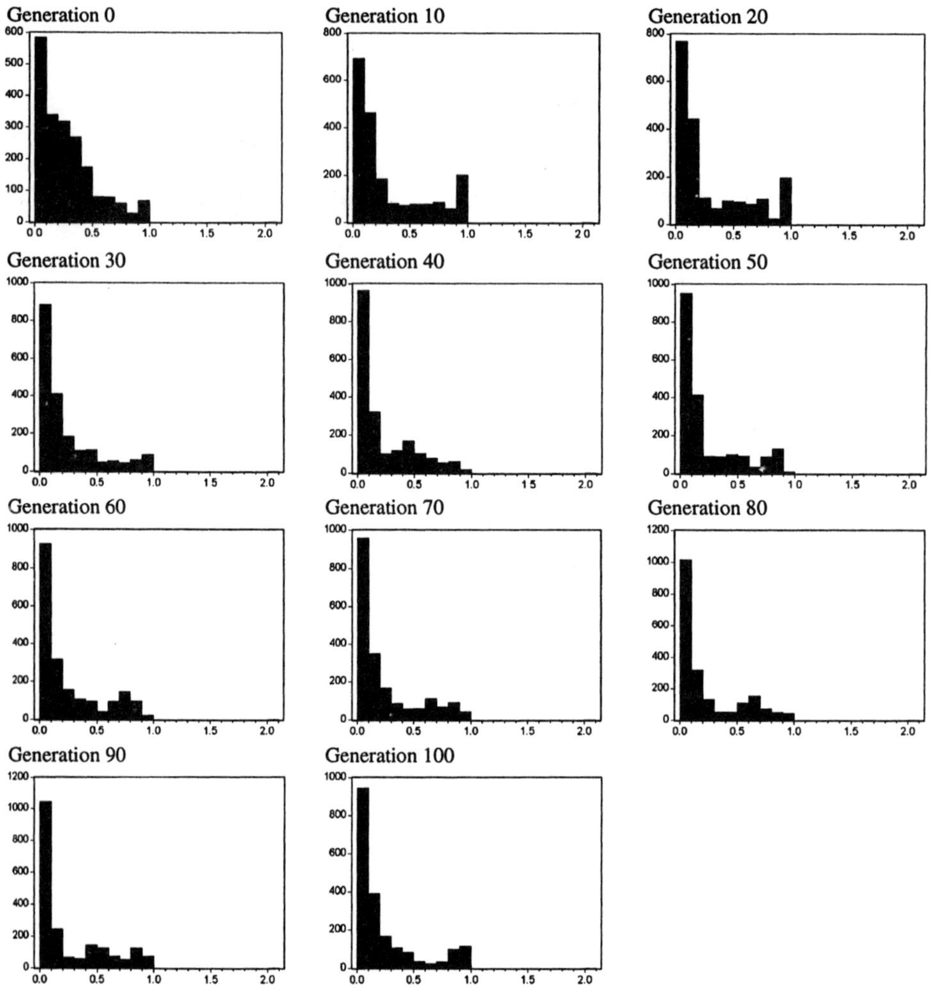

Figure 17.8. The Distribution of α Values: Experiment 2.

smarter agents (Experiment 1) tends to have a higher medium value at any point in time. This may help us answer the question posed above: *markets with smarter agents tend to be more fluctuating.* Hence, smarter agents play a *destabilizing* role for the market.

3.3. Beta Ratio

The β ratio, as defined in Equation 17.8, measures the allocative efficiency of a DA market. From the proceeding analysis above, we are

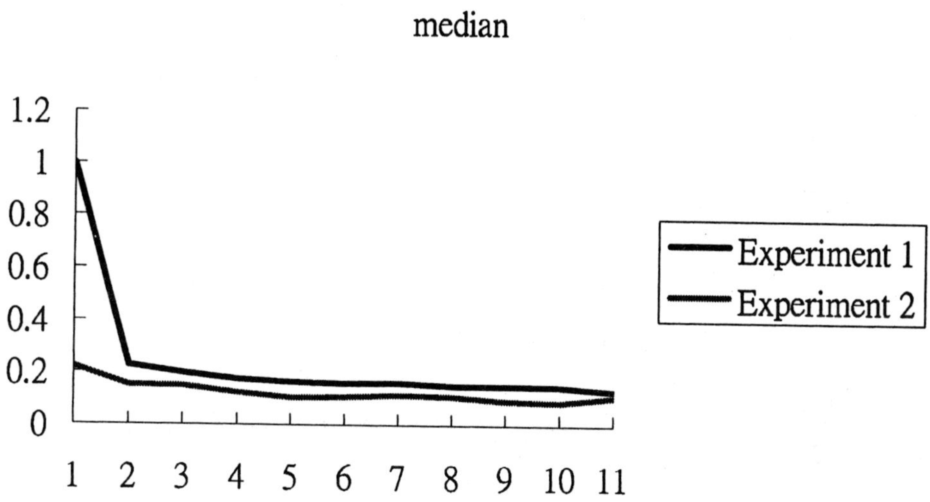

Figure 17.9. Time Series of the Median of the distribution of α Values.

interested in two things: first, the role of learning in allocative efficiency, and second, the role of smarter agents. Our first concern is shown in Figure 17.10. Figure 17.10 has three time series plots of the β ratio. The first time series plot shows that the β ratio increases with time. In this case, learning enhances the allocative efficiency, but, the second and the third do not. The third case even shows that allocative efficiency deteriorates after learning.

However, we have 20 markets for each experiment; therefore, we have numerous such time series plots. To economize our presentation, we use the *sup* function and the *inf* function to capture the essential characteristics of these three different time series plots.[11] Given an ordered series $\{x_t\}_{t=1}^{T}$, the *sup* function is defined as

$$sup_i(\{x_t\}_{t=1}^{T}) = \max\{x_t\}_{t=1}^{i}, \quad i = 1, ..., T, \tag{17.10}$$

and the *inf* function is[12]

$$inf_i(\{x_t\}_{t=1}^{T}) = \min\{x_t\}_{t=i}^{T}, \quad i = 1, ..., T. \tag{17.11}$$

The corresponding *sup* and *inf* functions of the three series are also drawn in Figure 17.10. As we can see from this diagram, if learning can enhance allocative efficiency, then these two functions should ideally display many upward jumps. On the contrary, if allocative efficiency deteriorates during the course of learning, then these two functions are

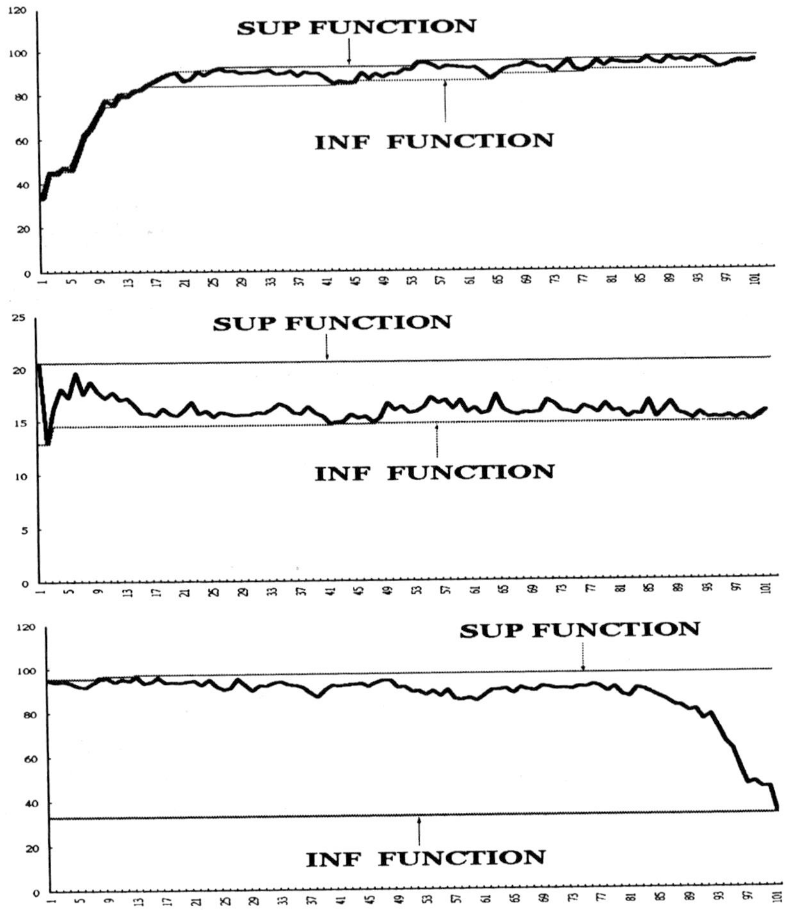

Figure 17.10. Three Possible Time Paths of Beta Ratio.

totally flat with no jumps at all. Therefore, by counting the number of jumps and the total jump size, one can effectively summarize the role of learning in allocative efficiency over different markets.

Figure 17.11 plots the number of jumps (on the y-axis) and jump size for the inf and sup function of the β values over 20 markets. Since each market may start with different initial ratios of β, we distinguish them by two different symbols, namely, diamond and box. Diamond stands for a lower initial ratio of β (below 0.4), whereas box stands for a higher initial ratio of β (between 0.4 and 0.6). Since jumps are prevalent for all markets, allocative efficiency becomes ameliorated as times goes on. Furthermore, for those markets with lower initial ratios of β, the

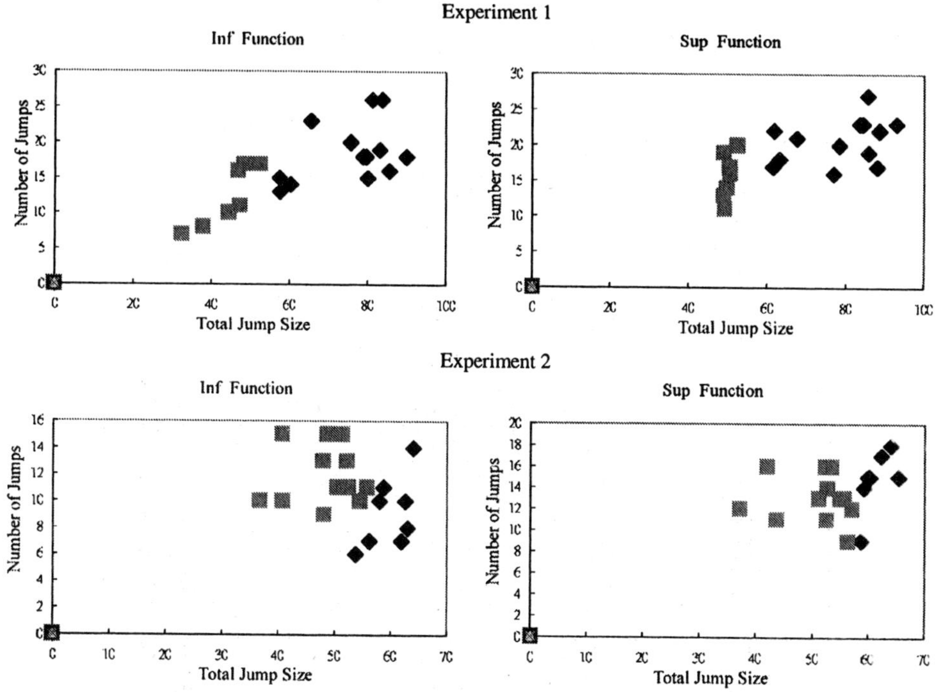

Figure 17.11. The Number of Jumps and Jump Size of Inf and Sup Functions.

amelioration degree is generally more significant than those with higher initial values.

To see the effect of smartness, the beta ratios of the initial generation and the last generation are shown in Table 17.3. The βs of Experiment 1 start with a range from 5% to 48%, and end up with a range from 79% to 98%. The βs of Experiment 2 are much higher: they start with a range from 33% to 59%, and end up with a range from 92% to 98%. If we compare these ratios pairwisely, then Experiment 1 is uniformly beaten by Experiment 2 in its resultant allocative efficiency, except for Market 9. Therefore, smarter traders do not enhance allocative efficiency.

4. Concluding Remarks

Our evidence is quite clear: smarter agents fail to enhance market performance.[13] They induce a relatively unstable price (higher alpha value) and lower allocative efficiency (lower β ratio). There is only thing left to address in this concluding section. *Why?*

Table 17.3. The Beta Ratio of the Initial and Last Generation

	Experiment 1		Experiment 2	
Market	Initial	End	Initial	End
1	29.09	94.67	56.93	97.56
2	32.09	85.60	44.01	96.14
3	45.56	83.44	41.19	97.00
4	8.59	90.91	45.32	93.37
5	11.11	92.28	33.75	96.78
6	46.66	89.28	46.74	94.73
7	6.41	86.45	59.69	96.29
8	48.32	96.63	46.72	97.25
9	46.09	98.31	38.50	97.23
10	9.92	79.33	42.12	94.41
11	33.42	90.91	55.26	96.00
12	45.50	92.62	39.66	97.66
13	4.47	80.07	34.59	98.59
14	8.57	84.27	33.78	96.34
15	48.42	95.18	36.86	98.78
16	6.44	89.19	46.69	98.00
17	34.14	89.10	43.62	98.11
18	3.80	89.37	38.82	92.55
19	47.26	79.40	38.02	94.26
20	5.46	89.03	45.80	94.51
Average		88.80		96.28

While it is not easy to provide a mathematical proof, we try to make our argument as plausible as possible. We shall build our argument based on the competition between *intra-marginal* and *extra-marginal* agents, who are the trading partners of *intra-marginal* and *extra-marginal* tokens. Intra-marginal and extra-marginal tokens are the tokens which are just inside or outside the equilibrium frontier. Without loss of generality, let us consider a very simple demand and supply schedule, as shown in Figure 17.12. Clearly, in this diagram only Seller 1 may have a successful trade. Let us also assume that Buyer 1 takes a random bid from the open interval (5,9). Given this bid and the imposition of the quote limit, Buyer 2 has no chance to beat Buyer 1 since his redemption value is only 5. In this case, Buyer 1 is an intra-marginal buyer, and Buyer 2 is the extra-marginal buyer.

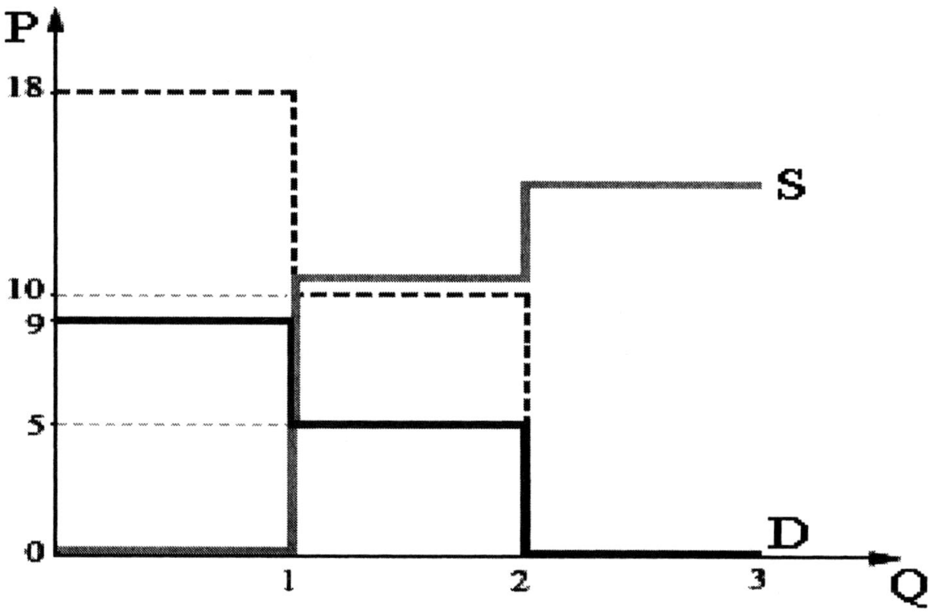

Figure 17.12. Removal of the Quote Limit and the Competition between Intra- and Extra-Marginal Agents.

Now consider the removal of the quote limit. Buyer 2 then *strategically* makes a bid up to 10 by not making a loss deal.[14] Supposing now that Buyer 1 still takes a random bid from (5, 9), the deal-winning chance for Buyer 2 will then jump from 0 to 0.2 if he takes a random bid from (5, 10). If Buyer 2 wins the deal, then the allocative efficiency derived becomes lower as opposed to the case where Buyer 1 wins the case. This gives us an explanation for why allocative efficiency will be deteriorated when the quote limit is removed.

One may next ask why Buyer 1 would not enlarge his bidding area, say up to 18. Yes, he will, if Buyer 2 continues to undertake the ambitious bidding. Nonetheless, when Buyer 2 is scared away, Buyer 1 may shift down his bidding area, and hence open the gate for the intruder again. Since we are using *multi-population genetic programming* to model our agents, it would be useful to make a distinction between *phenotype* and *genotype* in interpreting this dynamics. What we see from the outside is a competition between intra- and extra-marginal buyers, but what really happens inside (mentally) is the enduring competition between greedy strategies and cautious strategies. Greedy strategies nurse the growth of cautious strategies, which in turn do the same thing for the greedy strategies. The market just cannot settle its dynamics steadily. The in-and-out process is a generic property observed in many other agent-based markets.

Notes

1. Chen and Chie (2001) examined some *smart strategies* that evolved from their agent-based simulation of DA markets.

2. The number of bargaining strategies assigned to each bargaining agent is called the *population size*. AIE-DA Version 2 allows each agent to have at most 1000 bargaining strategies.

3. For the case when i is never selected, its mean profits would automatically be 0.

4. In other words, there totally are $H \times T = 10,000$ periods of trading for a single run of simulation.

5. Current bid refers to the highest bid at the current trading step, and current ask refers to the lowest ask. When CA is greater then CB, there shall be no match between buyers and sellers at the current step.

6. While the Aurora rules allow a random determination between CA and CB, we shall only consider the case by taking the average. See also Dawid (1999).

7. Notice that none of them in both series are born intelligent. They all have to learn and adapt to be intelligent. These terms only refer to the potentials which they may later develop.

8. To the best of our knowledge, Dawid (1999) is the only study that has drawn attention to this case.

9. Our results can be compared to Dawid (1999). Dawid (1999) found that price *always converges*, even though it did not necessary converge to the middle point of the demand and supply curves. However, there are several noticeable differences between our simulations and Dawid's. Firstly, Dawid did not use the Aurora Rule as the trading mechanism. Secondly, his market size is much bigger than us. Finally, his model of adaptive agents is also different from ours.

10. Since H is set to 100 in this paper, and there are 20 markets, there are 2,000 observations in each histogram.

11. This idea was first used in Chen, Kuo, and Lin (1996).

12. Actually, this is the inf function in reverse order.

13. Alternatively speaking, making everybody dumb by imposing a quote limit does not have a negative impact on market efficiency.

14. This upper limit is obtained by assuming that Seller 1 is a truth teller.

References

Chen, S.-H. (2000). "Toward an Agent-Based Computational Modeling of Bargaining Strategies in Double Auction Markets with Genetic Programming," in K.S. Leung, L.-W. Chan, and H Meng. (Eds.), *Intelligent Data Engineering and Automated Learning - IDEAL 2000: Data Mining, Financial Engineering, and Intelligent Agents*, Lecture Notes in Computer Sciences 1983, 517–531. Springer.

Chen, S.-H. (2001). "Evolving Bargaining Strategies with Genetic Programming: An Overview of AIE-DA Ver. 2, Part 1," in B. Verma, A. Namatame, X. Yao, H. Selvaraj, A. de Carvalho, and A. Ohuchi

(Eds.), *Proceedings of Fourth International Conference on Computational Intelligence and Multimedia Applications*, 48–54. IEEE Computer Society Press.

Chen, S.-H. and B.-T. Chie (2001). "The Schema Analysis of Emergent Bargaining Strategies in Agent-Based Double Auction Markets," in B. Verma, A. Namatame, X. Yao, H. Selvaraj, A. de Carvalho, and A. Ohuchi (Eds.), *Proceedings of Fourth International Conference on Computational Intelligence and Multimedia Applications*, 61–65. IEEE Computer Society Press.

Chen, S.-H., B.-T. Chie, and C.-C. Tai (2001). "Evolving Bargaining Strategies with Genetic Programming: An Overview of AIE-DA Ver. 2, Part 2," in B. Verma, A. Namatame, X. Yao, H. Selvaraj, A. de Carvalho, and A. Ohuchi (Eds.), *Proceedings of Fourth International Conference on Computational Intelligence and Multimedia Applications*, 55–60. IEEE Computer Society Press.

Chen, S.-H., J.-S. Kuo, and C.-C. Lin (1996). "From the Hayek Hypothesis to Animal Spirits: The Phase Transition based on Competitive Experimental Markets," in D. Vaz and K. Velupillai (Eds.), *Inflation, Institutions and Information: Essays in Honor of Axel Leijonhufvud*, 290–318. Macmillan.

Dawid, H. (1999). "On the Convergence of Genetic Learning in a Double Auction Market," *Journal of Economic Dynamics and Control*, 23, 1544–1567.

Gode, D. K. and S. Sunders (1993). "Allocative Efficiency of Market with Zero-Intelligence Trader: Market as a Partial Substitute for Individual Rationality," *Journal of Political Economy*, 101(1), 119–137.

Holland, J. H. and J. H. Miller (1991). "Artificial Adaptive Agents in Economic Theory," *American Economic Review: Papers and Proceedings*, 81(2), 365–370.

Smith, V. L. (1991). "Bidding and Auctioning Institutions: Experimental Results," *Papers in Experimental Economics*, 106–127, Cambridge: Cambridge University Press.

Chapter 18

A NUMERICAL STUDY ON THE EVOLUTION OF PORTFOLIO RULES

Is CAPM Fit for Nasdaq?

Guido Caldarelli
INFM - Unità di Roma 1 "La Sapienza", P. le A. Moro 2, 00185 - Roma, Italy
gcalda@pil.phys.uniroma1.it

Marina Piccioni
Abaxbank Corso Monforte, 34 20122 Milano
marina.piccioni@abaxbank.com

Emanuela Sciubba
Faculty of Economics and Politics, University of Cambridge
E.Sciubba@econ.cam.ac.uk

Abstract In this paper we test computationally the performance of *CAPM* in an evolutionary setting. In particular we study the stability of distribution of wealth in a financial market where some traders invest as prescribed by *CAPM* and others behave according to different portfolio rules. Our study is motivated by recent analytical results that show that, whenever a logarithmic utility maximiser enters the market, CAPM traders vanish in the long run. Our analysis provides further insights and extends these results. We simulate a sequence of trades in a financial market and: first, we address the issue of how long is the long run in different parametric settings; second, we study the effect of heterogeneous savings behaviour on asymptotic wealth shares. We find that *CAPM* is particularly "unfit" for highly risky environments.

JEL Classification C61, D81, G11

Keywords: Evolution, Portfolio Rules, *CAPM*, Kelly Criterion

Introduction

A major part of research in financial economics is aimed at improving our understanding of how investors make their portfolio decisions and hence of how asset prices are determined. In this paper we contribute to that research adopting a dynamic approach, where investors' wealth endowments and portfolio rules evolve in time.

We adopt here an evolutionary framework to address the issues of *fitness* and *survival* in financial markets, where traders differ in portfolio rules and/or savings behaviour. We build a simple framework that allows us to simulate a long sequence of trades in a competitive financial market and to test computationally the asymptotic wealth distribution of traders who follow heterogeneous portfolio rules. In particular we concentrate our analysis on the interaction between two types of traders: traders who use (the traditional version of) *CAPM* as a rule of thumb for their portfolio decisions and traders who make portfolio choices maximising a logarithmic utility function.

The choice of this particular focus of analysis is motivated by an old debate and a recent strand of literature. Despite the fact that its restrictive assumptions have often been criticised and its predictive power has been challenged by numerous contribution in empirical finance (see Fama and French 1996a, Fama and French 1996b), the mean variance approach is a standard in financial economics, and its main corollary in asset pricing, the Sharpe-Lintner-Mossin Capital Asset Pricing Model,[1] has been viewed as one of the "major contributions of academic research in the postwar era" (Jagannathan and Wang 1996, p.4). Since a seminal article by Kelly (1956), financial economists and applied probability theorists have been debating on the normative appeal of logarithmic utility maximisation as opposed to the mean-variance approach in financial markets. Several contributors have expressed their dissatisfaction with the mean-variance approach and argued that a rational investor with a long time horizon should instead maximise the expected rate of growth of his wealth.[2] The portfolio choice that this type of behaviour implies is equivalent to that prescribed by a logarithmic utility function (the so called "Kelly criterion"). Central to the debate was the claim that maximising a logarithmic utility function would be a "more rational" objective to follow for a trader with a long time horizon. This claim has been strongly opposed by Merton & Samuelson (1974) and Goldman (1974), who stressed the obvious contradiction that lies in arguing that rational traders should maximise a given utility function, possibly different from their own.

We believe that a useful contribution to this debate comes by the adoption of an evolutionary technique. By showing that logarithmic utility maximisers accumulate more wealth than $CAPM$-traders we certainly do not prove that they are more rational (or happier) than their mean-variance opponents. However we may gain a better understanding of the reasons for the support of logarithmic utility maximisation that originated the debate.

We do not attempt here a review on the literature on evolution and market behaviour. Instead we concentrate on the more self-contained strand of literature to which this paper specifically contributes. In particular, we focus on the literature that aims at studying long run financial market outcomes as the result of a process akin to natural selection. In a seminal paper (Blume and Easley 1992) develop an evolutionary model of a financial market and define notions of dominance, survival and extinction based on the asymptotic behaviour of traders' wealth shares. They provide us with the most general analytical result in this strand of literature. They show that, if all traders save at the same rate and under some uniform boundedness conditions on portfolios, there exists one globally fittest portfolio rule which is prescribed by logarithmic utility maximisation. Namely, if there is a logarithmic utility maximiser in the economy, he will dominate, determine asset prices asymptotically and drive to extinction any other trader that does not behave, at least in the long run, as a logarithmic utility maximiser.

In a different setting, where investors choose their investment rates endogenously (choosing an optimal consumption stream at the same time as an optimal sequence of portfolios), Sandroni (1999) provides analytical results that seem to weaken Blume and Easley's findings. He shows that, provided that agents' utilities satisfy Inada conditions (and that agents' beliefs are correct), then all traders survive. Clearly Inada conditions are crucial here as they require marginal utilities to go to infinity whenever the consumption level approaches zero. A rational trader that can also choose his investment intensity would avoid extinction by suitably modifying his consumption pattern.

Sciubba (1999) compares the relative fitness of logarithmic and mean-variance preferences. Mean-variance preferences do not satisfy Inada conditions and prescribe portfolio weights which do not necessarily display uniform boundedness properties, so that none of the two previous frameworks directly applies. She shows that when logarithmic utility maximisers invest at the same rate as mean-variance investors, logarithmic traders dominate, determine asset prices asymptotically and drive to extinction those agents who are either endowed with mean-variance preferences, ore use their theoretical predictions ($CAPM$) as a rule of thumb. One of the sections in (Sciubba 1999) is indeed devoted to the study of the role of heterogeneous savings rates in the dynamics of wealth accumulation. However, the stochastic process becomes so complex in this case, that the few results that are obtainable analytically are very weak. This is where we believe that numerical computation can provide us with more insights into the problem.

Our results show that, even when we allow for heterogeneity in savings rates, dominance of logarithmic utility maximisers proves robust, at least to a certain extent. In particular, we show that logarithmic traders can "afford" to save at a lower rate than $CAPM$ traders and still dominate financial markets. When logarithmic traders dominate, the wealth share of $CAPM$ traders converges to zero exponentially fast. In particular, we find that the more risk-averse $CAPM$ traders are, the faster they vanish. Clearly when $CAPM$ traders display a savings intensity that is much higher than their opponents, then — as one would expect — they indeed dominate and drive to extinction logarithmic utility maximisers. We identify the threshold in the savings rates differential such that this "fate reversal" occurs and find that it is increasing in the variance of the dividend stream. If the dividend stream is very volatile, then the advantage of logarithmic utility maximisers with respect to $CAPM$ traders in terms of portfolio selection is higher. Hence LOG traders can "afford" to save less and still dominate the market. Symmetrically, the fitness of $CAPM$ proves particularly low in environments characterised by high volatility. Should we trust what $CAPM$ prescribes when investing on Nasdaq? Probably not.

1. The Model

A detailed description of the model is to be found in (Sciubba 1999). Here we summarise its main features. Time is discrete and indexed by $t = 0, 1, 2, ...$ There are S states of the world indexed by $s = 1, 2, ..., S$, one of which will occur at each date. States follow an i.i.d. process with distribution $p = (p_1, p_2, ..., p_S)$ where $p_s > 0$ for all s. There is a finite set of assets, with the same dimensionality of the set of states of nature. With little abuse of notation we index both assets and states with the same letter. Asset $s \in \{1, 2, ..., S\}$ pays a strictly positive dividend $d_s > 0$ when state $s \in \{1, 2, ..., S\}$ occurs and 0 otherwise. At each date there is only one unit of each asset available, so that d_s will be the total wealth in the economy at date t if state s occurs. This wealth will be distributed among traders proportionately according to the share of asset s that each of them owns. Let ρ_{st} be the market price of (one unit of) asset s at date t. There is a heterogeneous population of long-lived agents in this economy, indexed by $i \in \{1, 2, ..., I\}$. Each agent is characterised by an investment rule and an initial wealth endowment. Agent i's investment rule is $\{\delta_t^i, \alpha_t^i\}_{t=0}^{\infty}$ where δ_t^i denotes agent i's investment rate at date t (i.e. the percentage of wealth endowment at date t, e_t^i, that i invests at date t) and α_t^i is a vector that describes agent i's portfolio choice at date t (i.e. the vector of portfolio weights in the S available assets $\{\alpha_{st}^i\}_{s=1}^{S}$ for agent i at date t). Agent i's initial wealth endowment is denoted by e_0^i. The tuple $\left(e_0^i, \{\delta_t^i, \alpha_t^i\}_{t=0}^{\infty}\right)$ constitutes a complete description of agent i. At each date t, agent i invests a portion δ_t^i of his wealth endowment e_t^i, in the S available assets.[3] We denote by w_t^i the total amount invested by trader i at date t :

$$w_t^i = \delta_t^i e_t^i \tag{18.1}$$

Aggregate total investment will be equal to aggregate expenditure in assets. Since there is only one unit of each asset available, then clearly:

$$\sum_{s=1}^{S} \rho_{st} = \sum_{i=1}^{I} w_t^i = w_t \tag{18.2}$$

Equation (18.2) provides us with a convenient normalisation for prices. We can, in fact, call π_{st} the normalised market price of asset s at date t and define it as follows:

$$\pi_{st} \equiv \frac{\rho_{st}}{\sum_{s=1}^{S} \rho_{st}} = \frac{\rho_{st}}{w_t} \tag{18.3}$$

Finally define: $\pi_t \equiv (\pi_{1t}, \pi_{2t}, ..., \pi_{St})$. In equilibrium, prices must be such that markets clear, i.e. total demand equals total supply. Agent i's demand for asset s will be equal to agent i's expenditure in asset s, $\alpha_{st}^i w_t^i$, divided by the market price of asset s, ρ_{st}. In the aggregate:

$$\sum_{i=1}^{I} \frac{\alpha_{st}^i w_t^i}{\rho_{st}} = 1 \tag{18.4}$$

Rewriting equation (18.4) we get:

$$\rho_{st} = \sum_{i=1}^{I} \alpha_{st}^i w_t^i \tag{18.5}$$

and using our price normalisation:

$$\pi_{st} = \sum_{i=1}^{I} \alpha_{st}^i \varepsilon_t^i \tag{18.6}$$

where ε_t^i denotes the investment share of agent i at date t:

$$\varepsilon_t^i = \frac{w_t^i}{w_t} \tag{18.7}$$

and also measures the "presence" of trader i in financial markets and his "importance" in determining asset prices.

1.1. The Dynamics

We now ask what is the period to period dynamics of trader i's investment share. This will allow us to follow the evolution of his "presence" in financial markets and of his "importance" in determining market outcomes, namely asset prices. Suppose that state s occurs at date t: asset s pays a dividend d_s while all other assets pay zero. The total wealth in the economy, d_s, gets distributed to traders according to the share of asset s that each of them owns. In particular, the investment income of trader i — that also constitutes his wealth endowment for period $t+1$ — is equal to the following:

$$e_{t+1}^i = \frac{\alpha_{st}^i w_t^i}{\rho_{st}} d_s \tag{18.8}$$

Aggregate wealth endowment at date $t+1$ is equal to the total payout of asset s:

$$e_{t+1} = \sum_{i=1}^{I} e_{t+1}^i = \sum_{i=1}^{I} \frac{\alpha_s^i w_t^i}{\rho_{st}} d_s = d_s \tag{18.9}$$

In period $t+1$ trader i invests a fraction δ_{t+1}^i of his wealth endowment:

$$w_{t+1}^i = \delta_{t+1}^i e_{t+1}^i = \delta_{t+1}^i \frac{\alpha_{st}^i d_s}{\rho_{st}} w_t^i \tag{18.10}$$

In the aggregate:

$$w_{t+1} = \sum_{i=1}^{I} w_{t+1}^i = \sum_{i=1}^{I} \delta_{t+1}^i \frac{\alpha_{st}^i w_t^i}{\rho_{st}} d_s = \sum_{i=1}^{I} \delta_{t+1}^i \frac{\alpha_{st}^i w_t^i}{\rho_{st}} e_{t+1} \tag{18.11}$$

In order to formulate (18.10) in terms of wealth shares, it is useful to define the market average investment rate δ_{t+1} as:

$$\delta_{t+1} = \sum_{i=1}^{I} \delta_{t+1}^i \frac{\alpha_{st}^i w_t^i}{\rho_{st}} \tag{18.12}$$

so that (18.11) can be rewritten as:

$$w_{t+1} = \delta_{t+1} e_{t+1} \tag{18.13}$$

Dividing (18.10) by (18.11) we obtain investment shares:

$$\varepsilon_{t+1}^{i} = \frac{w_{t+1}^{i}}{w_{t+1}} = \frac{\delta_{t+1}^{i}\alpha_{st}^{i}d_{s}}{\delta_{t+1}\rho_{st}e_{t+1}}w_{t}^{i} \tag{18.14}$$

Finally, using (18.9) and our price normalisation, we obtain:

$$\varepsilon_{t+1}^{i} = \frac{\delta_{t+1}^{i}\alpha_{st}^{i}}{\delta_{t+1}\pi_{st}}\varepsilon_{t}^{i} \tag{18.15}$$

Equation (18.15) describes the period to period dynamics of the investment share of trader i. It implies that the "impact" of trader i on financial markets follows a fitness-monotonic dynamic: trader i's investment share will increase if he invests more than the market on average and if he scores, with his investments, a payoff which is higher than the average population payoff. In fact, if state s occurs at date t, α_{st}^{i} and π_{st} give us a measure of trader i's payoff (his bet on the lucky asset) and of the average population payoff respectively. Following the definition given by Blume and Easley (1992), in order to establish whether trader i survives or vanishes, we consider the asymptotic value of his investment share and we check whether it is bounded away from zero or not. When trader i's investment share converges to zero,[4] his significance in determining market outcomes vanishes and he effectively disappears from the market. When trader i's investment share stays bounded away from zero, then he contributes to determine asset prices (also asymptotically) and we will say that, as a trader, he survives. The process described by Equation (18.15) is too complex to be studied analytically in a very general case.[5] In fact, even under the restriction that all shocks are i.i.d., asset prices and market savings rates depend on the whole sequence of past history, and the law of large numbers does not automatically help us with a solution. As we argued in the introduction, this is why we believe that numerical analysis can help to obtain further insights into the problem.

1.2. Types of Traders

As in (Sciubba 1999), we will consider the interaction between two different types of traders. The first type of agents is given by investors who believe in *CAPM* and use it as a rule of thumb. We think of them as of traders who have learned CAPM in their MBA finance courses and now apply its predictions to portfolio choice. In a CAPM economy where traders display mean-variance preferences, the well known result of two fund separation obtains. A CAPM trader's portfolio can be seen as the sum of two separate components: a part is invested in a portfolio of risky assets and the remainder is invested in the risk-free asset. The particular feature of a CAPM economy is that all rational agents invest in risky assets in the same proportions: as a result the risky component in every agent's portfolio can be interpreted as the market portfolio. Agents differ only in the weight that they assign to the risky component as opposed to the risk-free asset: they choose their preferred mix between these two components according to their individual degree of risk aversion. In the financial market that we model here there is no risk-free asset. However, since there are as many assets as states of nature (i.e. markets are complete), it is possible to construct a portfolio that yields the same return irrespective of the random draw over states of nature. Hence CAPM traders in our setting choose a mix of market and risk-free portfolios. In particular, at the beginning of each time period, they observe payoffs and market prices and work out the composition of the market and the risk-free portfolios. Finally, according to their degree of risk aversion they choose their preferred combination between the two. At date t, $CAPM$ investors choose $\gamma_t \in [0,1]$ and invest in asset s a portion α_{st}^{CAPM} of their portfolio such that:

$$\alpha_{st}^{CAPM} = \gamma_t \alpha_{st}^{F} + (1-\gamma_t)\,\alpha_{st}^{M} \tag{18.16}$$

where

$$\alpha_{st}^{F} \equiv \frac{\rho_{st}/d_s}{\sum_z \rho_{zt}/d_z} = \frac{\pi_{st}/d_s}{\sum_z \pi_{zt}/d_z} \tag{18.17}$$

and

$$\alpha_{st}^{M} \equiv \frac{\rho_{st}}{\sum_z \rho_{zt}} = \pi_{st} \tag{18.18}$$

The weights α_{st}^{F} and α_{st}^{M} are the weights on asset s in the risk-free and the market portfolio respectively. One can easily verify that the total return of the portfolio $\left\{\alpha_{st}^{F}\right\}_{s=1}^{S}$ if state s occurs at date t, does not depend on s:

$$w_{t+1}^{i} = \frac{\delta_{t+1}^{i}\alpha_{st}^{F}d_s}{\rho_{st}}w_t^i = \frac{\delta_{t+1}^{i}}{\sum_s \rho_{st}/d_s}w_t^i \tag{18.19}$$

Finally, the market portfolio can be defined as the portfolio where each asset is weighted by its relative price, precisely as in $\left\{\alpha_{st}^{M}\right\}_{s=1}^{S}$. The second type of traders is given by investors who are endowed with a logarithmic utility function (type LOG) and who actually maximise the growth rate of their wealth share and invest according to a "simple" and time invariant portfolio rule:

$$\alpha_{st}^{LOG} = p_s \tag{18.20}$$

More generally, at each date t, a rational trader i will choose $\{\alpha_{st}^i\}_{s=1}^S$ so as to maximise:

$$\sum_{s=1}^{S} p_s u^i \left(\frac{\alpha_{st}^i w_t^i}{\rho_{st}} d_s \right) \tag{18.21}$$

subject to the constraint that investment expenditure at each date is less than or equal to the amount of wealth saved in the previous period. If $u^i(\cdot)$ is logarithmic, it follows that $\alpha_{st}^L = p_s$.[6] We will denote by δ_t^{CAPM} and δ_t^{LOG} the investment rates of traders who believe in $CAPM$ and logarithmic utility maximisers respectively.

2. Computer Simulation and Numerical Results

Recall Equation (18.6) where normalised prices are expressed as a weighted average of portfolio rules. We assume that there are only two types of traders in this economy: CAPM and LOG traders. Finally we assume that there is a continuum of identical traders of each type and we denote by ε_t the investment share of LOG traders as a whole and by $(1 - \varepsilon_t)$ the investment share of CAPM traders as a whole. In this setting, Equation (18.6) becomes

$$\pi_{st} = \varepsilon_t \alpha_{st}^{LOG} + (1 - \varepsilon_t)\alpha_{st}^{CAPM} \tag{18.22}$$

where ε_t denotes the investment share of LOG traders and $(1 - \varepsilon_t)$ denotes the investment share of $CAPM$ traders. Also, for simplicity, we assume that the degree of risk aversion of the CAPM traders is constant over time, so that $\gamma_t = \gamma, \forall t \geq 0$. By substituting Equation (18.16) and Equation (18.20) in Equation (18.22) one has

$$\pi_{st} = \varepsilon_t p_s + (1 - \varepsilon_t)(\gamma \frac{\pi_{st}}{d_s} \frac{1}{\sum_{z=1,N} \frac{\pi_{zt}}{d_z}} + (1 - \gamma)\pi_{st}) \tag{18.23}$$

We solve the above equation through iteration by a numerical technique called "relaxation". Namely, we start from a trial value for π_{st}, we compute a new value through the above equation and then we iterate this procedure until a fixed point is reached. In other words, if we call τ the time of the iteration our numerical code computes the quantity

$$\pi_{st}^{\tau+1} = \varepsilon_t p_s + (1 - \varepsilon_t)(\gamma \frac{\pi_{st}^{\tau}}{d_s} \frac{1}{\sum_{z=1,N} \frac{\pi_{zt}^{\tau}}{d_z}} + (1 - \gamma)\pi_{st}^{\tau}) \tag{18.24}$$

and then it looks for the fixed point solution π_{st}^{∞} that is achieved for all the $\tau > \tau^*$ for which $|\pi_{st}^{\tau^*+1} - \pi_{st}^{\tau^*}|$ is less than a tolerance parameter fixed in advance for all the assets s. This fixed point is the numerical solution of Equation (18.24). Since it has been demonstrated that this solution exists and is unique (Sciubba 1999), it is possible to show that the above method gives a numerical value of the solution with a desired precision bounded only by computation time. In our simulations we stop the iteration when the relative precision of asset prices is lower than 10^{-5}. At this point we run two sets of simulations. The first set aims at detecting the time of convergence of the stochastic process given by the wealth shares. The second set of simulations aims at checking the robustness of *LOG* dominance results to heterogeneity in savings rates.

The first numerical check is devoted to study the average time one can expect *CAPM* traders to survive in a competition with the others. We run a Fortran code that simulates a real evolution in a market with *CAPM* traders and *LOG* traders. We consider a market where 100 assets are repeatedly traded, so that $S = 100$. We assume that the probability distribution over the states of nature is uniform, so that $p_s = 1/S$, $\forall s$. We assume that dividends are randomly drawn from a normal distribution $N(\mu, \sigma)$ with the mean to variance ratio large enough to guarantee that virtually every asset pays a positive payoff. We set the initial investment shares for the two types of traders to be equal. None of the qualitative results that we obtain in the simulation are driven by our choice of parameters and probability distributions. We run our first set of simulations under the assumption as in (Sciubba 1999 and Blume and Easley 1992) that both types of traders save at the same rate. This assumption will be relaxed later. Time step after time step the code generates pseudo-random numbers that identify states of nature. After each draw, prices are computed through "relaxation", the dividends are distributed and the investment shares of the traders are refreshed. Eventually the investment share of *CAPM* traders becomes lower than a certain threshold, sufficiently small such that we can conclude that *CAPM* traders are extinct. We then store in the computer the time at which this threshold is reached and we start again with a new realisation of the same market. Given the same initial conditions in the market, we collect up to 15000 different realisations of this competition between traders. The various realisations differ because of the randomness in the draw of the states of nature. In all the runs *CAPM* traders see their investment shares reduced until extinction takes place. This result confirms the theoretical findings. In (Sciubba 1999) where it is shown that the wealth share of CAPM traders converges almost surely to zero when a LOG trader enters the market. The simulation results add to the analysis in (Sciubba 1999) some interesting insights on the time needed for convergence. Through the extinction times recorded by the computer code, we are also able to measure the density function describing the probability that $CAPM$ traders survive up to a time t in a competition with LOG traders. The unit of time is given by the draw of a state of nature. That is, $t = 5$ means that after the initial condition 5 states of nature have been drawn in the market, and the 5 corresponding assets paid their dividends. The main result is that the density function is exponential. The probability for a $CAPM$ trader to survive decreases with time according to

$$P(t)dt = Ae^{-Ct}dt \tag{18.25}$$

We also run different simulations by changing the parameter γ of the model, that measures the risk-aversion of $CAPM$ traders. The result is that the more risk-averse the $CAPM$ traders are, the faster their wealth share converges to zero. The economic intuition for this result lies in the fact that a higher degree of risk-aversion implies that the portfolio of $CAPM$ is tilted towards the risk-free rather than towards the market portfolio, where the most successful trading rules are represented. In this setting, a $CAPM$ trader with an extremely low risk-aversion, and hence a value of γ approximately equal to zero, would indeed survive.

Our simulations show that exponential decay is robust with respect to the values of γ used. The functional form that can be hypothesised for such a decay is of the form

$$P(t) = Ae^{-Bt\gamma^2} \tag{18.26}$$

One can test this assumption by rescaling the different data. Namely in each simulation, if we multiply the value of time by the value of γ^2 used in that run, we should obtain different functions $P'(t')$ (where $t' = t\gamma^2$) all behaving in the same way. In particular these data obtained through simulations with different values of γ should collapse together. This technique note as "collapse plot" is shown in Figure 18.1. In the upper part of the figure we plot the data relative to $\gamma = 0.5, 0.7, 0.8, 0.9, 1$ testing the good validity of our assumption. We also run another set of simulations, by changing the threshold at which the $CAPM$ trader is considered extinct. By passing to a threshold of 0.5% of the total wealth shares from the 5% previously used, we noticed as expected a shift of the distribution to longer times. Nevertheless, qualitatively we obtain the same behaviour shown in the lower part of Figure 18.1. In both cases (above and below) the estimated exponential distribution is indicated by a dashed line.

In the second set of simulations we consider the situation in which the savings rates used by the two types of traders differ. In particular, we ask if LOG traders can survive and eventually dominate over

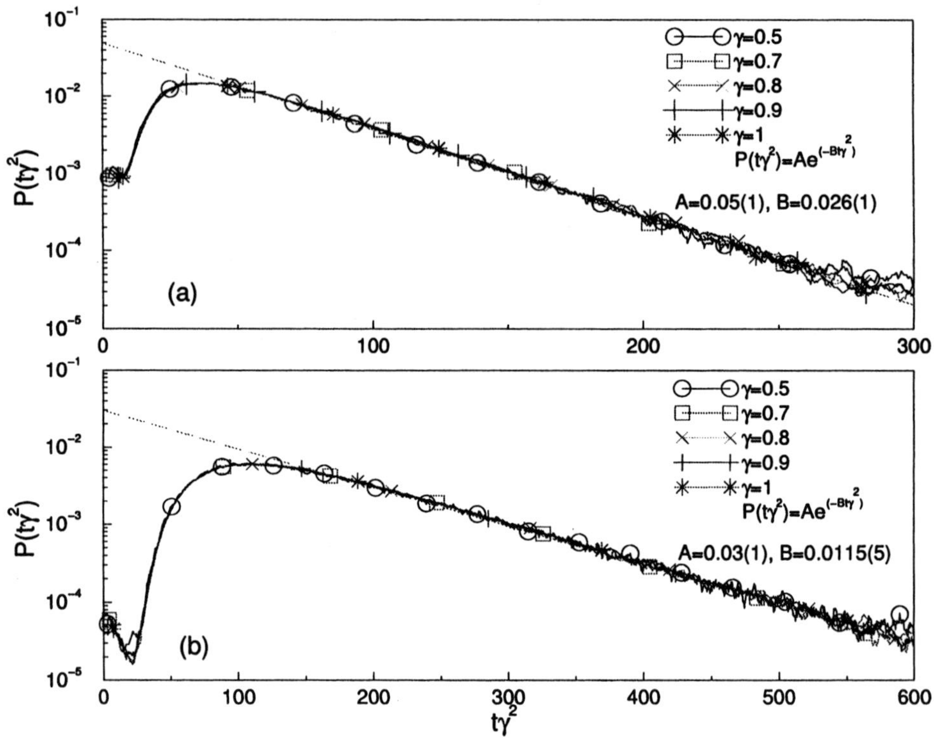

Figure 18.1. Different density functions for the survival probability for different values of γ. In the picture above a threshold of 5% of aggregate wealth has been adopted to claim traders' extinction. In the picture below we used a threshold ten times lower. This change affects only the scale of time and the parameters of the fit, but it does not affect the form of the distribution.

$CAPM$ traders that save at a higher intensity. We run simulations of the financial market described in the previous section, in this case under the assumption that traders have heterogeneous savings rates. Our aim is to check whether LOG dominance results are robust when $CAPM$ traders save at a higher rate than logarithmic utility maximisers. As in the previous set of simulations, we consider a market where 100 assets are repeatedly traded, such that $S = 100$. Again, we assume that the probability distribution over the states of nature is uniform, so that $p_s = 1/S$, $\forall s$. We assume that dividends are randomly drawn from a normal distribution $N(\mu, \sigma)$ with the mean to variance ratio large enough to guarantee that virtually every asset pays a positive payoff. We set the initial wealth shares for the two types of traders to be equal. Finally we normalise the savings rate of $CAPM$ traders to be equal to one. and we denote by Δ the difference between the two savings rates

$$\delta^{CAPM} = \delta^{LOG} + \Delta \tag{18.27}$$

such that for $\left(\delta^{CAPM} - \delta^{LOG}\right) > \Delta$, CAPM traders dominate and for $\left(\delta^{CAPM} - \delta^{LOG}\right) \leq \Delta$, LOG traders dominate.

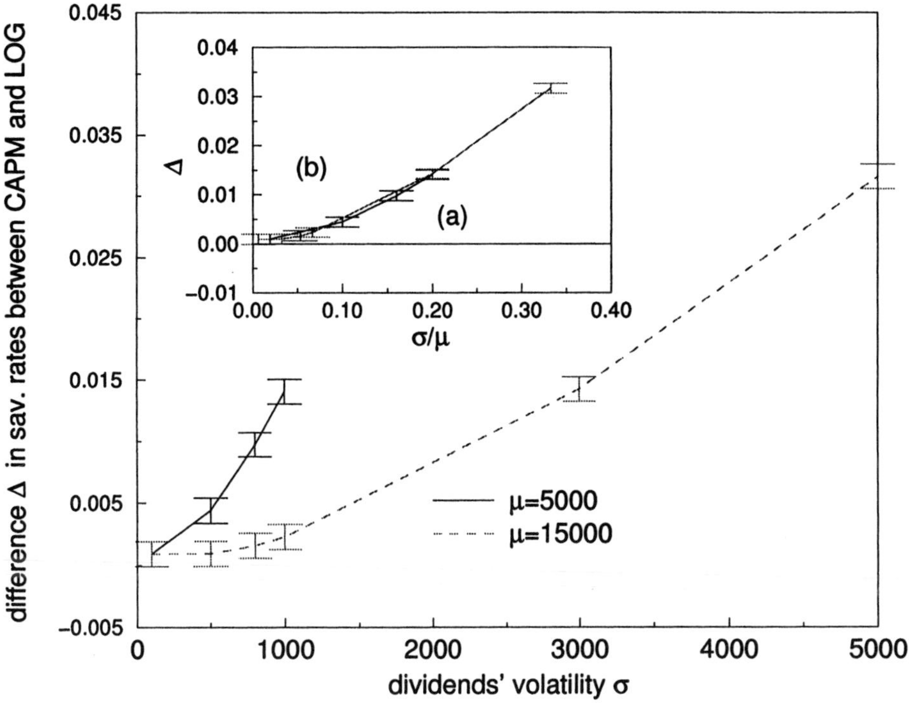

Figure 18.2. Values of Δ (the maximum difference between the savings rates of $CAPM$ and LOG traders that still allows the latter to dominate), with respect to the dividends' volatility σ.

Figure 18.2 shows the largest value of Δ that still allows for logarithmic utility maximisers' savings rate that still allows LOG traders to dominate (averaged across 1000 different simulations), for different values of the variance σ of the distribution of dividends, which measures the volatility of the dividend stream. Also in this case we apply the technique of the collapse plot in order to show that the threshold at which LOG traders commence to dominate depends on the ratio σ/μ. We can interpret the graph in Figure 18.2 as dividing the σ, Δ plane in two regions: in the area (a) below the graph, logarithmic utility maximisers dominate and drive to extinction $CAPM$ traders; in the area (b) above the graph, $CAPM$ traders dominate and drive to extinction logarithmic utility maximisers. From our results it appears that the dominance of logarithmic utility maximisers is robust to heterogeneity in savings rates, at least to a certain extent. In fact we obtain that LOG traders dominate and drive $CAPM$ traders to extinction even when the latter display a higher savings rate (provided that it is not much higher). This difference provides us with an immediate measure of the advantage of LOG traders with respect to $CAPM$ traders in terms of portfolio rules. We find that the higher the volatility of the dividend stream, the higher is the advantage of logarithmic utility maximisers over $CAPM$ traders. In particular, the threshold level for the LOG traders' savings rate appears to be a function of the mean to variance ratio of the probability distribution of the dividends that assets pay. The economic intuition of this result lies in the fact that, with a large variance of dividends, the behaviour prescribed by a logarithmic utility function differs greatly from the behaviour prescribed by CAPM. Think of the following example. Suppose that investors can place bets on two lotteries that pay positive prizes with equal probability: lottery "Rich" pays 1 million dollars with probability 1/2 and lottery "Poor" pays 1 single dollar whenever lottery "Rich" does not pay. A LOG investor would place equal bets on both lotteries. On the contrary, a CAPM trader (and any trader, in fact, that maximises the expected value of an increasing function of his wealth level, rather than the rate of growth of his wealth) would place a higher bet on the "Rich" lottery than on the "Poor" one. The divergence between his bets would be increasing in the difference between the two lotteries' prizes. As a result, the distance between the bets that LOG investors and CAPM investors would place is also increasing in the discrepancy between lotteries' payoffs. A higher variance of dividends implies that, in different states of nature, very high payoffs are paid as well as very low ones, and results in a higher divergence of portfolio choices between the two types of investors.

3. Conclusion

In this paper we test computationally the performance of $CAPM$ in an evolutionary setting. In particular we study the asymptotic wealth distribution across two types of traders that compete on financial markets: traders who invest as prescribed by $CAPM$ and traders who maximise the expected value of a logarithmic utility function of wealth. Our study provides further insights and extends some recent analytical results (see Sciubba 1999) that prove that the wealth share of $CAPM$ traders converges almost surely to zero, when a logarithmic utility maximiser with a savings rate at least as large as the savings rate of $CAPM$ traders, enters the market. We run two sets of simulation addressing two related, but separate, issues. First, we look at the time of convergence of the stochastic process given by the wealth share dynamics. We find that, when savings rates are identical across the two types of traders, the wealth share of $CAPM$ investors decreases exponentially fast towards zero. We also find that the degree of risk-aversion of $CAPM$ traders has a role in determining the speed of convergence: the more risk-averse the $CAPM$ traders, the faster their wealth share converges to zero. Second, we check the robustness of the analytical result in (Sciubba 1999) to heterogeneity in savings rates. We find that LOG investors dominate even when their savings rate is lower (but not too much lower) than the savings rate of $CAPM$ traders. We compute the maximum difference between the savings rates of the two types of investors that still allows LOG traders to dominate and drive to extinction all those who choose their portfolio according to what $CAPM$ prescribes. We argue that the difference between savings rates so computed can serve as a measure of the fitness of logarithmic utility maximisation with respect to $CAPM$ and we find that it is increasing in the variance of the dividend stream. Our results seem to suggest that, from an evolutionary perspective, if it is true that $CAPM$ could perform almost as satisfactorily as logarithmic utility maximisation in markets with low volatility, it proves particularly unfit for highly risky environments.

Notes

1. (See Sharpe 1964, Lintner 1965 and Mossin 1966).

2. For example: (Latane 1959 ,Breiman 1961, Hakansson 1971, Finkelstein and Whitley 1981, and Algoet and Cover 1988).

3. We are in fact assuming that agent i "eats" the rest of his endowment so that whatever is not invested, $\left(1-\delta_t^i\right) e_t^i$, is consumed by agent i.

4. Here and in what follows whenever we refer to convergence, we mean almost sure convergence.

5. I.e. without imposing some of the restrictions required, for example, by Blume & Easley (1992) and Sandroni (1999) or Sciubba (1999).

6. Note that we are assuming that traders know the probability distribution p over the state space S. In a more general framework, a trader who displays a logarithmic utility function bets his beliefs.

References

Algoet, P. H. and T. M. Cover (1988). "Asymptotic Optimality and Asymptotic Equipartition Properties of Log-Optimum Investment," *Annals of Probability*, 16, 876–898.

Blume, L. E., and D. Easley (1992). "Evolution and Market Behaviour," *Journal of Economic Theory*, 58, 9–40.

Breiman, L. (1961). "Optimal Gambling Systems for Favorable Games," in *Proceedings of the Fourth Berkeley Symposium*, University of California Press.

Fama, E. F. and K. R. French (1996a). "Multifactor Explanations of Asset Pricing Anomalies," *Journal of Finance*, 51, 55–88.

Fama, E. F. and K. R. French (1996b). "The *CAPM* is Wanted, Dead or Alive," *Journal of Finance*, 51, 1947–1958.

Finkelstein, M. and R. Whitley (1981). "Optimal Strategies for Repeated Games," *Advanced Applied Probability*, 13, 415–428.

Goldman, M. B. (1974). "A Negative Report on the 'Near Optimality' of the Max-Expected-Log Policy as Applied to Bounded Utilities for Long Lived Programs," *Journal of Financial Economics*, 1, 97–103.

Hakansson, N. H. (1971). "Multi-Period Mean-Variance Analysis: Toward a General Theory of Portfolio Choice," *Journal of Finance*, 26, 857–884.

Jagannathan, R. and Z. Wang (1996). "The Conditional *CAPM* and the Cross-Sections of Expected Returns," *Journal of Finance*, 51, 3–53.

Kelly, J. L. (1956). "A New Interpretation of Information Rate," *Bell System Technical Journal*, 35, 917–926.

Latane, H. A. (1959). "Criteria of Choice among Risky Ventures," *Journal of Political Economy*, 67, 144–155.

Lintner, J. (1965). "The Valuation of Risk Assets and the Selection of Risky Investments in Stock Portfolios and Capital Budgets," *Review of Economics and Statistics*, 43, 13–37.

Merton, R. C. and P. A. Samuelson (1974). "Fallacy of the Log-Normal Approximation to Optimal Portfolio Decision-Making over Many Periods," *Journal of Financial Economics*, 1, 67–94.

Mossin, J. (1966). "Equilibrium in a Capital Asset Market," *Econometrica*, 34, 768–783.

Sandroni, A. (1999). "Do Markets Favor Agents Able to Make Accurate Predictions," *Econometrica*, 68, 1303–1341.

Sciubba, E. (1999). "The Evolution of Portfolio Rules and the Capital Asset Pricing Model," DAE Working Paper n. 9909, University of Cambridge.

Sharpe, W. F. (1964). "Capital Asset Prices: A Theory of Market Equilibrium under Conditions of Risk," *Journal of Finance*, 19, 425–442.

Chapter 19

ADAPTIVE PORTFOLIO MANAGERS IN STOCK MARKETS: AN APPROACH USING GENETIC ALGORITHMS

Kwok Yip Szeto
Department of Physics
Hong Kong University of Science and Technology
Clear Water Bay, Hong Kong, China
phszeto@ust.hk

Abstract The problem of time series prediction is transformed into one of pattern recognition with rule extraction under the framework of genetic algorithm. The time series is digitized into an ordered sequence of alphabets representing different classes of fluctuation by standard signal processing techniques. The problem of forecasting financial time series is then mapped into one of classification of the daily rate of return. Results obtained on real as well as artificial time series indicate that genetic algorithm is a useful method for forecasting. These applications were also modeled in the context of multi-agent system, where each agent represents an adaptive portfolio manager who has a particular rule of prediction. These portfolio managers will be supplemented with certain human characters, such as a level of greed and fear, so that they will have different behaviors in their investment strategies. Group of investors form team that possesses certain average view on the market defined by the conventional wisdom of the team members. The effects of this collective view, which can be first taken as a source of random news, are incorporated via a model of herd effect to characterize the human nature of the investors in changing their original plan of investment when the news contradicts their prediction. Evolution of the net asset value of the teams, as a prototype of fund house, is monitored and discussed in terms of the general human characters assigned to the team members. The transactions taken by each team member are recorded and the decision process in the switching of assets is modelled with different level of sophistication. A universal feature that greedy and confident investors outperform others emerges from this study. A general discussion that relates the consistency of performance, rate of increase of net asset and the human nature in the decision process is made.

Keywords: Portfolio Management, Financial Time Series, Forecasting, Genetic Algorithm, Herd Effect

Introduction

Being one of the most complex systems with data readily available on the Web, stock market is an ideal system for investigations by scientists and financial analysts alike, while the spin-off is obvious for real profit taking by individuals or fund house. It is also an ideal complex system in that the laws underlying the dynamics are not even proven to exist, unlike the comparatively simpler system of physics where the laws are sometimes well known. The difficulty of analyzing stock data does not stop here, for even if the underlying laws of economics trends are known, there is no way to predict the rather elusive human behavior. Therefore, the aim of this exposition is not to provide any concrete solution to this complex problem. Instead, I hope to provide a framework for discussing this problem in the context of evolution computation. As the complex system evolves, the "underlying laws" should also evolve along, albeit at a slower time scale. The so called laws may be just a set of rules for prediction, but the platform on which they evolve and interact hopefully remains unchanged for a sufficiently long period to permit a sensible formulation. Indeed, to a large extent the questions posed by complexity arising out of the interacting rules of predictions can be answered by the simulation of the collective behaviors of group of individual with specific human character. It is this ultimate goal of bringing in the subjective and elusive psychological factors of the portfolio managers that makes the usage of evolution computation attractive. By formulating a simple architecture for the adaptive portfolio managers, I hope that I can arouse the interest of curious readers to build more complex models for the description of this fascinating subject.

In the analysis of stock market, often computer programs modelling various investment strategies are employed and statistical results on the yields of these strategies are used as a measurement for both the models and the decision process in the implementation of the strategies (Arthur 1995; Cutler, Poterba, and Summer 1989; Palmer et al. 1994; Friedman and Rust 1991). Usually the computer program is a given set of investment rules, extracted from historical data of the market, or rules based on fundamental analysis or news obtained from the inner circle of the trade. It is very difficult in real life to separate out the importance of technical analysis from other means of prediction, such as fundamental analysis with random news. Furthermore, the decision process of a trader with a given investment strategy or trading pattern may be very complex, as sudden news from fellow traders may alter the decision. Consequently, it is extremely difficult to combine artificial trading strategies with the complex psychological processes taking place in the actual decision of the trader to achieve a reasonable understanding of the global patterns of trading in stock markets (Palmer et al. 1994).

In order to initiate the investigation of the complex behaviours of traders in stock market, we break this problem into several simpler ones and hope that the solution of each will shed some light on the general pattern of traders. We first modify the traditional economic models where homogeneous agents operating in isolation are used by two simple steps. The first step is to incorporate a certain level of communication among traders so that their decision processes are affected by their interactions. The second step is to relax the condition that the agents are homogeneous. There are of course many ways to introduce heterogeneous agents, but our focus is on the individuality of the agent, with personal character that entails different psychological responses to other agents. These two steps are by nature not easy to model, as human interactions and psychological responses are themselves very challenging topics. Nevertheless, we initiate the investigation by considering some simple techniques used by physicists. The first problem of incorporating interactions can be modelled by the standard technique of mean field theory. This means that each trader will interact with the average trader of the market, who is representative of the general atmosphere of investment at the time. This will naturally incur some error as effects of fluctuation, or the volatility of the stock market, may not be adequately treated. Next, we like to consider the psychological response of individual trader. We can introduce simple quantitative parameters to measure individual characteristics of the trader, so that the response to the general atmosphere of the market is activated according to these parameters. A simple model that takes these considerations into account must therefore include heterogeneous agents who are represented by different rules of investment as well as different human characters. However, the interactions between agents are simplified into the interaction with a mean field, or the general atmosphere of the market, for which a sensible description must be first made. To this end, we model the general atmosphere of the market in a rather ad hoc manner. We do not deduce it from a model of microscopic interaction between agents, but rather by a source of random news that serves as a kind of external, uncontrollable stimulus to the market.

Before we discuss the model of the human psychological response to the "mean field" or the "general atmosphere" of the market, we must first model the individual investor. Indeed, the individual can be considered as an agent, belonging to a fund house that forms a team representing the interests of a particular population. In this context, the adaptive portfolio managers can be viewed as a multi-agent system. Now, the simplest model of the individual investor should at least include a rule of investment in the absence of noise. This should be made before one can discuss the architecture of the collective behavior of the adaptive portfolio managers. The common task of these agents is to analyze the market, which behavior is in term affected by their analysis and subsequent course of action. After we have built a model of the agent, we can supplement this agent with specific value and character, representing the human psychology of a particular subset of investors.

In order to model the agent or adaptive manager, one must first endow him/her with a particular skill of technical analysis. This skill is best modeled by a particular rule of prediction on the stock market, which the agent believe to be golden from his/her experience of the past. In order to obtain a golden rule, the agent must perform data mining of the past events. This then naturally leads towards the problem of prediction rule extraction from the past data. This becomes a problem of forecasting in financial time series.

Forecasting financial time series has been of great interest to investors due not only to its complex behavior, but also to the financial implications of a successful tool for prediction. Various models and methods (Campbell and MacKinlay 1997; Zhang 1999; Mantegna Stanley 1999) have been designed to give a good forecasting tool on stock prices and/or the collective behaviors of the investors. Here we discuss some examples of the successful application of genetic algorithm (Holland 1975 and Goldberg 1989) in forecasting patterns in financial time series. The results are usually better than many benchmark tests. The basic idea behind this successful approach is that prediction is transformed into a problem of pattern recognition (Szeto, Chan, and Cheung 1997 and Szeto and Cheung 1998). Instead of treating forecasting as a problem of dynamics of stochastic systems (Follmer and Schweizer 1993; Follmer 1995; Naik 1997; Lux 1997), the new approach treats it as one of pattern matching, learning, and generalization. Within the confines of the structure gained by limiting consideration to genetic algorithm for forecasting, we find the treatment to be fairly general. Furthermore, the flexibility inherent in this approach is guaranteed by the absence of any assumption on the underlying dynamics. Thus, we believe that the philosophy behind our approach in treating the forecast of financial time series as a pattern recognition problem can be more readily applied to other fields than the theory of chaos (Scheinkiman and LeBaron 1989; Brock and Hommes 1998; Hsieh 1991; Provenzale et al. 1992; Osborne and Provenzale 1989; Mantegna and Stanley 1996), in spite of its enormous success.

Pattern recognition problem can be investigated with many techniques, such as neural network, genetic algorithms and other heuristics (Weigend et al. 1997 and Froehlinghaus and Szeto 1996). Here we discuss genetic algorithm due to the relative flexibility of interpretation of the rules of prediction, in contrast to the more abstract approach of neural network. The raw data of a financial time series can be preprocessed using standard signal processing techniques before being treated by evolutionary computation. If one decides to digitize the signal, then the problem of forecasting can be further mapped into one of classification. In this special approach, many standard techniques in classification, such as CART, ID3, Bucket Brigade, etc. can be used. Our approach of extracting rules of prediction using genetic algorithm compares well with these techniques.

Since the common objective of forecasting financial time series is to optimize the rate of correct prediction, one of our formulations of genetic algorithm is to make use of an online update of the rate of correct prediction to map the problem of forecasting as one of simple optimization. The objective function being the rate of correct prediction, subjected to a set of constrains that is set by past patterns. This approach makes available for financial analysts a large bag of tricks in traditional optimization problem. Furthermore, our methodology of financial forecasting for single time series can be applied to multiple time series (Szeto and Cheung 1998). Of course, once multiple time series is addressed, the problem of portfolio management will have to be treated (Sharpe 1966 and Szeto and Fong 2000). This rather complex issue, which relates to scheduling and information theory, is outside the scope of the present exposition. Nevertheless, we find some universal behaviors in terms of the self-organization in the evolution of rules (Szeto and Luo 1999). These universal features provide useful insights into the collective behaviors without detailed fine-tuning of parameters.

Since the topics of discussion is on the collective behavior of adaptive portfolio managers, we will not go further into the more technical topics of forecasting financial time series. We will briefly describe the results of several forecasting tools developed by our group as example of the application of genetic algorithm in finance. The tools are basically models for individual agents. We understand that these simple models of stock market traders lack the generality of a realistic agent.

However, I emphasize that refinement can be introduced later. Once a specific model for the single agent has been adopted, we can begin the construction of a multi-agent system with interactions between agents. Further tuning in the heterogeneity of agents can be implemented for a closer approximation to reality. We hope that this new approach of modelling the microscopic agents is a first step towards building a more comprehensive model of the stock market and its complex patterns.

We discuss the data preprocessing techniques in section 1. We then discuss the performance of the forecasting tool for real time series as well as artificial time series with memory in section 2. In section 3, a simple model of greed and fear of the individual is built, and the collective behavior of a group of individual investors endowed with the human character of greed and fear will be tested. Finally, a brief outline for future research is made in section 4.

1. Data Preprocessing

Financial time series usually can be considered as the sum of a trend and noise. While the trend may be described by some complicated dynamics, the noise is usually assumed to be white. Since the trend is valuable for forecasting, one normally employ various filtering techniques common in signal processing to separate it from the noise. However this approach will be difficult to produce a generic tool for forecasting due to the general ignorance of the numerous factors that determine the trend. Indeed, the usual goal in application is to optimize the rate of correct prediction, not so much as to search for a fundamental theory behind the trend. Furthermore, a theory of the trend in a given time series cannot be too general, as the number of economic and political factors are numerous, not to mention the difficulty of their precise quantification. Consequently, the signal processing techniques used here is mainly for the purpose of noise reduction and not for prediction, and no attempt is made on the theory behind the trend.

For data preprocessing, we aim at transforming a time series of rational numbers into one of alphabets or integers. We use a vector quantization technique to encode the time series as a sequence of integers corresponding to q classes. For example, for $q = 2$, one can represent large fluctuation as class 1 and small fluctuation as class 0. In general, the fluctuation of the input $\{X'(t)\}$ is computed as the fractional change in each interval,

$$y(t) = \frac{X'(t) - X'(t-1)}{X'(t-1)}. \tag{19.1}$$

This is the daily rate of return. For $q = 2m + 2$, one can put down $2m + 1$ boundary values $\{y_{-m}, ..., y_0, ..., y_m\}$ for q levels of fluctuations. For example, if $m = 1$ and $q = 4$, we have four classes, with new data point $X(t) = -1$ when $y(t) \leq y_{-1}$, $X(t) = 0$ when $y_{-1} < y(t) \leq y_0$, $X(t) = 1$ when $y_0 < y(t) \leq y_1$, and finally $X(t) = 2$ when $y_1 < y(t)$. Note that for a given q, the original data set of N points will be divided into q sets, with $N_1, ..., N_q$ members respectively.

For a given q, *the choice of the boundary values* can be computed using standard statistical analysis on the original time series to satisfy several conditions. One criteria is to maximize the signal to noise ratio. Another is to set $N_1 = N_2 = ... = N_q$. This criterion imposes a strict constraint on the boundary values, but the results will ensure a more precise comparison on the performance of the prediction tools on each class. A third criterion on the boundary values is to adopt those normally used by traders on the daily rate of return to achieve immediate application. Since here we prefer to have a systematic comparison with other forecasting tools, we adopt the second criterion so that each class appears with almost equal probability. After obtaining the boundary values by analyzing the statistic, we convert a time series into an integer sequence of data $\{X(t)\}$ defined on the alphabets $A' \equiv \{-m, ..., 0, ..., m, m + 1\}$, for $q = 2m + 2$. Each string is called a message unit and is associated with an action unit which also belongs to A'. The problem of forecasting a time series is now transformed into a problem of *extraction of rules of prediction* based on a time-ordered sequence of integer data. To normalize the notation, we will rename the q classes with the alphabet $A = \{0, 1, ..., q - 1\}$.

Since rule extraction is inherently a pattern recognition problem, it is natural to divide the sequence $\{X(t)\}$ with N points into two parts: a training set and a test set. The training set consists of $M - L - 1$ strings of length equal to L digits, along with the known associated action unit of each string. The test set consists of $(N - (M - L - 1))$ sets of strings of length L, similar in structure to the training set, but the action unit should be used for performance evaluation. The choice of L is important and one method is to use *information entropy* to find the optimal M and L for major stock market (Cheung, Szeto, and Tam 1996; Diks et al. 1996; Granger 1969).

2. Results Of Genetic Algorithm As a Forecasting Tool

In testing genetic algorithms, we use data of major stock markets as well as controlled time series with short memory. Real data cannot provide a systematic study on the performance of the algorithms, but has the advantage of being immediately applicable. On the other hand, time series generated artificially with controlled parameters can be useful for a better understanding of the conditions under which the algorithms are most suitable.

2.1. Generation of Time Series

We generate time series with controllable memory measured by the autocorrelation function using the method of the *inverse whitening transformation* (Fukunaga 1990 and Fong and Szeto 2001b). First let Γ be a diagonalizable $n \times n$ matrix, with eigenvalue matrix $\Lambda = diag[\lambda_1, ..., \lambda_n]$, eigenvector matrix $\Phi = [\phi_1, ..., \phi_n]$, so that $\Gamma\Phi = \Phi\Lambda$. Using $T = \sqrt{\Lambda}\Phi^T$, the correlation matrix of $Y = TX$ can be shown to be the identity matrix if the covariance matrix of the random variable X is Γ. Our first step is to generate an independent, normally distributed random variable Y with zero mean and unit variance. Then we define the covariance matrix Γ, which is a Toeplitz matrix with entries $C_{ij} = C_{ji} = C(|i-j|)$. It is a real symmetric matrix with unit diagonal and the function $C(|i-j|)$ is related to the correlation function with given memory structure. If the required time series has a short memory, one assumes an exponentially decaying function for these elements $C_n = C(n) = exp(-n/\tau)$. Here τ is the range of correlation and n is the number of days in the past. The final result is a random variable $X = T^{-1}Y$ with correlation given by the covariance matrix Γ .

2.2. Forecasting of Artificial Time Series with Short Memory

To illustrate our method, three sets of short memory time series with 2000 data points are produced, with correlation functions $C(n)$ with τ=5, 10, and 15. The first 1000 data points form the training set and the next 1000 points for the test set. In the training set, cutoff values are used to put data points into five categories so that the number of data points of each class is approximately the same. We want to maximize the correct percentage as well as the guessing percentage. This difficult requirement is partially accomplished by including both specific rules and general rules in each generation. The former will try to increase the correct percentage while the latter will increase the number of guessing, N_g. In general the ratio of correct guess/total number of guess on the test set is around 50% to 60%. For benchmark comparison with random guess, it gets a maximum of 20% since there are 5 equally likely classes. Another benchmark is random walk. It uses the value of

Table 19.1. Result of Prediction

	Percentage of Correct Prediction		
Correlation Function	Genetic Algorithm	Random Guessing	Random Walk
$e^{-t/5}$	50	20	25
$e^{-t/10}$	58	22	27
$e^{-t/15}$	63	21	24

previous time unit to predict the present unit. This method gives 25% of correct predictions (See Table 19.1). We see that genetic algorithm performs best, especially in time series with long memory (larger time constant τ). This is quite reasonable since the rules must exploit the correlation within the time series. Furthermore, the method provides a set of interpretable rules for forecasting (Fong and Szeto 2001b). In this example, the number of points, 2000, is set for convenience. One can readily test the method with a larger set by performing the inverse whitening transformation on a larger covariance matrix Γ .

2.3. Forecasting of Real Financial Time Series

A second example (Szeto and Luo 1999) illustrates the application on real financial data. The blue chip data shown in Figure19.1 is divided into training set with 1712 data points and test set of 190 data points. Each experiment is started with different seed of random numbers and evolves for 2000 generations of the genetic optimizer. The experiment is repeated 15 times and the measurements are averaged.

Two benchmark tests of the same set of data are used. The first one is the random guess of two classes: stock prices go up and down with equal probability. The second one is the random guess of two classes based on past statistics, in which case the probability of choosing 1 is 0.5144 and choosing 0 is 0.4856. We summarize the results in the genetic optimizer in Table 19.2. The first row of data refers to the performance evaluation of the optimizer in the training set, and the second row to test set. The number C^i is the correct guess minus the wrong guess when i is the guess ($i = 0, 1$); G^j is the total number of guess made for j ($j = 0, 1$). Note that $G^0 + G^1$ is less than the number of examples in the training set, an indication that sometimes no guess is made. We also show in bracket the error bar of the number in each column. For comparison with the case of random guess, we look at the average number of correct guess minus the wrong guess, which is the sum of C^0 and C^1. This sum is a good measure of the performance of the predictor. The ideal result is that

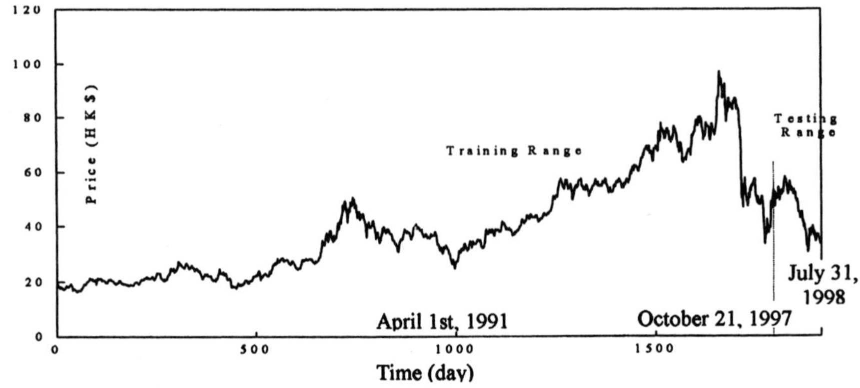

Figure 19.1. Stock price of a Blue Chip in Hong Kong Hang Seng index, before the Asian crisis and after.

C^1=80 and C^0=110, with a sum of 190. The worst case has C^1=-110 and C^0=-80, which sums to -190. Now, we have $C^0 + C^1$=+8.2 for the self-organized genetic optimizer, $C^0 + C^1$=-4.1 for random guess with equal probability, C^0+C^1=-5.6 for random guess using the probability of 1 (0.5144) and 0 (0.4856) from the training set. All tests are performed over the same set of data with 190 testing points. The conclusion in this simple comparison is that the genetic optimizer has some intelligence in that it beats random guessing in both cases. We can convert these numbers into probability of correct prediction. Let P^k_{test} and P^k_{train} be the probability of correctly predicting the class k for the testing set and the training set respectively. We have for the genetic optimizer, P^0_{test}=0.59, P^1_{test}=0.45, P^0_{train}=0.6, P^1_{train}=0.59. The sum of probability of making a correct guess for both classes in the test set is 0.59+0.45=1.03, greater than one, an upper limit of any scheme of random guess. This is a clear indication that the algorithm does learn from the training set and perform forecasting with some intelligence.

Table 19.2. Performance of Self-Organized Genetic Optimizer

Distance	C^0	C^1	G^0	G^1
Training Set	130.6(4.6)	169.4(4.7)	662.6(56.5)	942.3(54.8)
Test Set	17.3(3.7)	-9.1(8)	91.9(6.7)	94.4(6.4)

3. Portfolio Management

After building a model for the individual investor, we are in the position to discuss the collective behavior of many investors. We can regard them as a group of adaptive portfolio managers equipped with rules of prediction that are shown to be successful in the past. Though their collective view on the market constitutes the conventional wisdom, these investors can have very different "*personality*," reflected in their responses to the breaking news in the market. They may also have very different character, such as greed and prudence, which are manifested in their overall investment strategy. These complexities naturally demand a more detailed exposition on the psychology of the investors. Nevertheless, I hope that a simple model based on the "single particle approximation" in physics can be used to initiate the discussion of the collective behavior of agents. My approach is to simplify the psychology of the agents by considering a single agent with personality characterized by several parameters, and the complex interactions that exist in agents by the much simpler interaction of the single agent with breaking news. The news is poor man's approximation to the mean field, which is the standard trick of treating many-body interaction in statistical physics. As to the treatment of the psychological responses, I decide to model only two aspects of the personality of the investor: level of confidence and degree of greed. Though there are certainly many more aspects needed for the description of the character of the agent, I consider these two to be rather important. The level of confidence reflects the probability of change of the original strategy of investment, which is based on hard work on past data. The degree of greed reflects the relative portion of each asset involved in each transaction, which definitely affects the final outcome of the investment. We will see some interesting universal results generated from my mathematical treatment of these two aspects of the personality. Indeed, one can easily generalize my treatment below to other input parameters to the portfolio management problem. This flexibility inherent in my approach fulfills the key objective of my model, i.e., opens up the avenue for further investigation.

3.1. Level of Confidence and Response to News

Let's first consider the mathematical measure for the level of confidence of an investor. In order to find out how confident the investor is on his view on the market, we must test his response to the arrival of breaking news. Of course, a controlled experimentation will require a model for the breaking news, as the news in the real world is too complex and fundamentally uncontrollable.

The present work treats "news" as a randomly generated time series. Of course, this can be made more realistic by taking some kind of average of many real series of news as the input stimulus to the agent. One can also include a more detailed model of interaction of agents so that the input stimulus to the agent is neither an artificial time series, nor an external time series of news, but an internally generated series that reflect the dynamics of interacting agents. This will be studied in a future paper. For now we focus on a simple time series of news, which can be viewed as good or bad or neutral, by conventional wisdom.

Next, we consider the interaction of an individual agent with the randomly generated time series of news. The agent has to decide whether and how his action should be modified in view of the news. In making these judgements, the agent must anticipate certain probability of change, which reflects the level of confidence of the agent in his decision process. For example, when there is news that is good in conventional wisdom, the stock market price is generally expected to increase. An agent, who had originally forecasted a drop in the stock price tomorrow and planned to sell the stock at today's price by taking profit, may change his plan after the arrival of the "good" news, and halt his selling decision, or even convert selling into buying. This is a reversal of his original decision that is solely based on historical data analysis. Similarly, for an agent who originally wanted to buy the stock at today's price, as his forecast for tomorrow is a rise in stock price, may halt his buying action because "bad" news just arrives. Instead of buying today, he may sell or hold on to his cash, for fear of a crash. This kind of reversal of buying action may in reality trigger panic selling. These sometime irrational behaviours, in anticipation of the immediate effect of news on the stock market, will often reverse the original decision that is based on rational analysis on historical data through pattern matching.

To incorporate these realistic features of the markets, we introduce several numbers to model the market. First of all, we make use of the simplest model for the agent, since our objective here is to illustrate the method for measuring the level of confidence. The investor is a simple "rule of prediction." This rule is the result of education the agent received based on technical analysis of the past events. Furthermore, we consider only the simple case of binary prediction, in that the integer characterizing the class of prediction is either 1 for increases or 0 for decreases or unchanged stock price. To measure the level of confidence, I revert to a different concept of the level of fear first. I introduce a real number f to characterize the level of fear of the agent in his original decision. If f is 0.9, then the agent has 90% chance of changing his decision when news arrives that contradicts his original decision.

This denotes an insecure investor who easily changes his investment strategy. If f is 0.1, then there is only 10% chance of the agent of changing his original decision, a sign of a more confident investor. Thus, f is a measure of fear and the measure for the level of confidence is $1-f$.

Algorithmically, we first choose a random number c between 0 and 1. We choose the following rule: If $c > 0.5$, the news is good, otherwise it is bad. This model of random news may not be realistic, but will serve as a benchmark test on the effect of news on the agents by saying that there is equal chance of arrival of good news and bad news. There are four scenarios for the agent with news.

(1) News is good and he plans to sell.

(2) News is good and he plans to buy.

(3) News is bad and he plans to sell.

(4) News is bad and he plans to buy.

We note that there is no contradiction between the agent's original plan and the news for case (2) and (3). However, in case (1), the agent may want to reverse the selling action to buying action due to the good news, anticipating a rise in stock price in the future. Also, in case (4), the agent may decide to change his decision of buying to selling today, and buying stock in the future, as the news is bad and the stock price may fall in the future. Thus, in (1) and (4), the agent will re-evaluate his decision. He will first choose a random number p. If $p > f$, he will maintain his prediction, otherwise he reverses his prediction from 1 to 0 or from 0 to 1. Here we see that the parameter f is a threshold value for the probability above which he will maintain his prediction, and below which he will reverse his prediction. The bigger the value of the threshold f, the smaller the chance the random number p will be greater than f, implying that the smaller the chance he will maintain his original prediction. Therefore f is a measure of fear of the agent and $1 - f$ is the measure of confidence.

3.2. Level of Greed

In modelling the individual agents so that their collective behaviour observed in the stock market is simulated, one must consider many features associated with the human character of the investors. The level of confidence of the investor is revealed in the response to news. Another character is the level of greed of the investor. This human character is more intrinsic than the level of confidence, as the later is associated with the specific training the agent received as well as the news generated from the external world. Greed is an important aspect since it directly affects the size of investment once the investor decides on a particular course of action. For a greedy investor, he may be very aggressive in all his investment, while a prudent investor will be more conservative in his action.

Let's consider the more realistic situation of the action taken by the agent. Let's use a real number g to characterize the percentage of asset allocation in following a decision to buy or sell. If g is 0.9, it means that the agent will invest 90% of his asset in trading, a sign of a greedy gambler. On the other hand, if g is 0.1, the agent only invests 10% of his asset in trading, a sign of a prudent investor. Thus, g can be interpreted as a measure of greed. In the context of our simple model of stock market, the assets are either stock or cash. Thus, we can use g as a measure of the percentage of the amount of cash used for buying stock or the shares in selling stock. A large greed parameter g implies that the investor is a big gambler, who will invest a large fraction of his asset following the rules and the news, while a small g parameter characterizes prudent investors.

3.3. Portfolio Management in the Presence of News

For a given past pattern, a particular agent will first make a comparison of this data with his rule. If the pattern matches his rule, then the prediction according to the rule is made. Without the news, the prediction is definite, for the agent is supposed to execute the action suggested by the rule once the data is matched. However, in the presence of the news, the agent will have to re-evaluate his action, reflecting the changed circumstances implied by the news.

First of all, we define the forecasting problem by performing training, learning, and testing for a given time series, $x(t)$, with 2000 ($=N$) data points. We divide it into three parts. The first 800 ($=N_{Train}$) is the training set for extracting a rule using standard genetic algorithms. The next 100 ($=N_{Test}$) points form the test set, used for evaluating the performance of the set of rules obtained after training. The last 1100 ($=N_{News}$) points form the news set for investigating the performance of investors with different degree of greed and different level of indifference to the random news. In the training set, we make the usual assumption that at time t, the value $x(t)$ is a function of the value at $x(t-1), x(t-2), ..., x(t-k)$. Here k is set to 8. In the simplest linear model of time series forecasting, we can assume the following relation between the predicted value $\hat{x}(t)$ and its precedent:

$$\hat{x}(t) = \sum_{i=1}^{k} \beta_i x(t-i) \tag{19.2}$$

The objective is to find a set of $\{\beta_i\}$ to minimize the root mean square error in $\hat{x}$ compared with the true value x. Here, we do not perform any vector quantization on the series $\{x(t)\}$, so that each $x(t)$ is a real value. For the case of financial time series that we use for experimentation, these $x(t)$ values represent the daily rate of return of a chosen stock, with $|x(t)| \leq 1$. We also apply the same condition on the parameters β_i, so that $|\beta_i| \leq 1$, $i = 1, ..., k$. Since our objective is to look at performance of agents who buy and sell the particular stock, we limit ourselves to the simple sign detection of $\hat{x}$. What this means is that for $\hat{x}$ positive, the agent predicts an increase of the value of the stock and will act according to his specific strategy of trading. If $\hat{x}$ is non-positive, then he will predict either an unchanged stock price or a decrease, and will also act according to his specific strategy. We count the guess as a correct one if the sign of the guess value is the same as the actual value at that particular time, otherwise the guess is wrong. If the actual value is zero, it is not counted. The performance index P_c that measures the fitness value of the chromosome is designed as the ratio:

$$P_c = \frac{N_c}{N_c + N_w} \tag{19.3}$$

Here N_c is the number of correct guess and N_w is the number of wrong guess. Note that in this simple genetic algorithm, we do not worry about the absolute difference between x and $\hat{x}$. We only pay attention to their signs. This is acceptable in the context of our simple model and can be easily made more realistic by considering more refined classification of the quality of the prediction. Furthermore, while most investors make hard decision on buy and sell, the amount of asset involved can be a soft decision. Indeed, when we introduce the greed parameter for the agent as discussed below, the decision on the amount of asset involved in the trading of a stock can be considerably softened. For the purpose of the present work, the prediction based on signs will be sufficient to obtain general insights of the coupling between agents and news, and the effect of this coupling on the global behaviour. Finally, we should remark that agents do not predict when $\hat{x}$ is zero, corresponding to the situation of holding onto the asset.

We start with a set of 100 rules (chromosomes) represented by $\{\beta_i\}$. By comparing the performance of the chromosomes, the maximum fitness value is the result found by genetic algorithm using a modification of the Monte Carlo method that consists of selection, crossover and mutation operators (Szeto, Chan, and Cheung 1997; Szeto and Cheung 1998; Szeto and Luo 1999; Fong and Szeto 2001b). We will skip the technical details of the genetic algorithm. Here we just state the results. After several thousands of generations, we observe a saturated value of P_c, and we choose the chromosome corresponding to this P_c as the generic rule of prediction (Figure19.2). We simply make use of the adaptive nature of the different chromosomes in a Darwinian world to single out the best performing chromosome, or the fittest agent, to be the prototype of agents with different human characters. Now that we have the evolved agent, we can introduce two parameters to characterize different human nature of the best agent. These two parameters are the greediness "g" and level of fear "f". The final set of agents, all with the same chromosome (or rule), but with different parameters of greed g and fear f, will be used for the performance evaluation in their portfolio management in response to online news. In the news set (last 1100 data points), we assign identical initial asset to the agents. For instance, we give each rule (chromosome, portfolio manager, or agent) an initial cash amount of 10,000 USD and a number of shares =100. The value of f and g ranged from 0 to 0.96 in increment of 0.04 will be used to define a set of $25 \times 25 = 625$ different agents. We will then observe the net asset value of these 625 portfolios as they trade in the presence of news.

To perform portfolio evaluation, we choose a probation period of 10 days to examine the performance of each of the 625 agents. We use several sets of time series, including the real stock value of Microsoft, the long and short memory time series with controlled auto-correlation generated using the inverse whitening transformation. All these time series show similar behaviour. We will only illustrate the results on the Microsoft data. The initial net asset value of all 625 portfolios is \$19,900, corresponding to \$10,000 cash and 100 share of stock at \$99 a share. To characterise the effects of news on the performance of the agents, we illustrate the steady state values of the portfolio of the agents in a three-dimensional plot (Figure 19.2). The x-axis labels the degree of greed g of the investors, the y-axis the level of confidence of the agents and the z-axis the value of the portfolio. By steady state value, we mean the final value of the portfolio. We find that the net asset of the portfolio, (cash plus stock) reaches a steady value after more than 1000 responses to random news. From our numerical experiment with a ten days probation period, (we then have 110 evaluations on the news set), we observe to our surprise similar patterns to all the data set. In the three-dimensional plot of portfolio value vs. g and f shown in Figure 19.2, we observe a trend of the portfolio measured in net asset value in cash to raise at large g and small f. This is an interesting universal behaviour that

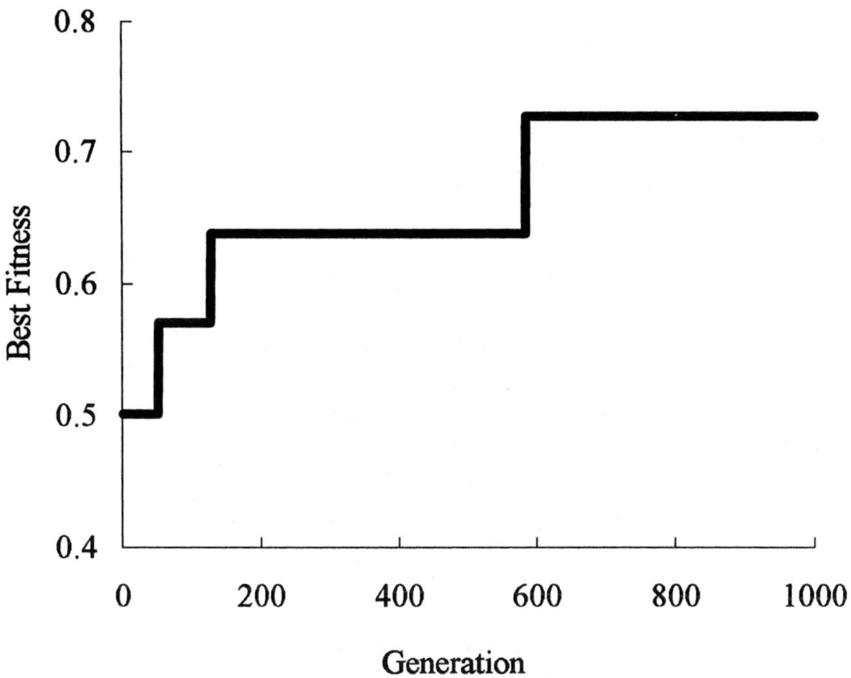

Figure 19.2. The fitness versus generation number for Microsoft.

demands an explanation. However, we do not attempt a detailed analysis of this observation or provide a model to explain the results. We just want to report this observation with the following interpretation. In a pool of agent that are trained by historical data, and endowed with individual characters like greed and fear in their investment exercises, the effects of news, generated randomly, show universal behaviour. This universal behaviour suggests that greedy (large g) and confident (small f) investors perform better.

4. Discussion

By transforming the problem of forecasting time series into a pattern recognition problem, we construct a learning classifier system based on genetic algorithm, which performs better than both the random guess and random walk method on artificial data as well as real data (Szeto,

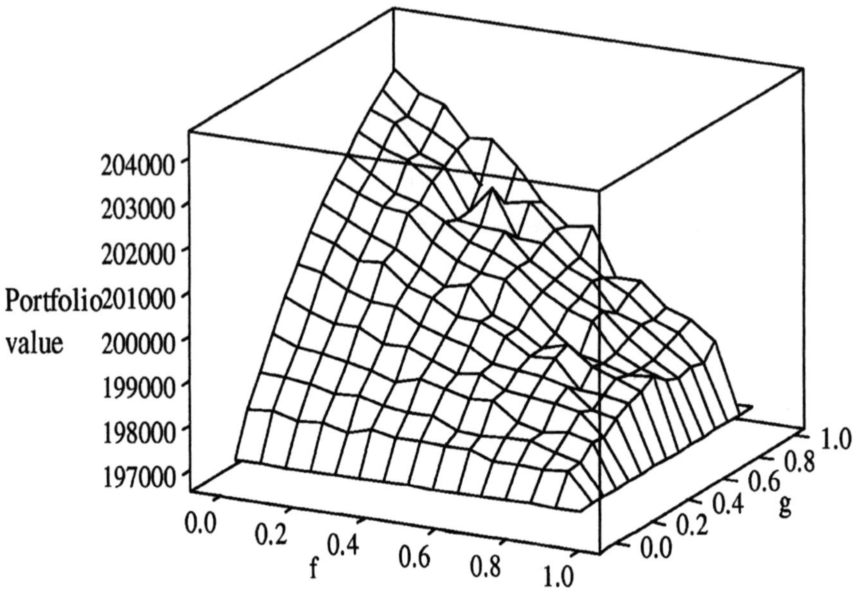

Figure 19.3. Final net asset values in cash of the portfolio of the 625 agents. Note that initially all agents have the same value at 19900. The time series for stock is Microsoft.

Chan, and Cheung 1997; Szeto and Cheung 1998; Szeto and Luo 1999; Fong and Szeto 2001b). Our method also provides a set of interpretable rules for forecasting (Fong and Szeto 2001b). In general, genetic algorithm performs better in time series with longer memory. A simple parameterization of fitness reveals an interesting self-organized behavior in the evolution of the population of prediction rules, with the ability to generalize with satisfactory diversity (Szeto and Luo 1999). This is superior to the use of Lagrange method for implementing constraints in the statistical properties of the rules (Szeto and Cheung 1998). Also, the tests based on real data as well as in other controlled data show evidences for cluster separation.

We have also investigated the interesting problem of interacting adaptive portfolio managers. The adaptive portfolio manager is a simple agent, or investor, who has to manage the two-assets portfolio by a given rule of prediction. This rule is obtained from the education on the technical analysis of time series, using past data as training. The best rule is a prototype for the agent. By endowing this agent with different set of human characters, we obtain a team of portfolio managers who have different level of confidence and greed. They form a team and make predictions on the market according to the technical rule, but have finite probability of changing the technical prediction when confronted with breaking news. Furthermore, they each have different level of greed, shown in the way they invest their asset in trading. This simple model has been tested on real data as well as artificial time series, and we obtain the interesting result that greedy and confident investors achieve a higher net asset value over a period of 1100 test points of prediction. The result is a simple illustration of the complex problem of adaptive portfolio management. Certainly many generalizations and better modeling with the inclusion of other human characters will make the collective behavior of the investors much more complex. This forms a basic platform for which genetic algorithm can be used as the first part in educating the agents by training with past data, whether real or artificial, and the subsequent analysis of the collective behavior an application of the result of genetic algorithm.

Our future work will focus on the effects of news and characters of portfolio managers, as modeled by specific set of rules for prediction. This opens up the studies of multi-agent dynamics (Szeto and Fong 2000 and Lor and Szeto 2000) involving herd effects as well as temporal optimization of the net asset value of the portfolio. In the analysis of stock market, statistical results on the yields of various strategies are often used as a measurement for the models and the decision process in the implementation of the strategies. Since in my framework investment strategies are simply rules extracted from historical data, they form a group of agents among which we can incorporate a controlled level of communication. The communication among agents will generally affect their decision, thus providing a feedback mechanism on the convention wisdom. Meanwhile, as it is easy to relax the condition that the agents are homogeneous, we can focus on the individuality of the agent, with personal character that entails different psychological responses to other agents. Consequently a natural direction of future research will be to look at interactions among heterogeneous agents. It will be very interesting if these agents manifest collective behavior found in SOC (Bergh 1996; Sneppen et al. 1995; Vandewalle and Ausloos 1997) and universal characteristics manifested in various scaling laws in physical systems (Takayasu, Sato, and Takayasu 1997 and Bouchaud and Sornette 1994). Finally, the agents in reality are not programs, but human, who are susceptible to the influence of others and personal mood. This certainly makes the entire decision process rather fuzzy. We have initiated the usage of fuzzy rules for forecasting and for making investment decision (Fong and Szeto 2001a). Further studies along these lines will be reported elsewhere.

Acknowledgement

K.Y. Szeto acknowledged that the work described in this paper was partially supported by a grant from the Research Grants Council of the Hong Kong Special Administrative Region, China (Project No. DAG grant 98/99.SC25, HKUST6123/98P, and HKUST6144/00P).

References

Arthur, W. B. (1995). " Complexity in Economic and Financial Markets," *Complexity, 1*,1995, 20-25

Bergh, J.C.J.M. van den (1996). *Ecological Economics and Sustainable Development: Theory, Methods, and Applications,* Cheltenham, UK: Edward Elgar Publishing Lirnited.

Bouchaud, J. P. and D. Sornette (1994). "The Black-Scholes Option Pricing Problem in Mathematical Finance: Generalization and Extensions for a Large Class of Stochastic Processes," *Journal de Physique I France*, 4, 863-881.

Brock, W. and C. Hommes (1998). "Heterogeneous Beliefs and Routes to Chaos in a Simple Asset Pricing Model," *Journal of Economic Dynamics and Control*, 22, 1235–1274.

Campbell, J. Y., A. W. Lo, and A. C. MacKinlay (1997). *The Econometrics of Financial Markets*, Princeton, NJ: Princeton University Press.

Cheung, K. H., K. Y. Szeto, and K. Y. Tam (1996). "Identifying Time Series Lag Structure for Developing Intelligent Forecasting Systems: A Maximum Entropy Approach," *International Journal of Computational Intelligence and Organization*, 1(1), 94.

Cutler, D. M., J. M. Poterba, and L. H. Summer (1989). "What Moves Stock Prices?" *Journal of Portfolio Management*, Spring, 4–12.

Diks, C., W. R. van Zwet, F. Takeiis, and J. DeCoede (1996). "Detecting Differences Between Delay Vector Distributions," *Physical Review E*, 53, 2169–2176.

Follmer, H. (1995). "Stock Price Fluctuations as a Diffusion in a Random Environment," in S. D. Howison, F. P. Kelly and P. Wilmott, Chapman and Hall (eds.), *Mathematical Models in Finance*, 27–33. London.

Follmer, H. and M. Schweizer (1993). "A Microeconomic Approach to Diffusion Models for Stock Prices," *Mathematical Finance*, 3, 1–23.

Fong, A. L. Y. and K. Y. Szeto (2001a). "Fuzzy Adaptive Rules in the Forecasting of Short Memory Time Series," *Proceeding of the Joint 9th IFSA (the International Fuzzy Systems Association) World Congress and 20-th NAFIPS (North American Fussy -1. Information Processing Society) International Conference*, 1, 598–603.

Fong, A. L. Y. and K. Y. Szeto (2001b). "Rules Extraction in Short Memory Time Series using Genetic Algorithms," *European Physics Journal B*, 20, 569–572.

Friedman, D. and J. Rust. (eds.) (1991). *The Double Auction Market: Institutions, Theories, and Evidence.* Proceeding Volume XIV, Santa Fe Institute Studies in the Science of Complexity. Menlo Park: Addison-Wesley.

Froehlinghaus, T. and K. Y. Szeto (1996). "Time Series Prediction with Hierarchical Radial Basis Function," *Proceedings of the International Conference on Neural Information Processing, ICONIP'96.* Springer-Verlag, 2, 799–802.

Fukunaga, K. (1990). *Introduction to Statistical Pattern Recognition.* Academic Press.

Goldberg, D. E. (1989). *Genetic Algorithms in Search, Optimization, and Machine Learning.* Addison Wesley.

Granger, C. (1969). "Investigating Causal Relations by Econometric Models and Cross Spectral Methods," *Econometrica*, 37, 424–438.

Granger, C. (1989). *Forecasting in Business Economics.* 2nd ed. San Diego: Academic Press.

Holland, J. H. (1975). *Adaptation in Natural and Artificial Systems.* Ann Arbor, MI: University of Michigan Press.

Hsieh, D. A. (1991). "Chaos and Nonlinear Dynamics: Application to Financial Markets," *Journal of Finance*, 46, 1839–1877.

Lee, C. Y. (1958). "Some Properties of Nonbinary Error-Correcting Codes," *IEEE Transactions on Information Theory IT-4*, 77–82.

Lor, W. K. F. and K. Y. Szeto (2000). "Switching Dynamics of Multi-Agent Systems: Soap froths Paradigm," *Proceeding of IEEE-INNS-ENNS International Joint Conference on Neural Networks, Como, Italy, 24-27 July 2000 ; Neural Computing: New Challenges and Perspectives for the New Millennium.*

Lux, T. (1997). "Time Variation of Second Moments from a Noise Trader / Infection Model," *Journal of Economic Dynamics and Control*, 22, 1–38.

Mantegna, R. N. and H. E. Stanley (1996). "Turbulence and Financial Markets," *Nature*, 383–587.

Mantegna, R. N. and H. E. Stanley (1999). *An Introduction to Econophysics: Correlations and Complexity in Finance.* Cambridge University Press. p.180.

Naik, N. Y. (1997). "On Aggregation of Information in Competitive Markets: The Dynamics Case," *Journal of Economic Dynamics and Control*, 21, 1199–1227.

Osborne, A. R. and A. Provenzale (1989). "Finite Correlation Dimension for Stochastic Systems with Power-Law Spectra," *Physica D*, 35, 357–381.

Palmer, R. G., W. B. Arthur, J. H. Holland, B. LeBaron, and P. Tayler (1994). *Artificial Economic Life: A Simple Model of a Stockmarket.* Physics D, 75, 264–274.

Provenzale, A., L. A. Smith, R. Vio, and G. Murante (1992). "Distinguishing Between Low-Dimensional Dynamics and Randomness in Measured Time Series," *Physica D*, 58, 31–49.

Scheinkman, J. A., B. LeBaron (1989). "Nonlinear Dynamics and Stock Returns," *Journal of Business*, 62(3), 311–337.

Sharpe W. F. (1966). "Mutual Fund Performance," *Journal of Business*, 119–138.

Sneppen, K., P. Bak, H. Flyvbjerg, and M. H. Jensen (1995). "Evolution as a Self-Organized Critical Phenomenon," *Proceedings of the National Academy of Sciences*, USA, 92, 5209–5213.

Szeto, K. Y. and K. H. Cheung (1998). "Multiple Time Series Prediction using Genetic Algorithms Optimizer," *Proceedings of the International Symposium on Intelligent Data Engineering and Learning, IDEAL'98*, 127–134.

Szeto, K. Y. and A. L. Y. Fong (2000). "How Adaptive Agents in Stock Market Perform in the Presence of Random News: A Genetic Algorithm Approach," in Kwong Sak Leung, Lai-Wan Chan, and Helen Meng (eds.), Lecture Notes in Computer Sciences Series, LNCS/LNAI, *Proceeding of the Second International Conference on Intelligent Data Engineering and Automated Learning- IDEAL 2000* 1983, 505–510, Heidelberg: Spriger-Verlag.

Szeto, K. Y. and P.X. Luo (1999). "Self-Organizing Behavior in Genetic Algorithm for the Forecasting of Financial Time Series," *Proceeding of the International Conference on Forecasting Financial Markets*, FFM99, CD-Rom.

Szeto, K. Y., K. O. Chan, K. H. Cheung (1997). "Application of genetic algorithms in stock market prediction," in A.S. Weigend, Y. Abu-Mostafa, and A.P.N. Refenes (eds.), *Proceedings of the Fourth International Conference on Neural Networks in the Capital Markets: Progress in Neural Processing Decision Technologies for Financial Engineering*, World Scientific, NNCM-96, 95–103.

Takayasu, H., A. H. Sato, and M. Takayasu (1997). "Stable Infinite Variance Fluctuations in Randomly Amplified Langevin Systems," *Physical Review Letters*, 79(6), 966–969.

Vandewalle, N. and M. Ausloos (1997). "Different Universality Classes for SOC in Models Driven by Extremal Dynamics," *Europhysics Letters*, 37(1), 1–6.

Weigend, A. S., Y. Abu-Mostafa, and A. P. N. Refenes (1997). "Decision Technologies for Financial Engineering," *Proceedings of the Fourth International Conference on Neural Networks in the Capital Markets: Progress in Neural Processing Decision Technologies for Financial Engineering, NNCM-96.* World Scientific.

Zhang, Y.-C. (1999). "Toward a Theory of Marginal Efficient Markets," *Physica A*, 269, 30–44.

Chapter 20

LEARNING AND CONVERGENCE TO PARETO OPTIMALITY

Chris R. Birchenhall
School of Economic Studies
University of Manchester, Oxford Road
Manchester M13 9PL, UK
chris.birchenhall@man.ac.uk

Jie-Shin Lin
School of Economic Studies
University of Manchester, Oxford Road
Manchester M13 9PL, UK

Abstract This paper investigates the performance of a number of variants of the GA within the context of the OLG model of Bullard and Duffy (1999), in which a GA is used to model a population of agents adjusting their heterogeneous beliefs about future prices. Bullard and Duffy found the population converged onto the Pareto superior equilibrium in the model. A number of experiments and analysis suggest this is a robust result. The paper also introduces a meta-learning model in which agents learn to learn. In this meta-learning model each agent uses an individual GA to adjust its own expectations; at the same time, the population of agents are selecting which form of the GA to use. Our results suggest that more exploitative variants of the GA have a greater probability of convergence and a greater probability of convergence to the Pareto superior equilibrium.

Keywords: Genetic Algorithms, Overlapping Generation Models, Pareto Optimal Equilibrium

Introduction

In recent years there has been an increasing interest in learning and adaptive behaviour including simulation models based on genetic algorithms; see for example (Arifovic 1994, Arifovic 1995, Arifovic 1996, Arifovic and Eaton 1995, Axelrod 1990, Birchenhall 1995, Birchenhall, Kastrinos and Metcalfe 1996, Bullard and Duffy 1999, Dawid 1994, Dawid 1996a, Dawid 1996b, Riechmann 1998 and Riechmann 1999). A common feature of these models is the use of a population of heterogeneous, adaptive agents in contrast to the models with a representative agent. One central question is whether such a population of adaptive agents can grope their way to any underlying "rational" equilibrium. Sargent (1993) offers a survey of the area and suggests genetic algorithms are one form of stochastic approximation.

We study a simple overlapping generation economy introduced in Bullard and Duffy (1999) in which agents have heterogeneous beliefs about the rate of inflation. After the actual inflation is realised the population is in a position to learn *i.e.* adjust beliefs. The learning is modelled as a genetic algorithm (GA). Bullard and Duffy (1999) found the population of beliefs converged on to the Pareto superior rational expectation equilibrium in contrast to the perfect foresight dynamic, but in line with other learning models *e.g.* Marcet and Sargent (1989). We raise two related questions. *How robust is this result to variation in the algorithm? What is it in the model that drives the population to this outcome?*

Our simulations suggest this result is robust, but that the rate and speed of convergence depends on the specific format of the algorithm. Arifovic and Eaton (1995) illustrates the fact that convergence to a Pareto optimal equilibrium is not guaranteed by the use of GA-like learning process. Such a process can get locked-in to an inferior outcome. So what is it about the current model that reduces the probability of lock-in? We offer a partial answer to this question by studying the dynamics of the population's mean expectations.

To investigate robustness we use a number of variants of the GA including Holland's standard GA (SGA), see Holland (1992) and Goldberg (1989), Arifovic's augmented GA (AGA), see Arifovic (1994), Birchenhall's selective transfer GA (STGA), see Birchenhall (1995), as well as Bullard and Duffy's own variant (BDGA), see Bullard and Duffy (1999). Furthermore, we compare population learning, in which a single GA models a population and agents are represented by a single string in the GA, against models of individual learning, in which each individual is modelled as a GA. We also introduce a model of meta-learning in what we call "*open learning*". This is an extension of individual learning in which agents adjust their learning algorithm *i.e.* the structure of their individual GA. The results on open learning re-confirm the robustness of Bullard and Duffy's result.

The rest of the chapter is organized as follows. In section 1 we outline the underlying overlapping generation model and the structure of our simulations. Section 2 describes the various forms of the GA used in our simulations. Section 3 presents the results of simulations. Section 4 sketches an analytical argument as to why the low inflation equilibrium tends to emerge from the GA process. Section 5 offers our conclusions.

1. The Overlapping Generation Economy

1.1. The Underlying Economic Model

We use a special case of the overlapping generation economy in which there is a single perishable commodity and an exogenous supply of fiat money in each period introduced by a government.[1] There are two co-existing populations in the economy. Each agent in the population only lives for two periods. Time is discrete with integer $t \in (-\infty, \infty)$. There is no growth of population, therefore, the whole population of agents at any date is $2 \times N$ where N is the number of agents in each generation. To keep things simple, we will assume that all agents born in generation t are endowed with an amount w_1 of the consumption good in the first period of life, and an amount w_2 of the consumption good in the second period of life, where $w_1 > w_2 > 0$. In the first period of life, agents may choose to simply consume their endowments, or they may choose to save a fraction of their first period endowment in order to increase consumption in the second period of life. Since the commodity is non-storable, agents in this economy can save only by trading a portion of their consumption good for fiat money. The only means of transferring wealth from young to old is by means of this fiat money.

We start by considering an individual's choice of consumption. Agent $i \in [1, N]$ born in period t solves the maximisation problem:

$$\max_{c_t^i(t), c_t^i(t+1)} U(c_t^i(t), c_t^i(t+1)) = \ln c_t^i(t) + \ln c_t^i(t+1), \tag{20.1}$$

subject to the budget constraint

$$c_t^i(t) + c_t^i(t+1)\beta_t^i \leq w_1 + w_2\beta_t^i, \tag{20.2}$$

where β_t^i denotes agent i's time t forecast of the gross inflation factor between dates t and $t+1$. Given their expectations on inflation β_t^i and current price level P_t, then the expected price in the next period P_{t+1}^i is given by:

$$P_{t+1}^i = \beta_t^i P_t. \tag{20.3}$$

In this model, the possibility of borrowing by agents is ruled out. When forecasts of the inflation factor are equal to or exceed an upper bound agents simply consume their endowments and save nothing.[2]

Solving the above optimisation problem gives us

$$c_t^i(t) = \frac{w_2}{2}[\lambda + \beta_t^i], \tag{20.4}$$

where $\lambda = w_1/w_2$. Therefore, individual agent i's saving decision at time t is given by:

$$s_t^i(t) = w_1 - c_t^i(t) = \frac{w_2}{2}[\lambda - \beta_t^i]. \tag{20.5}$$

The planned consumption in the second period is given by

$$c_t^i(t+1) = \frac{w_2}{2\beta_t^i}[\lambda + \beta_t^i]. \tag{20.6}$$

Actual consumption in the second period will depend on the realized actual rate of inflation β_t^* as follows.

$$c_t^i(t+1)^* = w_2 + \frac{s_t^i(t)}{\beta_t^*} = w_2 + \frac{w_2}{2\beta_t^*}[\lambda - \beta_t^i]. \tag{20.7}$$

Hence actual lifetime utility is, with a little manipulation, given by

$$U(c_t^i, c_{t+1}^i) = \ln[c_t^i(t) \times c_{t+1}^i(t+1)^*] = \ln \frac{w_2^2}{4\beta_t^*}[(\lambda + \beta_t^*)^2 - (\beta_t^i - \beta_t^*)^2]. \tag{20.8}$$

Given realized inflation β_t^*, an individual has a larger lifetime utility if its expected rate of inflation β_t^i is closer to the realized rate β_t^*.

Let $\bar{\beta}_t$ be the population average forecast of inflation in period t, so that

$$\bar{\beta}_t = \frac{1}{N}\sum_{i=1}^{N}\beta_t^i. \tag{20.9}$$

To find the relationship between β_t^* and $\bar{\beta}_t$ we need to look at equilibrium in the money market. At time t the government prints a quantity of fiat money M_t per capita and uses this money to purchase a fixed per capita amount g of the consumption good. Hence

$$M_t = M_{t-1} + gP_t. \tag{20.10}$$

It is assumed that these government purchases do not yield agents any additional utility. Since agents can save only by holding fiat money, the money market clearing condition is that average savings equal the per capita real money stock at every date t :

$$\bar{S}_t \equiv \frac{1}{N}\sum_{i=1}^{N} s_t^i(t) = \frac{M_t}{P_t} \tag{20.11}$$

Hence,

$$P_t\bar{S}_t = P_{t-1}\bar{S}_{t-1} + gP_t \tag{20.12}$$

and

$$\beta_t^* = \frac{P_t}{P_{t-1}} = \frac{\bar{S}_{t-1}}{\bar{S}_t - g}. \tag{20.13}$$

Yet

$$\bar{S}_t = \frac{1}{N}\sum_{i=1}^{N} s_t^i(t) = \frac{1}{N}\sum \frac{w_2}{2}(\lambda - \beta_t^i) = \frac{w_2}{2}(\lambda - \bar{\beta}_t) \tag{20.14}$$

and thus

$$\beta_t^* = \frac{\frac{w_2}{2}(\lambda - \bar{\beta}_{t-1})}{\frac{w_2}{2}(\lambda - \bar{\beta}_t) - g} = \frac{\lambda - \bar{\beta}_{t-1}}{\lambda - \frac{2g}{w_2} - \bar{\beta}_t} \tag{20.15}$$

1.2. Modelling Learning using Genetic Algorithms

Excellent introductions to GAs are available elsewhere including Holland's original exposition Holland (1992), Goldberg's classic tutorial

Goldberg (1989), Michalewicz (1996) general survey of evolutionary programs and the popular Mitchell (1996). A brief summary is presented in Birchenhall (1995). Much of this literature has concentrated on the GA as a problem solving and/or optimization tool, though Mitchell (1996) does pay some attention to the use of GAs to model biological evolution. Dawid (1996b) is a good starting point when interpreting a GA as a model of learning. We assume the reader has some familiarity with the GA.

Our simulations have the following structure.

i. Randomly initialise the beliefs of the first young and old generations.

ii. Calculate consumption, savings and actual inflation as in Equations 20.4, 20.5 and 20.15.

iii. Apply the GA operators to form a set of beliefs for the new generation. The fitness of a belief is taken to be the life time utility the individual would have achieved if he had implemented that belief and actual inflation was unchanged.

iv. Check for convergence and stopping rules. If there is no convergence and the stopping rules does not apply then return to 2, otherwise stop.

Coding of inflation forecast into bit strings is the same as that used by Bullard and Duffy (1999). If the individual elements of the bit string are represented by b_i, so that $b_i = 0$ or $b_i = 1$, the bit string is converted to an inflation forecast β as follows.

$$\beta = \left(\frac{\Sigma b_i 2^i}{\Sigma 2^i} \right) \beta_{\text{max}},$$

where the sum run over the length of the bit string and β_{max} is the maximum value of the inflation forecast.

The best interpretation of the models presented in this paper is that they model the evolution of *belief structures* and not the learning of individual agents. It is the belief structure that replicates across generations in our models. In that respect, we could think of a belief structure as a form of meme.[3] The selection operator, together with selective forms of crossover and mutation, can be interpreted as imitation. Learning by imitation can be seen to be analogous to the evolutionary process in nature insofar as imitation has a bias to select the more successful ideas or beliefs and thus replicate the more successful ideas faster.

Selection alone is not an adequate model of learning. What is appealing in the GA is that crossover and mutation combine to create new options. A belief structure can be divided into components, such that individuals adopt different combinations of these components. New generations remix the belief structures of the older generations and introduce new variants of those structures. Crossover and mutation mimic these processes.

In this vein we can consider belief structures as consisting of a number of belief schema in the manner of Holland (1992) or, as pointed out by Sargent (1993), we can consider a belief structure as the intersection of equivalence classes. Holland's analysis of the GA—the Schema Theorem—suggests we see belief schema being the essential replicators in our models. This perspective is useful if the schema naturally embody complex interdependencies between components, that is the epistatic structure of the underlying search space. This is often difficult to discern and makes the application of Holland's analysis a non-trivial task.

2. Genetic Algorithm Variants

There are many variants of the genetic algorithm. In this work the main variants we consider are the standard genetic algorithm (SGA), augmented genetic algorithm (AGA), Bullard and Duffy GA (BDGA), and selective transfer genetic algorithm (STGA). We outline the differences by looking at their implementation of selection, crossover and mutation.

2.1. Selection Operator

A common form of selection generates a new population P' from the old population P element by element, in the manner of a biased *roulette wheel.* Each member of the new population is selected at random from among the elements of the old population P, where the probability of selecting a member of the old population is proportional to its "fitness." This selection operator can be seen to be a stochastic form of the replicator dynamic.

A second form of selection is *probability selection.* Here each member of the new population is selected at random without any bias *i.e.* all old members have the same chance of been selected.

A third form is selection is *Top 50 selection.* Each time the top 50% of the old population P are selected. The other half of the new population are created by random initialisation, that is to say are random selections from the space of all possible members rather than random selection from the set of existing members.

A fourth form is *section selection.* The first step is to select the top 50% of the old population. The second step is to select members from the third one-fourth part of the old population and it will produce 17.5% of the new population. The third step is to select members from fourth one-fourth part of the old population and will produce 7.5% of the new population. Then, the remaining 25% of the new population is created by random initialisation.

A fifth form of selection is *tournament selection.* For each member of the new population, two members of the old population are selected at random and the one with the highest fitness value is selected.

The SGA, AGA, and STGA use roulette wheel selection. The GA in Bullard and Duffy (1999) used tournament selection.

2.2. Crossover Operator

Crossover is the GA's distinctive feature. Viewing the strings in the GA as representation of a combination of schema, the crossover operator aims to create new combinations from the existing population. Crossover invariably involves creating two "child" strings from complementary parts of two "parent" strings. The simplest method is to cut both parents at a common random point and exchange the elements to the "left" of the cut point between the two parents. The left part of the first parent becomes the left part of the second parent and vice-versa. A crossover probability controls the frequency of the operator. If the crossover probability is p then for each pair of parents the probability that the children are generated by crossover is p and the probability that the children are identical to the parents is $1 - p$. The current experiments allowed 16 different crossover probabilities, varying from 0.25 to 1.00 in increments of 0.05, see Grefenstette (1986).

There are variants on the standard crossover; for example, we can choose more crossover points. Birchenhall, Kastrinos and Metcalfe's STGA used two-point cuts. Here the two points cut the parent strings in to three connected subsets – left, centre and right. Crossover would involve exchanging the central subset between the parents.

A more radical change in the STGA is the replacement of crossover with *selective transfer* that involves a "one-way transfer" of subsets, not an exchange. Part of one parent is used to replace the corresponding part of a second parent to generate a single child. This child is seen to replace the second parent and the first parent is left unchanged. Note in this transfer the original part of the second parent is destroyed rather than transferred to the first parent. At the same time, STGA filters out transfers that would lower the "estimated" fitness of the second parent. This operator is motivated by the following thought experiment about imitation. Consider two people meeting and one is wearing a blue sweater that the second person admires. After the meeting, the second person imitates the first by obtaining a similar blue sweater. It would seem unnatural to insist that the imitation involve the first person exchanging the blue sweater for the corresponding item of clothing worn by the second person. Insofar as we interpret crossover as modelling imitation, it has to be suggested that it lacks credibility. We can note that Birchenhall, Kastrinos and Metcalfe (1996) had individuals co-evolving models of fitness and the estimated fitness a potential transfer was generated by these models and replaced both selection and crossover by selective transfer. This is not pursued in this paper.

2.3. Mutation Operator

After crossover is performed, mutation is applied. Mutation applies random changes to the new offspring. When using binary strings, as we do in these experiments, the standard form of mutation involves passing through all bits in the string and randomly flipping the bit from 1 to 0 or from 0 to 1. The probability of a flip is the mutation probability. Mutation is usually seen to be a safeguard in that allows lost schema to be recreated. The current experiments allowed eight values for the mutation rate, increasing exponentially from 0.0 to 1.0, see Grefenstette (1986).

2.4. Election Operator

The election operator discards the products of crossover and/or mutation if their potential fitness is less than the original or parent strings. With the "*one-to-one*" election, a child string replaces a parent string only if the potential fitness of that child is greater than the fitness of that parent. Another version is the "*best two*" election. Once crossover and/or mutation are completed, the election operator then chooses the best two strings out of the four strings (two children and two parents). The election operator has an important impact on convergence of the population. Arifovic (1994) found the GA did not converge without the election operator. Also, Birchenhall (1995) suggested that the presence of election operator is important if population convergence is to be a feature of models. Population convergence is not significant for optimisation problems, but may be an important feature when using the GA to model a population of adaptive agents. In the current application convergence of agent's beliefs to a common value seems a reasonable outcome. Furthermore, the idea that even adaptive agents attempt to filter out "bad" changes is not unreasonable. On both counts the use of something along the lines of the election operator seems apt. A weakness of the election operator is that fitness values need to be known before a string is realized, which is problematic when there is no well defined fitness function or it is complex. As noted above, in Birchenhall, Kastrinos and Metcalfe (1996) agents coevolved a model of fitness and this model was used to filter potential changes. In the experiments reported here, the augmented GA and STGA use the "*one-to-one*" election operator. The "*best two*" election operator is applied in Bullard and Duffy's GA. Table 20.1 lists the GA variants used in this study.

Table 20.1. GA Types

GA	Name	Operators
SGA	Simple GA	Roulette Wheel Selection, Single Point Crossover, Mutation
AGA	Augmented GA	Roulette Wheel Selection, Single Point Crossover, Mutation, One-to-one election
STGA	Selective Transfer GA	Roulette Wheel Selection, Selective Transfer, Mutation
MSGA	Modified SGA	Tournament Selection, Single Point Crossover, Mutation
MAGA	Modified AGA	Tournament Selection, Single Point Crossover, Mutation, One-to-one election
MSTGA	Modified STGA	Tournament Selection, Selective Transfer, Mutation
BDGA	Bullard-Duffy GA	Tournament Selection, Single Point Crossover Mutation, Best-two election

3. The Simulation Results

This section presents summary results on a large number of simulations. Extended results will be presented in Jie-shin Lin's PhD thesis and can be obtained from the authors on request.

3.1. Population Learning

Table 20.2 summarises the key results from a large number simulations on population learning. In these simulations, there is a single GA and each string in the GA represents the belief of an individual agent. In the context of the overlapping generation model, the GA is to be

Table 20.2. Population Learning

GA Type	SOC*	SOL	SLC	Mean	Std Dev
SGA	0.065	0.048	0.734	520	241
AGA	0.689	0.617	0.896	60.4	56.6
STGA	0.971	0.919	0.946	68.1	77.8
MSGA	0.298	0.258	0.868	107.1	134.7
MAGA	0.687	0.593	0.864	25.6	20.8
MSTGA	0.997	0.963	0.966	25.0	18.3
BDGA	0.984	0.906	0.920	36.4	71.7

* SOC is the proportion of runs that converge, while SOL is the proportion of runs that converge to the low inflation equilibrium. SLC is the ratio SOL/SOC. Mean is the mean number of periods required to attain convergence, while Std Dev is the standard deviation of these convergence times.

interpreted as a model of intergenerational learning with individuals in each generation adopting beliefs based on the experience of the previous generation. The GA can be interpreted as follows: each individual in the new generation makes a biased and mutated combination of beliefs held by members of the previous generation.

Each row of Table 20.2 summaries the results from 100 runs for one of the seven GA types listed in Table 20.1. Consider the first row that reports results using the Simple GA (SGA). The first column (SOC) indicates the proportion of runs that converged, in the sense that all agents held the same beliefs within the maximum number of periods (1000). We can see about 7% of the runs based on the SGA converged in this sense. The second column (SOL) gives the proportion of runs that converged onto the low inflation equilibrium, so that some 5% of the SGA runs converged to this equilibrium. The third column (SLC) is the ratio of SOL to SOC and is thus the proportion of convergent runs that end up at the low inflation equilibrium. In the case of SGA about 70% of the convergent runs lead to the low inflation equilibrium. The fourth column (Mean) reports the mean time taken to achieve convergence, while the fifth column (Std Dev) reports the standard deviation of these convergence times. Of the 10% of SGA runs that converged, the average number of cycles needed to obtain convergence was 520, while the standard deviation of this convergence time was 241.

Consider first the SOC column and note the difference in the rates of convergence between SGA, AGA, STGA and BDGA. Almost 70% of the runs of the AGA converged, compared with 7% for the SGA. Introduction of the election operator significantly increases the rate of convergence. Election strongly reduces the explorative nature of the GA and makes the process more of a hill-climbing algorithm. Some care needs to be used when comparing the SGA and the other variants. It has to be noted that each cycle of the AGA requires a significant increase in fitness calculations and computation. In reality the fitness calculations underlying the election operator would have to be replaced by a more or less accurate estimate; insofar as there are significant errors in these estimates the ensuing GA would retain much of its explorative nature. By applying the election operator we are implicitly assuming the agents have "good" models of their environment. The election operator is the adaptive agents equivalent of rational expectations.

Turning to the STGA and BDGA, it can be seen that these variants further enhance the rate of convergence. Consider the similarity in performance of these seemingly unrelated variants. The difference between SGA and STGA is the latter's use of selective transfer, while BDGA uses tournament selection and best-two election. When comparing BDGA and SGA we have to suggest that tournament selection is not the main difference. While the results for MSGA suggest that the move from roulette wheel selection to tournament selection is significant, the comparison of MAGA with AGA and MSTGA with STGA suggests it has a secondary role. Rather, it is the use of selective transfer (as in STGA) or best-two selection (as in BDGA) that has a higher order impact.

Our results suggest the difference in the convergence performances of BDGA and STGA are not significant. Both enhance the exploitative and play down the explorative nature of the GA. Insofar as calculating fitness values dominates other calculations, then BDGA and STGA have similar computational costs, for example two fitness estimates for every pair of strings affected by transfer or crossover. Nevertheless, STGA seems a more natural expression of component-wise imitation.

3.2. Individual Learning

Table 20.3 summaries the results of our simulations using individual learning. The values reported are the averages arising from four sets of experiments for each GA type. The sets differed in the number of agents in each generation, namely 30 or 60, and the number of strings used in the GAs, namely 30 or 60. In these models there is a GA associated with each individual agent. The strings in each GA represent a set of potential belief structures. By applying the GA operators on this pool of beliefs each individual agent can search the belief space for fitter options. In each period the best belief in the GA is chosen and is the basis of

Table 20.3. Individual Learning

GA Type*	SOC	SOL	SLC	Mean	Std Dev
STGA	0.675	0.675	1.0	142.1	179.7
BDGA	0.708	0.708	1.0	120.1	180.5

*SOC is the proportion of runs that converge, while SOL is the proportion of runs that converge to the low inflation equilibrium. SLC is the ratio SOL/SOC. Mean is the mean number of periods required to attain convergence, while Std Dev is the standard deviation of these convergence times.

the agent's consumption plan. Note we assume all agents can correctly estimate the current fitness of all of its alternative beliefs even for those beliefs that has not been implemented.

In the context of the overlapping generation model each GA can be seen as a bundle of knowledge been passed from old to young generations, say within a family or close community. It is to be noted the GAs are independent of each other, there is no exchange of strings between the GAs. Due to the large computational cost of individual learning only two variants were considered, namely STGA and BDGA.

Despite the computational cost of individual learning the rate of convergence (SOC) in these models is lower than observed in population learning. Convergence means that all agents implement the same belief, and does not require the individual GAs to be identical. At the same time, all the convergent runs lead to the lower inflation equilibrium. As individuals do no communication directly, the coordination of beliefs can only be achieved indirectly. Coordination or convergence of beliefs arises from the fact that all agents are trying to identify the same best belief. This indirectness explains the lower convergence rates.

3.3. Open Learning

Table 20.4. Open Learning

Model*	SGA	AGA	STGA	BDGA	Others
1	3.2%	6.4%	28.2%	28.1%	34.1%
2	3.1%	6.4%	28.0%	28.4%	34.1%

*Model 1 and 2 differ in the level of Government finance. In Model 2 the two equilibria are closer together. The percentages show the relative numbers of the GA types in the population. For example, 3.2% of the agents adopted a SGA in Model 1. Convergence to the lower inflation equilibrium occurred in all runs.

Table 20.4 summaries the results of our simulations on open learning. As in individual learning each agent uses a belief-GA to search the belief space. Unlike individual learning the structure of the belief-GA is not fixed but is adjusted in light of experience. Each individual belief-GA has a specification string. This string specifies the structure of the belief-GA, in that it specifies which forms of selection, crossover and mutation operators are to be implemented in the individual's belief-GA. For the simulations reported here, the set of GA specification strings form the basis of a population wide specification-GA. Essentially, the GA specification strings evolve through population learning modelled as a SGA. Each specification string is associated with one of the individual belief-GAs and the fitness of the specification string is, in any given period, the average fitness of the belief strings in the associated belief-GA.

What is of interest here is the types of belief-GA that emerged from the specification-GA. There were 32 types of belief-GA that an individual could have adopted, as there were 4 possible forms of selection, 4 forms of crossover and 2 forms of mutation. Of these 24 alternatives 4 were variants of the SGA, 4 variants of AGA, 4 were variants of STGA and 4 were variants of BDGA — leaving 8 "others." In Table 20.4 we report summary information on the types of belief-GA that emerged from the specification-GA for cases. The difference between Model 1 and Model 2 is that the low and high inflation equilibria are closer together in Model 2. The results are very similar for both Models. Consider the results for Model 1. You will see the variants of STGA and BDGA emerge quite strongly from this process. While some 10% of the population are in the SGA/AGA group, 56% are in the STGA/BDGA group and 34% are in the "others" group. The specification GA seems to be selecting the exploitative forms e.g. STGA and BDGA.

All runs of the open learning model converged — all agents eventually adopted the same beliefs — and all converged to the low inflation equilibrium, which is significantly different from the results for individual learning. It would seem some diversity in the structure of the belief-GAs assists convergence. Here the population as a whole will exhibit both explorative and exploitative features; some agents continue to explore and search with explorative versions of the belief-GA while the exploitative belief-GAs of others agents will tend to converge. Exploitative versions of the belief-GA that manage to find a good belief structure will tend to have higher average fitness values and thus tend to be selected in the specification-GA. It is suggested there is something akin to simulated annealing here; the presence of explorative belief-GAs continue to disturb the overall population beliefs, but over time their influence is reduced and the disturbances are reduced, but do not disappear in our simulations.

4. Analysis of Selection

In this section we offer an argument that suggests two conclusions on the workings of the GA in the current model. First, we argue that selection in the GA will tend to generate more accurate forecasts, that is forecasts that are closer to the realized inflation. Second, we offer an argument that helps to explain why the GA tends to converge to the low inflation equilibrium. Bullard and Duffy (1999) used simulation results to suggest agents adopting a low inflation belief would be better off than adopting a high inflation belief. Our analysis looks at the dynamics of the population's mean belief to argue that this mean will converge to the lower equilibrium.

4.1. Accurate Forecasts

Given realized inflation β_t^*, an individual has a larger lifetime utility if its expected rate of inflation β_t^i is closer to the realized rate β_t^*, see Equation (20.8). Hence, all other things been the same, individuals with a more accurate forecast will have a higher probability of been selected in a population GA. In this way, the utility driven selection operator in the GA will tend to propagate accurate forecasts and discard relatively poor forecasts. It has to be noted that election-free crossover and mutation will reduce this tendency. A more exploitative GA, with some form of election, such as AGA, BDGA and STGA will propagate accurate beliefs across the population and thus lead to convergence. A more explorative GA, such as SGA, will continue to generate more speculative and potentially inaccurate beliefs and have greater difficulty in converging.

In the context of individual learning the argument that more accurate beliefs will have a higher probability of selection applies to all individual GAs and the more accurate belief in the individual GA will be implemented by the individual. Furthermore, in open learning, the specification-GA will favour those GAs whose current beliefs are on average more accurate.

As all individuals are otherwise identical and experience the realized inflation this pressure for accuracy will lead all beliefs to the common realized inflation and thus to convergence.

4.2. Convergence to Low Inflation

While the previous analysis helps to explain why an exploitative GA tends to generate accurate forecasts and thus induce convergence, it does not explain why the low inflation equilibrium tends to emerge from the GA. To address this issue we offer the following argument. Insofar as the GA selects more accurate forecasts, then in each period the population average forecast will tend to move toward the realized inflation. Hence, if the average inflation forecast is greater than the realized inflation then the average forecast will tend to fall in the next period and if the average inflation forecast is lower than the realized inflation then the average forecast will tend to rise in the next period. This specifies a qualitative dynamic on the average forecast. We now argue that under this qualitative dynamic the average forecast will converge to the lower inflation forecasts. Combined with the pressure to convergence, this leads to the emergence of the lower inflation equilibrium.

The key to our argument is the relationship between average forecasts and the ensuing realized inflation given in Equation (20.15).
Let $f_t = \beta_t^* - \bar{\beta}_t$ be the difference between the realized inflation and the mean level of beliefs. From (20.15) we get

$$f_t = \frac{\lambda - \bar{\beta}_{t-1}}{b^* - \bar{\beta}_t} - \bar{\beta}_t \tag{20.16}$$

where $b^* = \lambda - \frac{2g}{w_2}$. We assume $g < w_1/2$ so that $b^* > 0$.
If $\bar{\beta}_t < b^*$ then

$$\operatorname{sgn}(f_t) = \operatorname{sgn}\left(\lambda - \bar{\beta}_{t-1} - \bar{\beta}_t(b^* - \bar{\beta}_t)\right) = \operatorname{sgn}(q_t) \tag{20.17}$$

where

$$q_t = \bar{\beta}_t^2 - b^*\bar{\beta}_t + (\lambda - \bar{\beta}_{t-1}). \tag{20.18}$$

Given $\bar{\beta}_{t-1}$ the sign of f_t is the same as the sign of this positive quadratic in $\bar{\beta}_t$. This quadratic has zeros $b^{L,H}$ at

$$b^{L,H} = \frac{1}{2}\left[b^* \mp \sqrt{b^{*2} - 4(\lambda - \bar{\beta}_{t-1})}\right]. \tag{20.19}$$

It follows that $f_t = \beta_t^* - \bar{\beta}_t > 0$ if $\bar{\beta}_t < b^L$ or $b^H < \bar{\beta}_t$ and in these regions $\bar{\beta}_t$ will tend to rise toward β_t^*. If $b^L < \bar{\beta}_t < b^H < b^*$ then $f_t = \beta_t^* - \bar{\beta}_t < 0$ and $\bar{\beta}_t$ will tend to fall toward β_t^*. Hence as long as $\bar{\beta}_t < b^H$ then $\bar{\beta}_t$ will tend to b^L. It is to be noted at this juncture that the larger the value of $\bar{\beta}_{t-1}$ the larger is the basin of attraction of b^L.

In our simulations $w_1 = 4$, $w_2 = 1$, $g = 0.333$ so that $\lambda = 4$ and $b^* = 3.333$. In most simulations expectations were restricted to lie between 0 and 4, while in some the range was from 0 to 5. With random initialisation of expectations the population will in the most part lie near 2 and 2.5 for these simulations respectively. Hence initial expectations will typically be less than b^*. If the initial average expectation is 2 then $b^H \approx 2.54$ and the realized inflation will be 1.5 and we can expect average expectations to fall. If the initial average expectation is 2.5 then $b^H \approx 2.8$ and the realized inflation will be 1.8 and we can expect average expectations to fall. In this way, we expect a random initial average expectation to fall within the attractor of b^L. While this analysis is only suggestive, it does suggest that the lower equilibrium is the more probable outcome of an exploitative GA for which this analysis is a closer approximation.

5. Conclusion

In this paper we reconsidered the overlapping generation model investigated in Bullard and Duffy (1999); these authors found the model's Pareto superior low inflation equilibrium emerged from their GA-based learning process. Our simulations suggests their conclusion is robust to the nature of the GA, at least as long as the GA involves some form of Arifovic's election operator. In particular, the selective transfer GA introduced in Birchenhall (1995) has similar behaviour to the GA used in Bullard and Duffy (1999). An analysis of the dynamics of the mean population beliefs goes some way to explain the robustness of this result. Finally, we can note that a form of meta-learning, which we called open learning, illustrated how the structure of the GA can be made endogenous. Open learning exhibited interesting convergence properties and promises to allow us to take GA modelling to a higher level.

Notes

1. The model used here is identical to the version of Bullard and Duffy (1999).

2. Following Bullard and Duffy (1999), the highest inflation factor, λ is set equal to the ratio of endowments w_1/w_2.

3. Dawkins (1989) introduced the term meme to refer to a cultural replicator *i.e.* an idea or behaviourial pattern that replicates across human agents. The concept has generated a good deal of debate and we make no attempt to enter into that debate here. We simplify wish to stress that in our models belief structures replicate across individuals and generations.

References

Arifovic, J. (1994). "Genetic Algorithm Learning and the Cobweb Model," *Journal of Economic Dynamics and Control*, 18, 3–28.

Arifovic, J. (1995). "Genetic Algorithms and Inflationary Economies," *Journal of Monetary Economics*, 36, 219–243.

Arifovic, J. (1996). "The Behavior of the Exchange Rate in the Genetic Algorithm and Experimental Economics," *Journal of Political Economy*, 104(3), 510–541.

Arifovic, J. and C. Eaton (1995). "Coordination via Genetic Learning," *Computational Economics*, 8, 181–203.

Axelrod, R. (1990). *The Evolution of Cooperation.* Penguin.

Birchenhall, C. R. (1995). "Technical Change and Genetic Algorithms," *Computational Economics*, 8, 223–253.

Birchenhall, C. R., N. Kastrinos, and S. Metcalfe (1996). "Genetic Algorithms in Evolutionary Modeling," *Journal of Evolutionary Economics*, 7, 375–393.

Bullard, J. and J. Duffy (1999). "Using Genetic Algorithms to Model the Evolution of Heterogeneous Beliefs," *Computational Economics*, 13, 41–60.

Dawid, H. (1994). "A Markov Chain Analysis of Genetic Algorithms with a State Depenent Fitness Function," *Complex System*, 8, 497–417.

Dawid, H. (1996a). "Learning of Cycles and Sunspot Equilibria by Genetic Algorithms," *Journal of Evolutionary Economics*, 6, 361–373.

Dawid, H. (1996b). "Genetic algorithms as a model of adaptive learning in economic systems," *Central European Journal for Operations Research and Economics*, 4(1), 7–23.

Dawkins, R. (1989). *The Selfish Gene.* Oxford University Press.

Goldberg, D. E. (1989). *Genetic Algorithms in Search, Optimization and Machine Learning.* Addison-Wesley.

Grefenstette, J. J. (1986). *Optimization of control Parameters for Genetic Algorithms*, IEEE Transactions on Systems, Man, and Cybernetics, Smc-16(1), January/February.

Holland, J. H. (1992). *Adaptation in Natural and Artificial Systems*, A Bradford Book. MIT Press.

Marcet, A. and T. J. Sargent (1989). "Least-Squares Learning and the Dynamics of Hyperinflation," in W.A. Barnett, J. Geweke, and K. Shell (eds.), *Economic Complexity: Chaos, Sunspot, Bubbles and Nonlinearity*, Cambridge, MA: Cambridge University Press.

Michalewicz, Z. (1996). *Genetic Algorithms + Data Structures = Evolution Programs.* Springer.

Mitchell, M. (1996). *An Introduction to Genetic Algorithms.* MIT Press.

Riechmann, T. (1998). "Genetic Algorithms and Economic Evolution," Discussion Paper No. 219, University Hannover.

Riechmann, T. (1999). "Learning and Behavioral Stability," *Journal of Evolutionary Economics*, 9, 225–242.

Sargent, T. J., (1993). *Bounded Rationality in Macroeconomics*, Oxford: Clarendon Press.

VI

RETROSPECT AND PROSPECT

Chapter 21

THE NEW EVOLUTIONARY COMPUTATIONAL PARADIGM OF COMPLEX ADAPTIVE SYSTEMS

Challenges and Prospects for Economics and Finance

Sheri M. Markose

Economics Department and Institute of Studies in Finance (ISF)

University of Essex, UK.

scher@essex.ac.uk

Abstract The new evolutionary computational paradigm of market systems views these as complex adaptive systems. The major premise of 18th century classical political economy was that order in market systems is spontaneous or emergent, in that it is the result of "human action but not of human design." This early observation on the disjunction between system wide outcomes and capabilities of micro level rational calculation marks the provenance of modern evolutionary thought. However, it will take a powerful confluence of *two* 20th century epochal developments for the new evolutionary computational paradigm to rise to the challenge of providing long awaited explanations of what has remained anomalies or outside the ambit of traditional economic analysis. The first of these is the Gödel-Turing-Post results on incompleteness and algorithmically unsolvable problems that delimit formalist calculation or deductive methods. The second is the Anderson-Holland-Arthur heterogeneous adaptive agent theory and models for inductive search, emergence and self-organized criticality which can crucially show and explicitly study the processes underpinning the emergence of ordered complexity. Multi agent model simulation of asset price formation and the innovation based structure changing dynamics of capitalist growth are singled out for analysis of this disjunction between non-anticipating global outcomes and computational micro rationality.

Keywords: Complex Adaptive Systems, Emergence, Self-Organized Criticality, Algorithmic Unsolvability, Inductive Search, Innovation, Market Efficiency

Introduction

As with the contents of this book and with growing contributions of the thinkers of the Santa Fe Institute (SFI) and others, notably, Nicolis and Prigogine (1987), Chen and Day (1993), Dosi and Nelson (1994), Epstein and Axtell (1996), Krugman (1996), Albin (1998) and Velupillai (2000), there is clearly a resurgence of interest among economists in a world view that market systems are *complex adaptive* systems. However, as pointed out by Arthur (1993a), thought habits of economists dominated by a deductive/formalist methodology have in part been a barrier to understanding the significance of the development of inductive/evolutionary models of economic systems. In his strongest diatribe, Simon (1981) is dismissive of the neoclassical economists' lack of concern with the procedural lacunae of rationality: "rules of substantive rationality that are not backed by executable algorithms are worthless currency" (*ibid.*, p. 43).

In so far as an intuitive notion of an algorithm or calculation can be formalized, one of the major 20th century intellectual achievements is the Church-Turing thesis. By this thesis all finitely encodable sets of instructions defining algorithms implementable in a number of equivalent ways, including that of the Turing machine, can be formalized by the class of general recursive functions. The units of modern adaptive models is what Arthur (1991) describes as "parameterized decision algorithm" or units whose behaviour is algorithmic and hence brought about by finitely codifiable programs. However, in Church's Theorem we have the basic result on algorithmic unsolvability generically referred to as the halting problem (see, Cutland 1980). In other words, dynamical system outcomes produced by algorithmic agents need not be computable. Typically, as the set of all (countable infinite) partial recursive functions is co-extensive with the set of all Turing machines, problems for which no partial recursive function or universal Turing machine guarantees a solution are called undecidable or incomplete.[1]

The purpose of this review is to highlight that a schism between the so called formalist/deductive school and the inductive/evolutionary school is an outmoded framework of scientific discourse which has unfortunately prevailed far too long. In *Table 21.1*, I have summarized the framework of discourse to assess the challenges and prospects of the new evolutionary computational paradigm for economics. I will argue that a powerful confluence of the 'new logic' of the limits of formalistic calculation (column II, *Table 21.1*) with the adaptive algorithms of inductive search (column III, *Table 21.1*) is required for the new evolutionary paradigm to rise to the challenge of providing long awaited explanations of what have remained anomalies or outside the ambit of traditional economic analysis. Indeed, in a discipline where its elites give pride of place to axiomatic and formal analysis, it is unfortunate that they remain for the most part ignorant of the epochal results in formalist mathematics of Gödel (1931), Turing (1936) and Post (1944) on the limits of the formalist/deductive methodology.

Table 21.1. The New Framework of Scientific Discourse

I . Formalist / Deductive Inference	**II. The 'New Logic'**	**III. Inductive Inference**
♯Axiomatic Proof Theory ♯Model Theory	♯Gödel(1931) Self-reference Undecidability and Incompleteness ♯Church-Turing-Post Thesis Algorithmic Unsolvability And Limits of Formalist Calculus	♯Theory of Emergence And Self-Organizing Complex Systems ♯Non-recursive Tiling Problems
Formalistic Methods	**Computability Methods**	**Methods of Adaptive Computing**
*Predicate and Propositional Calculus *Classical methods of optimization * Classical Probability Models	*Recursion Function/ Computability Theory *Algorithmic and Stochastic Complexity Theory	*Cellular Automata *Classifier Systems *Genetic Algorithms and Genetic Programs *Neural Networks *Numerical Multi-Agent Simulations

In Gödel (1931) we have for the first time, a proof of an impossibility result that strongly self-referential system wide properties such as a formal requirement of order in terms of internal consistency is not one that can be established by an algorithmic decision procedure by an internal observer who operates on codifiable information.[2] The undecidable proposition which is known to be true to an internal observer is without any unique recursive procedure in its derivation and hence contradictory inferences can be drawn from the same information. This also clearly sets a limit to knowledge that can be transferred in a codifiable form. Gödel (1931) axiomatically derived a class of algorithmically unsolvable decision problems, viz. the diophantine equations or polynomial equations with integer solutions. With the formal conditions for the incompleteness of predicate calculus also being identical for geometrical patterns arising in tiling problems, Roger Penrose (1989), one of the active proponents of a theory of patterns that do not have any recursive implementation, has emphasized that non-recursive outcomes rather than being a curiosum of science have an objective and pervasive existence that theoretical and empirical investigators should explicitly take on board as the probable explanation for many anomalies in the domain of their sciences.

Thus, a formalist cognizant of the limits of formalist calculation will be open to the necessity of trial and error style inductive inference while in turn the latter can be fully justified by the existence of decision problems that are algorithmically unsolvable. When search has to proceed in domains that cannot be recursively enumerable or when patterns emerge for which no recursive implementation exists, the mathematical and the methodological framework given in column III is radically different to traditional methods in column I of *Table 21.1*. The efficacy of classical optimization algorithms requires a recursive bijective mapping between actions and outcomes and this fails when the outcomes cannot be enumerated in advance.[3] For algorithmically unsolvable problems, in the absence of an unique decision procedure, the hallmark of inductive inference is that a multiplicity of decision procedures have to be considered, with issues of algorithmic unsolvability governing the search for which decision procedures to include, how to alter existing procedures and when to stop searching and so on.

As *Table 21.1* shows the paradigm shift from traditional constraint optimization methods of column I to adaptive methods of inductive inference in column III is symbiotically related to the powerful axiomatic limitative results on deduction and calculation given in column II. In other words, the full recognition that there are problems, indeed all non-trivial market related problems may be such, for which methods in column I in Table 21.1 are of limited use, is slow in coming. The whole thrust of adaptive computing methods in the form of Classifier Systems, Genetic Algorithms (**GA**s) and Genetic Programs (**GP**s) pioneered by John Holland (1975, 1992) and John Koza (1992), respectively, is to evolve computer programs that can solve a problem from elementary bit strings or units of several different programs. In the classical methods, known solution algorithms are tried out sequentially or even in parallel but they cannot be grown over time from simpler units as in the adaptive methods such as **GA**s and **GP**s where the programs coevolve as the problems change. In an evolutionary framework, the domain of decision rules contain implementable solution structures or objects which are a subset of an uncountably infinite set of total computable functions. As this set has no algorithmic decision procedure, the outcome of an inductive search may often be far from a global optimum. Holland (1975) pioneered the evolutionary principle for the selection of decision rules. The latter are selected in proportion to their fitness relative to the average fitness of the population of decision rules. In multi-species environments, the rates of cross-over and mutation operators that can improve fitness of decision rules may arise endogenously. As we will see, oppositional structures that arise between species can encourage strategic innovation and prevent entrapment in local optima.

The adaptive computing method of neural networks (see *Table 21.1*, column III) involve universal function approximators. Thus, while it is the case that for any function there is a neural network able to approximate it, there is, however, no general way in which such a neural network can be approximated in terms of the weights of the links and the threshold values. Again the networks have to learn to recognize patterns in a supervised or unsupervised way, by trial and error, Hertz *et al.* (1991). But, the complexity of the task that can be learnt is unlimited and can exceed the human capacity to specify the task in logical terms. The theory of validation in adaptive inductive inference when the data generating process is unknown is still in its infancy. Algorithmic and stochastic complexity theory of column II in *Table 21.1* has developed some necessary principles, while in column III, characteristic features are now well known for macro systems with large numbers of interacting agents at a micro level.

Building on the cellular automata with the same recursive power of a Turing machine,[4] John von Neumann pioneered the theory of self-organizing and complex systems, Burks (1957). It is now well known from the Wolfram-Chomsky scheme (see, Wolfram 1984, Dawid 1999, Foley in Albin 1998, pp. 42-55, Markose 2001a) that on varying the computational capabilities of agents, different system wide or global dynamics can be generated. Finite automata produce **Type 1** dynamics with unique limit points; push down automata produce **Type 2** dynamics with limit cycles; linear bounded automata produce **Type 3** chaotic output trajectories with strange attractors. However, it takes agents with full powers of Turing machines capable of simulating other Turing machines, a property called *computational universality*, to produce the **Type 4** irregular innovation based structure changing dynamics associated with capitalist growth. While in general, the computational agents associated with higher types of dynamics can compute/learn lower type of dynamics, **Type 3** and **Type 4** dynamics pose computational problems. **Type 3** dynamics in the Wolfram-Chomsky schema though algorithmically solvable, in principle, could pose problems of computational intractability. In the latter case, the computational cost of solving a problem optimally may place a barrier to the determination of global optima leading to the decision rule being based on arbitrary local conditions and/or beingtemporally myopic. **Type 4** dynamics is algorithmically unsolvable. The central issues that arise here are many. We need to understand how computational agents produce such complex global dynamics which pose either computational intractability or algorithmic unsolvability in terms of their individual decision making. How is order then brought about? How does such complex global dynamics impinge on agents' capability to learn, cope and operate effectively? How do agents acquire computational universality, Langton (1992, p. 69)? The modern theory of emergent and complex phenomena deals with this.

The 1977 Nobel laureate in Physics, Phillip Anderson, is considered to be the father of emergent phenomenon. The seminal contributions of Holland (1975,1992), Koza (1992), Goldberg (1989), Arthur (1993b), Kaufmann (1993) and others have made it possible to simulate heterogeneous computationally intelligent **(CI)** agents in adaptive settings to give a material counterpart in computer or virtual environments of the otherwise elusive phenomena of emergence and self-organization. It is increasingly becoming a methodological tool that will be used as a means of understanding complex social and other environments. The theory of emergent phenomena for intelligent adaptive agents identifies at least five following characteristics of complexity.

(i). First, there are adaptive agents. Adaptive agents have algorithmic capabilities that are not pre-programmed to respond in a fixed fashion to changes in global states but have a non-linear feedback loop that enable them to change rules of behaviour which in turn change global properties to produce coevolving local and global systems.

(ii). The global properties of a heterogeneous adaptive set of agents is not the scaled up version of the purposive behaviour or program of any agent or group of agents in the system.

(iii). Knowledge of the local programs of agents gives no clue to the global outcomes. The non-anticipating nature of global outcomes causes and retains heterogeneity in agents despite local self-organizing tendencies and homogeneity characterized by convergence to attractors. It must be noted that self-organized order as in the classic Schelling (1978) result on racially segregated neighbourhoods does not necessarily correlate with what many regard to be desirable. Self-organizing and emergent systems produce typical statistical 'signatures' in the macro dynamical data such as the power laws,[5] logistic curves, self-similar structures or fractals and chains of interrelations that manifest long memory. Power laws have known to characterize statistical distributions in in natural, social and artificial systems ranging from earthquakes, size of cities Krugman(1996), and income distribution.

(iv). With adaptive learning in complex environments canalization and lock ins can stabilize certain categories of behaviour within shared operational schemes that may sometimes become maladaptive (see, Ackley and Littman 1992, Kaufmann 1993). With the emergence of canalization or lock ins, learning becomes instinctive or habitual and atrophied into skillful behaviour. The Ackley and Littman (1992) Artificial Life simulation where agents have neural network brains show that when the adaptive operational schema emerge they free up neural networks to do other things.

(v). As first postulated in the Wolfram-Chomsky schema, the **Type 4** irregular structure changing dynamics, over and above those characterized solely by convergence to attractors, arise only in the case of agents with powers of a universal Turing machine. Langton (1992) makes an important observation that physical systems capable of complexity, experience a critical slowing down at the phase transition between order (analogous to halting computations) and chaos, as they are in principle involved in a non-terminating computational loop that characterizes an undecidable global ordering problem at that juncture. Langton concludes that physical dynamical systems "are bound by the same *in principle* limitations as computing devices" (*ibid.* p. 82). Thus, equilibria associated with the famous Langton (1992) thesis on "life at the edge of chaos" correspond to recursively inseparable sets first discovered in the context of algorithmically unsolvable problems, Post (1944). Further, in system simulations with agents it has been found that cooperative and competitive structures develop. Competitors soon learn, in what can only be described as paradoxical conditions, the advantages of adopting surprise or innovative strategies that their rivals cannot predict (see, Ray 1992, Hillis 1992). This can move the system in unpredictable structure changing directions which are error driven and highly dissipative of the old order. In this context, of coevolving species of computationally intelligent agents each trying to enhance its fitness relative to others is the evolutionary principle called the Red Queen, Ray (1992) which exacerbates rivalrous behaviour in agents competing for scarce resources.

Clearly, the **ACE** (Adaptive Computational Economist, see, Testfasion 1998) are extricating us from decades of methodological entrainment of neoclassical economics that simply failed to see that the domain of economic analysis cannot be restricted to column I of *Table 21.1.* However, my preamble here aims to guard against a cavalier attitude that emergent phenomena is what ever that comes out of an **ACE** computer simulation or that non-trivial mathematical issues of non-anticipating or 'surprise' outcomes remain unaddressed.

The principal aspects of the new science of emergent complex adaptive systems **ECAS**, briefly outlined above raises issues on —

(a)Rationality of agents, viz. when 'zero' intelligence and the highest level of intelligence with full powers of Turing machines are appropriate.

(b)'New' evolutionary equilibrium concepts such as the Red Queen principle and the Langton (1992) hypothesis of equilibria governed by undecidable dynamics.

The extant well established concept of equilibria with evolutionary stable strategies Weibull (1996) is useful to explain outcomes relating to lock-ins and canalization in (iv) above that are resistant to mutant strategies. However, as the *sine qua non* of complex systems with computationally intelligent agents is the capacity to produce innovations or surprises and system outcomes of greater algorithmic complexity than the computational capabilities of any of the agents within it (see, Casti, 1994, pp. 143-149), it is clear that the mathematics of algorithmic unsolvability and incompleteness along with algorithmic and stochastic complexity theory of column II in *Table 21.1* is needed to give formal underpinnings for this. To date, proponents of agent based models such as Axtell (2000) have raised such foundational issues only in passing. Further, it is plausible that the principle of the Red Queen that arises in evolutionary biology and immunology is also appropriate as a model of competition in markets. The Red Queen based on the observation made to Alice by the Red Queen in Lewis Caroll's Through the Looking Glass: "in this place it takes all the running you can do, to keep in the same place," can manifest in areas such as the arms race in product innovation by firms to maintain status quo in market shares, regulatory arbitrage and financial market efficiency where returns on portfolio has to match the market index. I will provide some discussion with regard to the latter two.

The rest of this chapter is organized as follows. Section 1 briefly outlines the classical 18th century provenance of evolutionary theories of market systems and the thesis of spontaneous order or order without design to show how the challenges of an evolutionary agenda for economics remain perennial and wide ranging even though the domain of scientific discourse has altered greatly. As decentralized heterogeneous agents are crucial in the agenda of **ECAS** adaptive agent models, in Section 1.2, I briefly review the well articulated thesis on this from classical political economy and assess also the logical/methodological impasse that the neoclassical theory of decentralization ran into in the 1970s. Section 2 focuses on price formation in stock/asset markets. It is an area in which the following issues abound: the algorithmic unsolvability of rational expectations equilibria, Spear (1989); the necessity of heterogeneous inductive models for trading modelled in artificial stock markets (Arthur *et al.* 1997, Chen and Yeh 2000, Challet and Zhang 1998, Lux and Marchesi 1999, etc.)and other cited properties of complex systems, Friedman and Rust (1993). In Section 2.4, I will discuss the theoretical desirata of a Emergent Efficient Market Hypothesis, **EEMH**, to contrast it with the traditional Efficient Market Hypothesis, **EMH**. Firstly, as in studies by Solomon (2000), Solomon and Levy (1996), power laws in investment wealth distribution are upheld as a manifestation of self-organized complexity in micro activity. Here, the principle of the Red Queen is proposed as a testable hypothesis on the emergence of market efficiency and power laws in artificial stock market models. Section 3 surveys the ubiquitous rivalrous structure of capitalism that has produced unprecedented novelty to the system. In no other area has the classical methods of constraint optimization (column I, *Table 21.1*) failed economic analysis more than in the explanation of irregular structure changing innovative growth in market systems. Again I hope to show how point (v) on the emergence of surprises or novelty from complex system simulations with adaptive computational agents has an analogue in formally undecidable systems. In particular, I will sketch an analytical proof as to why computationally universal agents with full powers of deductive inference, including that of self-reference, are necessary for innovation to arise from rational strategic necessity in the Nash equilibrium of a game rather than from random mutation. This is followed by a brief concluding section.

1. The Classical and other Precursors of Spontaneous or Emergent Order: The Challenges Outlined

1.1. The Classical Legacy

The provenance of evolutionary theories of society that goes back to the classical forebears of Economics with the Scottish Enlightenment is known to predate even Darwin's evolutionary thesis on the origin of species, Hodgson (1993). On observing the unfolding of early western capitalism and the libertarian market forms, the classical Scottish thinkers were led to conclude that the elaborate structures of society ranging from language, civil society, monetary exchange, *laissez faire* and economic progress did not appear as if they were the product of human execution of a human plan. These outcomes were not produced by intentional design. Spontaneity of the order or pattern, therefore, refers to the absence of direct intentionality of a designing mind in the emergence of such observable outcomes. The following epigrams were coined: nations "*stumble* on establishments that are the result of *human action and not the execution of any human design*" (Adam Fergusson, circa, 1767, italics added); or civil society is 'the unintended consequence' of actions of individuals pursuing some other proximate objective which in the hands of Bernard Mandeville may seem like private vices. Adam Smith's famous "invisible hand" explanation refers to the elusive nature of the ordering principle manifesting entirely as observable outcomes of individuals' actions especially in the case of equilibrium in multiple markets, rather than the proximate objective of anybody within the system implemented by rational calculation.

There is a direct parallel here on the spontaneous development of a system of legal and moral rules governing cooperation and competition in interactions between individuals capable of producing stable outcomes in society with what is found in Smith's arbitrageur in the economic realm. It is in the work of David Hume that we have the clearest statements on the non-constructivist view of reason's role in the development of the moral and legal rules of liberty and just society. Thus, "the rules of morality are not the conclusions of our reason," Hume (1888). Further, Hume couches the non-consequentialist nature of the abstract and general structure of the rules of justice with their independence from satisfying any particular desires of agents: "these rules are not derived from any utility or advantage which either the particular person or public may reap from his enjoyment of any particular good," Hume (1888). Kant (1965) is known to have given a formal characterization of the rules of just society in that coercively applied rules are end neutral and by their operation do not bring about predetermined outcomes in society. Absenting knowledge of the utility that rules produce in particular instances is seen by many a liberal theorist (see, O'Neill 1989) as a useful methodological ploy often called the 'veil of ignorance' in the construction of rules of justice so that they are not instruments satisfying the invidious interests of elites or special groups[6]. Nevertheless, it is still not fully understood how or why libertarian market systems evolved rules which possess the formal quality of their end neutrality and specifically why their capacity to produce non-anticipating outcomes in society is upheld as an important normative property in liberal Kantian political philosophy. Likewise, the Humean position that no *a priori* rationalism, can succeed in designing these system of rules, *de novo*, that would produce the desired outcome of liberty, to this day, remains one of the most baffling tenets of classical liberalism (see, Suzumura 1990).

Modern theorists of emergent phenomena will no doubt see a parallel between the late 20th century theory and the classical thesis on spontaneous order. The issues covered in an evolutionary agenda on spontaneous order are very far reaching indeed: it includes explanation of autonomy, decentralization, market institutions and even libertarian morality in its ambit. However, despite its venerable origins, the classical tradition of spontaneous order in the two centuries that followed, more often than not, fell prey to the ultra-rationalist riposte that the patterns of spontaneous order "look to be the product of someone's intentional design," Nozick (1974, p. 19) or "appear to be a product of some omniscient designing mind," Barry (1982, p. 8). As the feeble foundations of the evolutionary tradition made arguments to the contrary seem theoretically unconvincing (see, Ullmann-Margalit 1978), the 19th and early 20th century political economy was dominated by experiments to exert rational and centralized control on society. In this context, Ullmann-

Margalit's statement that "it took the powerful minds — and all the logical arsenal at their disposal — of Hume and Kant, as well as the works of Darwin and Mill, to explode the logic of this Argument from Design," (*ibid*, p. 268) as if to say that the task at hand was successfully accomplished by the said luminaries, is hopelessly optimistic. In other words, it was not well into the 20th century that the twin theoretical pillars for an evolutionary thesis on markets were in place. The first pillar is the *contra* Argument from Design which requires an impossibility result on why no agent or agents in a system can bring about determinate systemic outcomes when all agents have computational capabilities of simulating other such agents. Indeed, not till Gödel (1931) and undecidability of the Church-Turing halting problem could the mathematical principle of non-recursiveness or algorithmic unsolvability, namely a logical impossibility result on the limits of finitary procedures, be brought to bear on the anti-creationist principle at the heart of evolutionary systems.[7] As observed earlier, Penrose (1988) can be credited to be the first to have explicitly mentioned the connection between non-recursive or non-algorithmic implementation as the *sine qua non* of emergent patterns in nature and in artificial settings.

The second major pillar of the evolutionary thesis is the demonstration of emergent phenomena in the absence of a blueprint or of an encoding for system wide outcomes. Not till the seminal work of von Neumann in the 1950's on cellular automata and then of Holland, Koza and thinkers of the SFI, did we have the necessary tools to pin down the elusive and central tenet of evolution that is of its spontaneous and emergent properties specifically in terms of orderliness and patterns. The system has to run its course as there is a lack of algorithmic predetermination of outcomes. To be in a position at the end of the 20th century to recreate dynamical systems of heterogeneous agents with computational intelligence of varying degrees that can evolve complexity and self-organize in virtual environments of the computer is an outstanding achievement.

1.2. F. A. Hayek: Decentralization and Autonomy As Emergent Phenomena

For most part in equilibrium theory of decentralized markets, decentralization refers to both the units of decision making in terms of their autonomy of action and the informational setting that guides their decisions. In the formal equilibrium theory of markets of Arrow and Hahn (1971) it was thought that the burden of proof placed by the invisible hand type argument involved the establishment of a *formal possibility* or *an existence result* on an equilibrium in a decentralized economy where individuals motivated by self interest and guided by price signals alone will result in a resource allocation that not only satisfies the consistency of the economic plans of the individuals but also one that is most efficient. It was never in the domain of discourse that the emergence of decentralized market systems with the dynamical properties described by the classical economists also placed a burden of proof along the lines of what John Rust (1987) has called *computational decentralization*, viz. the impossibility or the non-existence of an overarching program that can control programs of all agents in the system in the determination of system wide outcomes.[8]

The main bridgehead between the classical thesis on spontaneous or emergent order and the modern one is F. A. Hayek. Hayek (1945) cites the two most commonplace but singularly intractable informational constrains in society as being part of the rationale for decentralized systems. First, information in society is found in a dispersed form subject to time and place matrices and it is perceived by individuals in a subjective fashion. Second, it is impossible to centralize all information by communication alone as the knowledge needed to make decisions is tacit and not in a codifiable form. Hayek's much quoted observation on this is that: "We cannot expect that this problem will be solved by first communicating all this information to a central board which, after integrating all knowledge, issues its orders .... *the problem is to show how a solution is produced by the interaction of people each of whom has partial knowledge*," Hayek (1945, italics added).

Hayek's *third* postulate on markets is remarkable in that he had well before 1950 made the connection that many economists have yet to do so, that is, market institutions that have coevolved with human reason enable us to solve problems that is impossible to do so by direct computation. He called the latter the limits of constructivist reason and on why we failed for so long to acknowledge these limits he relates back to Cartesian rationalism, Hayek (1967).

Hayek's fortuitous Viennese connection with Kurt Gödel led him to see that there was a logical impossibility result on why algorithmic centralized control is impossible. There is ample evidence that Hayek's rampant evolutionary thinking on morals to markets to price formation and human reason itself arose as he traversed the modern Gödelian route from column II to III in *Table 21.1.* There is explicit reference to the Gödel incompleteness result and its implications for his work on *The*

Theory of Complex Phenomena, Hayek (1982) and also in his work on cognition and the brain as complex and incomplete phenomena in *The Sensory Order* (see, Hayek 1953 and Weimar 1982). Hayek saw that complexity and incompleteness are two sides of the same coin and that as such domains make problem solving by direct rational calculation impossible he was led to the necessity of evolutionary solutions.

Not withstanding problems of localized information, Hayek clearly saw that autonomy of action is also necessary on account of tacit knowledge or cognitive incompleteness. The latter which is a by product of evolved complexity of the brain (Hayek 1953) permits the agent no algorithmic access to rules of inference which prompt action. Much tacit knowledge will be lost to the world unless it can be directly precipitated in action in a manner that warrants no verbal or algorithmic justification. Thus, Hayek's work (Hayek 1967, 1953) does lend itself to the interpretation that classes of tacit knowledge can fall into two respective categories of (i) emergent phenomena and (ii) the products of canalization from the evolution of the species or from learnt behaviour in the process of socialization. Hayek has provided a cogent bridgehead to the classical liberal tradition and particularly to the Kantian normative injunctions on the end neutrality of coercive rules of the state which then give scope to emergent phenomena with autonomous actors. Interestingly, no thinker is more struck by the irony of evolutionary selection especially in reference to libertarian values on which the West owes its unprecedented material success : these are best served when they operate as category (iv) in the **ECAS** schema (see, Introduction) as habitual and atrophied skilful behaviour. With these provisos in place, Hayek may well qualify to be an **ACE**, as Vriend (2000) puts it.

Decentralization in society, thus, necessarily presupposes the granting of some degree of autonomy to the individual decision maker. In the mainstream literature on decentralized decision making initiated by Marschak (1959), the optimal level of decentralization determining the extent or degree to which decision making entities should be given autonomy to make decisions on the basis of private information, requires direct and rational calculations of the relative speeds and costs involved in the alternative systems of communication and control. The logical and analytical impasse such an approach to decentralization posed was soon detected by Hurwicz (1960). Hurwicz's seminal development of the notion of incentive compatibility indicates that the reporting protocol is open to abuse as agents find it in their interest to misrepresent their endowments and preferences. He concluded that "it is the characteristic of the current state of the literature on decentralization, that one may be provided with a definition of what it means to have a more or less centralized *command* (italics added) economy;" virtually nothing is known about the decentralizing processes of the market system. A consistent paradigm for the existence of decentralized systems requires in particular that the process that determines the level of decentralization in the system should itself conform not only with the informational constraint but with what Rust (1997) calls decentralized computational constraints. How do market systems actually put in place a self-enforcing structure of rules that brings about non-anticipating global outcomes ?

In Markose-Cherian (1991) the political economy of expanding market societies of Europe is studied to glean insights into how in fact a larger market order is formed. In the creation of a larger market order with the European Economic Community, end neutral or end independent rules can be observed to emerge from a negative selection process of rule elimination. Through a decentralized litigious process initiated by individual litigants who challenge rules for their inability for general implementation leading to equal treatment; rules are progressively eliminated as unjust as they cannot be 'universalized' over what is now a larger territory and peoples. In Section 3, an even more powerful and commonplace hypothesis is put forward for the development of legal rules that favour *laissez faire* type arbitrage free non-anticipating systemic outcomes. Regulation that aims to bring about specific predetermined outcomes can be rendered dead letter by regulatory arbitrage when private individuals find it profitable for them to contravene these rules. As individuals break rules and attempt to exit from given regulatory systems, it is rationally strategic for them to innovate as they step out. Many policy rules whose outcomes are predictable may in fact suffer elimination as they fail to be Nash implementable in the face of such contrarian strategic behaviour on part of regulatees. This is part of a very large literature on the regulatory dialectic that has recently enjoyed great resurgence of interest as a process of institutional innovation (e.g. Miller 1986, Schanze 1995, and others). However, what is interesting, for our purposes, is that we can relate the above issues on autonomy and decentralization in markets to the **Type 4** structure changing dynamics in the Wolfram-Chomsky schema produced only by agents with computational universality.

2. Emergence of Efficiency in Asset Markets and Individual Rationality

In the previous section the emphasis was on the open ended structure of the constitutional rules of market systems which appears to permit the evolution of complexity and emergent outcomes. In this section, the focus is on a specific class of markets, viz. financial markets, in which controversy and hence a lack of understanding of the nexus between the non-anticipating nature of global outcomes relative to individual agent programs or rationality has featured in a big way. Here I will address two issues: (a) Why does asset price formation cause problems of inductive inference? (b) Is asset market efficiency an emergent outcome?

2.1. Inductive Rationality and Heterogeneous Beliefs

The problem with the traditional approach to efficient market hypothesis (**EMH**) is that an unspecified and uniform capacity of rationality is postulated for agents which allegedly enables them to systematically extract patterns leaving only residual white noise. Typically, from the famous Samuelson (1965) view that 'proper' anticipation of prices implies taking conditional expectations of the price P_{t+1} conditional on information H_t, $E(P_{t+1} | H_t)$, random fluctuations follow by tautology.[9] Clearly, taking conditional expectations is without unique procedural content. Further, this uniform capacity for rationality implies that all agents are random walk believers. This implies the specious no trade results on non-existence of speculative trading for pecuniary gains based on price expectations. We have the paradox that with the cessation of trade, the price at t+1, P_{t+1}, never gets determined.

Arthur *et al.* (1997) make a case for heterogeneous multi-agent models where each agent uses genetic algorithms to arrive at price predictions. "Agents, in facing the problem of choosing appropriate predictive models, face the same problem that statisticians face when choosing appropriate predictive models given a specific data set, but *no objective means by which to choose a functional form.* The expectational models investors choose affect the price sequence, so that our statisticians very choices of model affect their data and so their choices of model" (*ibid.* p.305, italics added).

The initial Santa-Fe Institute micro architecture of artificial stock markets of Arthur *et al.* (1997) has influenced a number contributions such as that of Chen and Yeh (2000), Brock and Holmes (1997), LeBaron

(2001), etc. These models typically determine the traded market price for a single risky asset which has a fundamental value for the dividends determined by an AR(1) process. The decision to buy or sell shares is governed by a standard portfolio choice result given by the Sharpe ratio defined as the expected excess return divided by a homogeneous constant sample variance of the stock returns (see, Arthur *et al.* 1997). In contrast, there are minority game stock market models first developed by Arthur (1994) as the El Farol game and further formalized by Challet and Zhang (1998), Savit *et al.* (1999). The minority game yields a useful prototype of speculative markets where price dynamics are driven solely by endogenous trading behaviour rather than by an objective value determining process and investors are rewarded for being contrarian, viz. selling when the majority are buying and vice versa. Significantly, in all variants of the SFI stock market model as well as the minority stock market games, agents are identical except in how they make inductive inference with regard to future stock returns. There is a finite set of forecast rules at the start. Agents are either randomly allotted a subset of these or have access to all of them with selection of and innovation to forecast rules proceeding according to evolutionary methods given in column II in Table 21.1. The initial forecast rules typically fall into categories of fundamental value determining or technical/trend following ones.

Following the pioneering work of Arifovic (1994), the question posed of artificial stock market models with coevolving populations of forecast rules based on genetic programs is whether these converge to a homogeneous rational expectations equilibrium **HREE** with noise/technical trading rules being eventually driven out. Briefly, the following has been observed. As the market environment is maintained relatively stationary with a low intensity of search by **GAs** or **GPs**, the unique homogenous rational expectations result with a slow down of trade followed in the Arthur et al. (1997). In contrast when the rate of **GPs** exploration for 'better' predictors was speeded up, the stock market prices begin to show historically observed properties of volatility clustering with little tendency for convergence to **HREE**. Note, that this crucial criterion on the intensity of search leading to forecast performance of agents applies in an *ad hoc* manner in some of these models. The insight from the minority game structure with multi agents, Savit *et al.* is both different and important in that there is a critical degree of heterogeneity of forecast rules that is required to obtain the optimal social gains from trade. The nature of emergence of the outcome rests on the fact that agents adaptively and in a decentralized way adopt this critical degree of heterogeneity in forecast rules and this outcome improves on the Nash equilibrium one with random/ mixed strategies to buy or sell. Thus, in a framework of emergence of stock market efficiency, one first needs a justification for the existence of agents with heterogenous forecast models of stock prices.

2.2. Algorithmic Unsolvability of REE Asset Prices

To pin down the algorithmically unsolvable nature of a rational expectations price or the absence of an unique objective decision procedure for agents to compute the functional fixed point mapping, I will outline the issues on inductive inference learning first raised in Spear (1989). The model is tailored to suit an asset market equilibrium for a single asset with fixed total supply that can be bought and sold in standardized units. Time is discrete and denoted by $t = 0, 1, 2, ..., T$. There are N agents indexed by $i = 1, 2, 3, ..., N$. Agents can choose to buy (B_i) or sell (S_j) one unit of the asset, $\mathrm{B}_i + \mathrm{S}_j \leq N$, $i \neq j$.

In a one period ahead forecast horizon for agents, agents execute their trades at t based on price forecasts, $P(t+1)$, conditional on price information up to t, $H_t = \{P_{t-s}\}$, $s = 0, 1, 2, ..., t$. In the absence of any fundamental value determining factors for the asset price, as in the minority game, spot price dynamics is entirely generated by endogenous uncertainty arising from speculative trading strategies of agents.

For computational agents, their decision procedures and forecast rules are computable functions.

Definition 1: Following a well known notational convention, Cutland (1980), a single valued computable function as follows

$$f(x) \cong \phi_a(x) = q. \tag{21.1}$$

That is, the value of a computable function $f(x)$ when computed using the program/Turing machine with index a is equal to an integer $\phi_a(x) = q$, if $\phi_a(x)$ is defined or halts (denoted as $\phi_a(x) \downarrow$) or the function $f(x)$ is undefined ($\sim$) when $\phi_a(x)$ does not halt (denoted as $\phi_a(x) \uparrow$). See also footnote 1.

Each agent's prediction function of price at $t+1$ given H_t is given by $\hat{f_i}$; it is a mapping from the information set to set of spot prices for the asset,

$$f_i^{\hat{}} : H_t \to \{P_{t+1}\}. \tag{21.2}$$

Note, there are three pure strategies to buy, sell or not to trade, {B, S, ♯}. Agents' decision procedure in pure strategies is defined as

$$b_i : (H_t, f_i^{\hat{}}) \to \begin{cases} \text{If } f_i^{\hat{}} - r < 0 & \text{then buy, } \mathrm{B}_i & (a) \\ \text{If } f_i^{\hat{}} - r > 0 & \text{then sell, } \mathrm{S}_i & (b) \\ \text{If } f_i^{\hat{}} = r & \text{then no trade, } \sharp_i. & (c) \end{cases} \tag{21.3}$$

Above, r is the reservation price which is the same for all agents. The decision rule for speculation is simple: agent i buys, B_i (sells, S_i) if the spot price at $t+1$ is predicted to make i a winner and does not trade if the agent is a random walk believer, $f_i^{\sharp}$ in (21.3c), and the expected return is equal to r. In other words, it is plausible to assume that a speculator will not trade unless he is expects to win.

The spot market prices are determined by a total computable function g which is a mapping from agents' decision rules based on their forecast functions to the set of admissible spot prices,

$$g : (b_{it}, f_{it}^{\hat{}}, \forall_i, i = 1, 2, ..., N) \to \{P_{t+1}\}.$$

The spot market price function g is given by

$$g = \begin{cases} (\{P^S\} | P_{t+1} > r + \varepsilon), & \text{if } \sum\limits_i B_i > \sum\limits_j S_j, \ \ i \neq j, \quad (a) \\ (\{P^B\} | P_{t+1} < r + \varepsilon), & \text{if } \sum\limits_i S_i > \sum\limits_j B_j, \ \ i \neq j \\ & \text{or } \ \forall_i, f_i^{\hat{}} = f_i^{\sharp}, P_{t+1} = 0, \quad (b) \\ (\{P^{\sharp}\} | P_{t+1} = r + \varepsilon), & \text{if } \sum\limits_i B_i = \sum\limits_j S_j, \ \ i \neq j. \quad (c) \end{cases} \tag{21.4}$$

Here, ε is a white noise term, ε ~N(0,1). The market price function g precisely determines the payoff for the minority speculative market game. In (21.4a) it yields the set of prices $\{P^S\}$ which leads sellers to win given they are in the minority as stated on the **RHS** of (21.4a). Likewise, in (21.4b) the set of prices $\{P^B\}$ makes it a win situation for buyers given they are in the minority. The extreme case of $P_{t+1} = 0$ is obtained when no trade occurs and when all agents become random walk believers, viz. $\forall_i, f_i^{\hat{}} = f_i^{\sharp}$.

Theorem 1: The spot market price function g is a total computable function given in (21.4). By the Second Recursion Theorem (Cutland 1980), for any total computable function g and for a fixed enumeration of partial computable functions ϕ_0, ϕ_1, ϕ_2, ..., g has a fixed point in the sense that there exist computable functions ϕ_a such that $\phi_{g(a)} = \phi_a$.

Definition 2: In a rational expectations equilibrium (**REE**) there exists some computable forecast function $f^{\hat{}} = \phi_a$ such that

$$\phi_{g(a)} \cong \phi_a, \tag{21.5}$$

then a is a fixed point of the market price function.[10]

Note, a is the encoding of the algorithm or program that computes that output of the market game when the market price function g that determines the outcome is consistent with agents' prediction functions for $\{P_{t+1}\}$ and strategies defined in (21.3). In the absence of perfect information on the population distribution of forecast rules, in principle an agent has to find a meta forecast rule as on the **RHS** of (21.5). That is, the agent has to enumerate a proper subset of the set of all partial computable functions, ϕ_0, ϕ_1, ϕ_2, ..., such that only the fixed points of the total computable function g are identified, viz.

$$\{m \mid \phi_{g(m)} = \phi_m\}. \tag{21.6}$$

Theorem 2: By Rice's Theorem (see, Spear 1989) there is no recursive/algorithmic procedure to identify the set of indices in (21.6) and hence to learn the **REE** of the market price function g.

Inductive trial and error processes specified in Column III of *Table 21.1* have to be used with search beginning from an arbitrary subset of forecast rules. It is useful here to formalize why a homogeneous computable **REE** cannot exist in market games that resemble the minority game.

Definition 3: We say agent i has rational expectations of the spot price if $f_i^{\hat{}} = \phi_a$ and a is the fixed point in (21.5).

Definition 4: The **REE** of the minority market game has a computable fixed point if (21.5) is computable and $\phi_{g(a)} = \phi_a = \{P_{t+1}\}$.

Then the **RHS**, ϕ_a, and the **LHS** $\phi_{g(a)}$ of (21.5) must produce the same outcome here in the classes of $\{P^S\}$, $\{P^B\}$, $\{P^\sharp\}$ and hence which is the appropriate pure strategy for $t+1$ is predictable at t.

Definition 4: A homogenous **REE (HREE)** is one in which, $\forall_i, i = 1, 2, 3, ..., N$, there exists a $f_i^\wedge$, such that $f_i^\wedge = \phi_a$.

Theorem 3: Given that the total computable market price function g always has a fixed point, there is no computable homogeneous **REE** for the minority speculative market game in the three pure strategies buy, sell or no trade {B, S, ♯}. There is no algorithmic decision procedure to determine optimal winning strategies in (21.3) or $\{P_{t+1}\}$ is not predictable.

Proof: We consider two cases. Assume that (21.5) is computable and for $\forall_i, i = 1, 2, 3, ..., N$ there exists a $f_i^\wedge$ such that $f_i^\wedge = \phi_a$.

Case 1: *For the pure strategies to sell or buy*: By ϕ_a on the **RHS** of (21.5) let $\{P^S\}$ be the predicted outcome. This results in the decision rule (21.3b) to follow for all agents. Hence, all agents become sellers which, however, results in $\phi_{g(a)}$ to output $\{P^B\}$ in (21.4b). As this leads to the **RHS** and the **LHS** in (21.5) to yield contradictory outcomes $\{P^S\}$ and $\{P^B\}$, we conclude that (21.5) is not computable.

Case 2: *For pure strategy not to trade*: The proof is analogous to Case 1 and is left to the reader.

The upshot of this is the following. From **Theorem 2**, learning **REE** is in general non-recursive, viz. with no unique decision procedure. With no computable fixed point as in **Theorem 3**, homogeneity of agents' forecast rules is impossible in stock market models where contrarian actions are reinforced as borne out in Savit *et al.* (1999). The algorithmic unsolvability of **REE** of prices in **Theorem 2** makes it tenable that even agents with the full powers of Turing Machines and the same information sets must agree to disagree with regard to price predictions. This goes against the grain of no trade theorems that rational agents cannot agree to disagree. The next section briefly reviews how algorithmic unsolvability of a problem may prompt an evolutionary solution along the lines of an adaptive operating schema or institution as in category (iv) of **ECAS** in the Introduction.

2.3. Price Formation in Double Auction (DA) Markets: Why Zero Intelligence?

The rules of double auction (**DA**) that determine transactions prices in goods with standardized units are known to have been adopted in markets that we now call organized markets at least as early as the advent of the London Stock Exchange in the 17th century. In continuous **DA** agents simultaneously post bids (the price at which they will buy) and offers (the price at which they will sell). The success of the **DA** in achieving transactions prices close to equilibrium market clearing prices rests on the simple institutional rule that traders can get their bids/offers accepted only if they are the best prices at a given time. The best price rule requires that the bid is the highest quoted and the ask is the lowest at time of transaction.

The problem as critiqued by papers in Friedman and Rust (1993) of the institution free analysis of price formation is that in the absence of institutions, agents have to show extra ordinary powers of computation to work out the Bayesian Nash equilibrium of a continuous double auction. The latter has no analytical solution and specifically "nobody has yet been able to calculate the exact timing and size of bids and offers" (*ibid.* p. xxi). One of the important insights from multi-agent model simulations such as by Ackley and Littman (1992) is the inverse relationship between the need for explicit calculation and learning and the evolutionary development of system wide well adapted operational schema or institutions. Once the latter are in place agents need to exercise very little intelligence to get things 'right.' What explicit learning was initially needed when fully adapted becomes instinct and atrophied into skilled behaviour. It is precisely and only in this framework that one must interpret Hayek's early observation (Hayek 1945) on how market institutions in the context of equilibrium price setting facilitate the emergence of the latter with agents trading on the basis of limited information, calculation and explicit learning of their environment. The bounded rational behaviour observed in contexts such as of double auctions can be interpreted to be the product of evolution.

The recent experimental behavioural studies surrounding the variants of **DA** show that the market in action can produce easy convergence to competitive market clearing prices with very few traders who use a minimum of strategic behaviour at that. Easley and Ledyard (1993) led the way by showing that the Bayesian game against nature strategies, played by agents who ignore the impact of their decisions on that of their opponents, successfully generate competitive equilibrium trajectories. In a now celebrated paper, Gode and Sunder (1993) show that continuous **DA** markets populated by *zero-intelligent* agents are highly efficient in extracting gains from trade and price trajectories converge to competitive equilibrium prices. Zero intelligence corresponds to simple computer programs that generate random bids (or asks) subject to a no loss constraint. The latter means that traders cannot buy above their redemption values or sell below their costs and no attempt is made to maximize profits. This was sufficient to obtain 98% of the gains from trade. It is highly conceivable that successful market making in specialist markets bolstered by non-disclosure rules on bloc trades does not involve strategies that are more complicated than the zero intelligence one. Thus, the best price rule of execution in continuous auctions is a simple but powerful device to obtain competitive outcomes with the great economies of computation and information that Hayek emphasized.

2.4. Emergent Efficient Market Hypothesis (EEMH), Power Laws and The Red Queen

Shiller (1981) provides early evidence that the volatility of traded stock returns does not support the **HREE** model that has dominated Finance for over four decades. Chen and Yeh (2000) who pioneered the notion of **EEMH** (Emergent Efficient Market Hypothesis), show that agents who are non-random walk believers are not deluded. Some significant proportion of them at any one time make profitable trades contrary to the view that technical trading is irrational and will in time be driven out of the market. The simple random walk model of asset prices has given way to a current consensus which favours the following class for asset market prices. Stock market returns are serially uncorrelated but their variance is long memory fractally integrated process rather than **GARCH** (Generalized Autoregressive Conditional Heteroscedasticity). The latter has an autocorrelation function that dies down far too fast at an exponential rate while long memory in volatility of stock returns shows persistance and follows a power law decay. The fractal Pareto-Levy Stable distribution with (infinite variance) but independent increments with rich volatility dynamics as a possible contender of asset returns was first proposed by Benoit Mandelbroit (1966). The professed goal of a growing number of papers in the area of the multiagent stock markets has been to see what type of microbehaviour (Kirman and Teyserrie 2001, Lux and Marchesi 1999) can generate in the simulated asset prices the properties of fat tails, volatility clustering, crashes and the like that have been observed in traded asset market prices.

In keeping with the recent literature on complex adaptive systems which upholds the manifestation of power law distributions as evidence of self organized complexity in large ensembles of micro interacting agents, I will follow the work of of Solomon (2000) and Solomon and Levy (1996). They use a micro architecture similar to SFI stock market model and attempt to give a micro level explanation for power law in investment wealth distribution. Indeed, they give the necessary analytical condition for this but curiously fail to see its significance of it within an evolutionary framework. I will argue that what is involved here is none other than the principle of the Red Queen which has as yet not been fully articulated in the literature of stock price dynamics and market efficiency. Technically, as will be seen the power law distribution of investor wealth from a population of coevolving trading strategies of agents modelled as genetic programs requires that each satisfies a time varying constraint embodying the Red Queen principle that each agent must 'keep up' with the average performance of the market.

A. Steps in the Test for the Red Queen Effect

In the Solomon-Levy thesis on endogenous emergence of power law in stock returns, the power law is defined in terms of the distribution of investor wealth in a large microagent system of N agents. The probability distribution, or the proportion of individuals in a population with wealth of size w, is given as

$$P(w) \sim w^{-1-\alpha}, \alpha > 0. \tag{21.7}$$

Here, w (integer valued)[11] is a certain value of wealth w in the population. The total wealth is generated from the N sub/micro agent systems is

$$W(t) = w_1(t) + w_2(t) + ... + w_N(t). \tag{21.8}$$

Solomon and Levy (1996) discovered that dynamics characterized by generalized Lotka Voltera equations for each micro system can under certain conditions bring about the power law distribution in (21.7). The simplest form of this is

$$w_i(t+1) = \lambda_i(t) w_i(t), i = 1, 2, ..., N. \tag{21.9}$$

Here, $\lambda_i(t)$ is the random multiplicative wealth generating factor which incorporates the performance of each agent's forecast based decision rule to buy, sell or not to trade and also the market's generation of the asset returns which is self reflexive on all traders' strategies.

The power law in (21.7) follows if and only if the multiplicative coefficient $\lambda_i(t)$ in (21.9) on agent's wealth becomes independent of agent i factors and all agents' payoffs from strategies must be drawn from a uniform probability distribution. Note, short term returns distribution on stocks satisfies the truncated Levy stable distribution. This follows because the returns appropriately defined is the sum of the N agents' trades at time t *if* each $w_i(t)$ satisfies the power law in (21.7). How does the latter come about?

B. Constraint Satisfaction and The Power Law

The steps in the proof of the result requires that the α parameter in (21.8) has to be positive. For this there has to be a lower bound, $w_{min}(t)$, which dictates that the central limit theorem no longer applies at large t and log normal distributions do not follow for $w_i(t)$. The lower bound $w_{min}(t)$ is specified as

$$w_{min}(t) = q\bar{w}(t), \tag{21.10}$$

where $\bar{w}(t) = W(t)/N$, i.e. $\bar{w}(t)$ is the mean income at time t.

On placing the lower bound constraint regarding minimum wealth given above, the wealth dynamics equation in (21.9) is defined as

$$w_i(t+1) = \lambda_i(t)w_i(t) \geq q\bar{w}(t). \tag{21.11}$$

When the power law in (21.8) holds for the system dynamics in (21.11), for given N and q with q in range of $1 > q > 1/lnN$, the exponent α in (21.8) is given by

$$\alpha = 1/(1-q). \tag{21.12}$$

Typically financial stock market data gives $\alpha \sim 3/2$ and $q \sim 1/3$ with q being interpreted as the reciprocal of the long term market impact factor.

C. Constraint Enhanced GPs and The Red Queen

To underscore the importance of the role of constraint satisfaction and the how the Red Queen Effect works, the lower bound condition in (21.11)(taking the case of equality) is expressed as

$$\lambda_i(t) = q\frac{\bar{w}(t)}{w_i(t)}, 1 > q > \frac{1}{lnN}. \tag{21.13}$$

As the first term (21.13) is common for all agents, the implication of (21.13) is that $\lambda_i(t)$ has to be proportionately greater (smaller) when the mean wealth of traders $\bar{w}(t)$ exceeds (is less than) that of the ith agent. In other words, agents are constrained to improve their forecast/investment performance when it is below average and vice versa. Note, that the economist's invisible hand metaphor applies to the Red Queen principle in the emergence of market efficiency in that it occurs amongst traders who are each trying to find rules to 'beat' or 'keep up' with the market and assiduously select 'good' forecast rules by generic evolutionary fitness criteria of rewarding those rules that increase investor wealth shares, $w_i(t)/W(t)$. Technically, agents are constrained enhanced **GPs**. They bring about an outcome not of their original intent: they find it increasingly harder to enhance their fitness (here the relative wealth shares) relative to the rest of the population. This is the point of i-independence of $\lambda_i(t)$ in (21.9). Thus, **EEMH** is sustained in microsimulation models under circumstances very different from what is traditionally associated with trader rationality. Note, under **EEMH** at each t a range of strategies/forecast rules are used so that investor wealth distribution far from being egalitarian or equal on average over time (as might be the case if agents were making random choices), satisfies the power law distribution in (21.7). As per the latter, the emergent efficient market does reward a small proportion of investors with large returns while many have to make do with very little.

While artificial stock market models have included some criteria governing search intensity for better forecast/investment performance, often in an *ad hoc* way, what has not been precisely stated as above nor tested is how the emergence of the power law in investor wealth distribution follows as a population of N agents apply the constraint in (21.13) in their selection of trading strategies. Specifically, in Solomon and Levy (1996) the Red Queen principle is emphatically not the explanation they give for the required lower bound constraint in (21.10). The failure to explicitly identify and test the Red Queen principle of evolutionary competition for the emergence of power laws in multi-agent stock market models has led to vague conclusions such as "almost every realistic microscopic market model we have studied in the past shares this characteristic of i-independent λ distribution," Solomon (2000).

3. The Ubiquitous Structure of Opposition, Emergence of Innovation and Irregular Structure Changing Dynamics

The ubiquitous structure of opposition that necessitates secrecy and emergence of innovation though intuitively familiar has not received formal attention in economic models. Recent work in this direction by Ray's Tierra (1992) and Hillis (1992) show how in complex system simulations both cooperative and competitive structures develop. When some agents learn that others are parasitic on them, they begin to adopt secrecy or surprise/innovative strategies that their rivals cannot predict. This can move the system in unpredictable structure changing directions which are highly dissipative of the old order.

In extant game theory whether eductive or evolutionary there is no notion of innovation being a Nash equilibrium strategy let alone one that is necessitated as a best response by a structure of opposition.[12] Innovation is either brought about by random mutation or is an ad hoc addition in the form of trend growth.

Goldberg (1995) claims that the mystery shrouding innovation can be dispelled "...by a heavy dose of *mechanism.* Many of the difficulties in the social sciences comes from a lack of a *computational theory* of actor innovation. population oriented systems are dominated by what economists call the law of unintended consequences (which is itself largely the result of the innovative capability of the actors) and interacting with **GA**s provides hands-on experience in understanding what for most people is counterintuitive behaviour," (*ibid.* p. 28). In this section I will briefly show using the mathematics of incompleteness and non-recursiveness why novelty producing behaviour can only be brought about by agents with full powers of Turing machines.[13]

3.1. A Computational Theory of Actor Innovation

Despite Binmore's (1987) seminal work that introduced to game theory the requisite dose of mechanism with players with powers of Turing machines, and along with it 'the spectre of Gödel', the computational theory of actor innovation did not follow. Binmore's critique of traditional game theory is that it cannot accommodate a generic model of a rule breaker. The centre piece of Gödel (1931, p. 19) is a formal analogue of the Liar which leads to the proof of the limits of calculation in self-referential structures.[14] The Liar strategy in a two person game is defined as being capable of systematically contradicting the mutually predicted outcomes of the other person's strategy.[15] This structure of opposition is universal as in Ray's Tierra (1992) and *ipso facto* renders the transparent or predictable rule inoptimal. I will proceed to show that when there is mutual knowledge of the Liar qua rule breaker, viz. at the fixed point with the Liar, we are at Gödel's famous uncomputable fixed point. Technically, from the latter the only total computable best response function is one that maps into a domain of the player's strategy set that cannot be algorithmically enumerated. Further, we can show that this surprise strategy function can implement a new action/institution outside extant action sets. Formally, as the surprise or innovative strategy involves a total computable response function that corresponds to the productive function, the encoding of which provides an ever extendible set of explicit 'witnesses' for incompleteness in set theoretic proofs of the 1931 Gödel result developed by Post (1943) (see, also Cutland 1980). The analogue of this in the formalisation of a game with computational agents is that when they have mutually identified the Liar or the structure of opposition, then not only does secrecy become paramount for their objectives, but it also follows that the only Nash equilibrium strategies thereof are surprises or innovations implemented by the productive function.

A major implication of computational agents is that all meta - information with regard to the outcomes of the game for any given set of state variables, $s \in S$, can be effectively organized by the so called prediction function $\phi_{\sigma(x,y)}(s)$ in an infinite chequer board like matrix Ξ of the enumeration of all partial computable functions, given in Figure 21.1, Cutland (1980). The tuple (x, y) identifies the row and column of this matrix Ξ whose rows are denoted as Ξ_i, $i = 0, 1, 2, ...$

The function $\phi_{\sigma(x,y)}(s)$ if defined at a given state s and $\sigma(x, y)$ yields

$$\phi_{\sigma(x,y)}(s) = q. \tag{21.14}$$

Here in (21.14), q in some code, is the vector of state variables determining the outcome of the game. Note, $\sigma(x, y)$ is the index of the program for this function ϕ that produces the output of the game when

$$
\begin{array}{lllllll}
\Xi_1 & \phi_{\sigma(0,0)} & \phi_{\sigma(0,1)} & \phi_{\sigma(0,2)} & \phi_{\sigma(0,3)} & \phi_{\sigma(0,y)} & \cdots \\
\Xi_2 & \phi_{\sigma(1,0)} & \phi_{\sigma(1,1)} & \phi_{\sigma(1,2)} & \phi_{\sigma(1,3)} & \phi_{\sigma(1,y)} & \cdots \\
\Xi_3 & \phi_{\sigma(2,0)} & \phi_{\sigma(2,1)} & \phi_{\sigma(2,2)} & \phi_{\sigma(2,3)} & \phi_{\sigma(2,y)} & \cdots \\
\cdot & \cdot & \cdot & \cdot & \cdot & \cdot & \cdots \\
\cdot & \cdot & \cdot & \cdot & \cdot & \cdot & \cdots \\
\cdot & \cdot & \cdot & \cdot & \cdot & \cdot & \cdots \\
\Xi_x & \phi_{\sigma(x,0)} & \phi_{\sigma(x,1)} & \phi_{\sigma(x,2)} & \phi_{\sigma(x,3)} & \phi_{\sigma(x,y)} & \phi_{\sigma(x,x)} \quad \cdots
\end{array}
$$

Figure 21.1. **Prediction function in an infinite chequer board.**

the first player plays strategy x and the second player plays a strategy *that is consistent with his belief that the first player has used strategy y.*

We will now adopt this generic framework for the analysis of a two person game with computational agents. The game can best be interpreted as one of regulatory arbitrage where an oppositional structure can arise between the players.

The policy game is played by the authorities (g) and the private sector (p) with their respective objective functions defined by computable functions Π_i , $i \in (p, g)$. Each player optimizes using strategies denoted by (β_p^*, β_g^*) which involve computable functions that specify the optimal response functions (f_p, f_g) which incorporate elements from the respective action sets $A = (A_p, A_g)$ *and* given mutual beliefs of one another's optimal strategy.[16] Here, $\hat{b_p}$ denotes (codes of) p's meta -representations of g's optimal strategy and $\hat{b_g}$ denotes (codes of) g's meta-representations of p's optimal strategy. Further, (g, p), (A_p, A_g) and all archival information of past and current state variables (where s is a given vector of state variables) are assumed to be in the public domain. All meta calculations on the strategies played and capable of being played in the game are based on this information and recorded in the matrix Ξ.

The determination of Nash equilibrium strategies involve the use of total computable best response functions (f_p, f_g) which operate directly on points such as $\sigma(x, x)$ to effect computable transformations of the system from one row to another of matrix Ξ with special reference to its diagonal array, see, Figure 21.1. Thus,

$$\phi_{f_i \sigma(x,x)}, i \in (p, g). \tag{21.15}$$

Again, proceeding informally, all fixed points and Nash equilibria have to be elements along the diagonal array of this matrix. A typical Nash equilibrium are at points defined by $\sigma(x,x)$, viz. player p plays x and g correctly identifies this. Off diagonal elements along any row defined by strategy, say y, employed by the private sector, cannot be Nash equilibria, as these off diagonal terms imply that authorities are choosing their strategy assuming the wrong meta representation of p's play. Consistent alignment of beliefs by which we have $\sigma(x,x)$ is a necessary condition of a Nash equilibrium as is the condition that there is rational expectations and both agents choose their optimal Nash equilibrium strategies such that each identifies the same function as producing the outcome of the game.

The best response functions f_i, $i \in (p,g)$, that are total computable functions can belong to one of the following classes —

$$f_i = \begin{cases} 1(\text{Identity Function}) & - \quad \text{Rule Abiding} \\ f_i^{+} & - \quad \text{Rule Bending} \\ f_i^{\neg} & - \quad \text{Rule Breaking (Liar Strategy)} \\ f_i! & - \quad \text{Surprise} \end{cases} \tag{21.16}$$

such that the codes of f_i are contained in set $\Re$,

$$\Re = \{m | f_i = \phi_m, \phi_m \text{ is total computable}\}. \tag{21.17}$$

The set $\Re$ which is the set of all total computable functions is not recursively enumerable. The proof of this is standard, Cutland (1980).

As will be clear, (21.17) draws attention to issues on how innovative actions/institutions can be constructed from existing action sets. The remarkable nature of the set $\Re$ is that potentially there is an uncountable infinite number of ways in which 'new' institutions can be constructed from extant action sets A. In standard rational choice models of game theory, the optimization calculus in the choice of best response requires choice to be restricted to given actions sets. Hence, strategy functions map from a relevant tuple that encodes meta information of the game into given action sets

$$\beta_i(f_i\sigma(x,x)), s, \mathrm{A}) \rightarrow \mathrm{A} \text{ and } f_i = \phi_m, m \in \mathrm{A}, i \in (p,g). \tag{21.18}$$

Unless this is the case, as the set $\Re$ is not recursively enumerable there is in general no computable decision procedure that enables a players to determine the other player's response functions. However, in principle, a strategic decision procedure (β_g, β_p) for choice of best response, $f_i = \phi_m, m \in \Re, i \in (p, g)$, can map into $\Re - A$, implying that an innovative action not previously in given action sets is used. We define the surprise/innovation producing strategy as follows:

$$\beta_i(f_i\sigma(x,x)), s, \mathrm{A}) \rightarrow \Re - \mathrm{A} \text{ and } f_i = f_i^! = \phi_m, m \in \Re - \mathrm{A}, i \in (p, g). \tag{21.19}$$

It has been noted in passing by Anderlini and Sabourian (1995, p. 1351), based on the work of Holland (1975), that heterogeneity in forms do not arise primarily by random mutation but by algorithmic recombinations that operate on existing patterns. However, a number of preconceptions from traditional game theory such as the 'givenness' of actions sets prevent Anderlini and Sabourian (1995) from positing that players who as in (21.19), equipped with the wherewithal for algorithmic recombinations of existing actions, do indeed innovate from strategic necessity rather than by random mutation. The innovation *per se* is emergent phenomena, but the strategic necessity for it is fully deducible. Hence, it is adduced that in the Wolfram-Chomsky schema on dynamical systems with computationally intelligent agents, agents with full deductive powers of Turing machines capable of self-referential calculations are necessary to bring about innovation based structure changing dynamics. It only remains to show the specific structure of opposition that logically and strategically necessitates surprise strategies in the Nash equilibrium of the game.

Consider the state of affairs given by

$$\phi_{\sigma(b_a,b_a)}(s) = q, \tag{21.20}$$

where outcomes of a policy rule a is predictable and q is the desired outcome that g wants in state variables when applying this policy rule a. Here, in our two place notation $\sigma(b_a, b_a)$, the **RHS** b_a is g's belief that p is rule abiding and the **LHS** b_a confirms that this is true. By rule abiding is meant that p will leave the system unchanged in terms of the row b_a of matrix Ξ. It is convenient to assume that policy rule a is optimal for g *if* the private sector is rule abiding.

However, for player p, for the given (a, s) it is optimal for p to apply the Liar strategy, $f_p^{\neg}\ \sigma(b_a, b_a)$, the code of which is, say, $b_a^{\neg}$. Formally, the Liar strategy has the following generic structure. For any state s when the rule a applies,

$$\phi_{f_p^{\neg}\sigma(b_a,b_a)}(s) = q^{\sim}, q^{\sim} \notin E_{\sigma_{b_a}} \leftrightarrow \phi_{\sigma(b_a,b_a)}(s) = q, q \in E_{\sigma_{b_a}}. \qquad (21.21)$$

For all s when policy rule a does not apply,

$$f_p^{\neg} = 0, \text{viz. do nothing.} \qquad (21.22)$$

The Liar can successfully subvert with certainty in (21.21) if and only if ($\leftrightarrow$) the policy rule is transparent with predictable outcomes and $f_p^{\neg}$ itself is total computable. Also, $f_p^{\neg} = \phi_m$, m $\in A_p$, must include a codified description of an action rule if undertaken by the Liar can subvert the predictable outcomes of the policy rule a. Formally, if q is predicted then the application of $f_p^{\neg}$ to $\sigma(b_a, b_a)$ will bring about an outcome $q^{\sim}$ $\notin E_{\sigma_{b_a}}$ which belongs to a set disjoint from the set that contains the desired output of rule a for all s for which rule a applies, viz. $E_{\sigma_{b_a}} \bigcap E_{\sigma_{b_a\neg}} = \emptyset$. The outcomes ($q^{\sim}$, q) can be zero sum but in general we refer to property $q^{\sim} \notin E_{\sigma_{b_a}}$ in (21.21) as being oppositional or subversive. This underpins the intuition behind Ray's Tierra simulation where agents recognize the necessity for secrecy. This is also well known from the Lucas (1972) postulate on policy ineffectiveness in the case of fully anticipated policy and the wisdom behind the panacea that to forestall subversion, the policy rule must be undefined and fraught with ambiguity.

Thus, we come to the point as why agents who precipitate the Wolfram-Chomsky **Type 4** dynamics with innovation have to have powers of self-referential calculation. Firstly, g acknowledges the identity of the Liar in (21.21) and understands that the transparent rule a cannot be implemented rationally as the outcome defined by $\phi_{\sigma(b_a^{\neg},b_a)} = q^{\sim}$ follows.[17] The latter outcome is the opposite what is desired by g. Player g updates beliefs so that formally we obtain the fixed point involving the Liar which is $\sigma(b_a^{\neg}, b_a^{\neg})$ where $b_a^{\neg}$ is the code for the Liar strategy in (21.21).[18] Now, the Liar, p, knows that g knows that p is the Liar. The prediction function indexed by the fixed point of the Liar/rule breaker best response function $f_p^{\neg}$ in (21.23) is not computable and corresponds to the famous Gödel uncomputable fixed point.

$$\phi_{f_p^{\neg}\sigma(b_a^{\neg},b_a^{\neg})}(s) = \phi_{\sigma(b_a^{\neg},b_a^{\neg})}(s). \tag{21.23}$$

The proof is standard.[19]

There is no paradox in stating that as both players can prove the non-computability of (21.24) they will have mutual knowledge that the only Nash equilibrium strategies for both players that is consistent with meta information in the fixed point in (21.23), is one that involves strategies that elude prediction from within the system. On substituting the fixed point $\sigma(b_a^{\neg}, b_a^{\neg})$ in (21.23) for $\sigma(x, x)$ in (21.19), g's (p's) Nash equilibrium strategy β_g^E (β_p^E) implemented by an appropriate total computable function must be such that, for $i = (p, g)$,

$$\beta_i^E(f_i\sigma(b_a^{\neg}, b_a^{\neg}), s, \mathrm{A}) \rightarrow \Re - \mathrm{A} \text{ and } f_i = f_i! = \phi_m, m \in \Re - \mathrm{A}. \tag{21.24}$$

That is, $f_i!$ implements an innovation and or action outside extant actions sets for $i = (p, g)$.

The intuition here is that from the non-computable fixed point with the Liar, the total computable best response function implementing the Nash equilibrium strategies can only map as above into domains of the action and strategy sets of the players that cannot be algorithmically enumerated in advance.[20]

As in Ray's Tierra and the Hilles simulation models, once computational agents have enough capabilities to detect rivalrous behaviour that is inimical to them, they learn to use secrecy and surprises. In a two person game with computational agents, this can be fully formalized using the Gödelian result on incompleteness. To show how with parallel computing agents, we have cooperation and competition not simply as in Prisoners Dilemma, but with the use of periodic adoption of new institutions outside of extant action sets, we need the new technology of virtual models of emergent phenomena. The above framework supports the new regulatory arbitrage theory of institutional innovation (see Miller 1986, Schanze 1995, etc.) the latter is brought about by rule breaking regulatees who in their attempt to contravene regulation with predictable outcomes do so by 'exit' and innovation. Schanze (1995) calls such innovative behavior *regulatory bifurcation* for which the constitutional structure of rules may have to be modified to accommodate innovation. The vulnerability of formalist control of society to the Liar (see, Koons 1992, p. 151) and innovative behaviour makes unwritten constitutions, deontic virtue, truth, meaning and such tacit values the predominant guides to spontaneous order.

4. Concluding Remarks

The science of complex adaptive systems has been a truly interdisciplinary venture being spearheaded by physicists, biologists, computer scientists, mathematicians, economists and others. In the case of the evolution of complexity in intelligent social systems, it is not far off the mark to say that the modern evolutionary agenda on spontaneous or emergent order was clearly identified at the provenance of Economics in the 18th century. A survey of classical political economy has been given here to redress the balance in the growing contributions by physicists and mathematicians on the evolution of complexity. The *sine qua non* of a system capable of complexity is manifested in the disjunction between system wide outcomes and the micro level computational capabilities and also in the non-anticipating or surprise producing features of the system. The normative necessity of a structure of constitutional rules that permits this was clearly articulated in classical liberal tradition even if its implications were and are not yet fully understood. Over two centuries have had to elapse till we have the methodological tools necessary to counter, on grounds of computational impossibility, the ultra rationalist position that sees no distinction between what can be constructed by calculation and what can only emerge. Scientifically, the latter category of events has had a tenuous status till recently. The advances in the mathematics of non-linear dynamics, emergence and evolutionary computation has made it possible to pin down the elusive properties of non-computable dynamics produced by computationally intelligent agents.

I started with the premise that when the domain of a decision problem is non-recursive, the loss of a categoric decision procedure prompts a multiplicity of models for inductive inference. Indeed, a critical degree of competing heterogeneous 'world views' are needed to achieve the emergent coordination and enhanced social welfare even in the simple minority game. In Section 2, the Red Queen principle which manifests as a constraint on the performance of agents to 'keep up' with the average performance of the market was given as a testable hypothesis for the the emergence of market efficiency and of power laws in asset prices. Section 3 sketched a computational theory of actor innovation based on the mathematics of Gödel incompleteness which permits computational agents to exit and innovate when caught up in an oppositional structure. This ubiquitous dynamic behind innovation has been observed in **Type 4** dynamics of Wolfram-Chomsky cellular automata and more recently been seen in the Ray's Tierra (1992) and Hillis (1992) Artificial Life simulations. Despite efforts in this direction as in Easley and Rustichini (1999), Arthur (1993b) and others, it is clear that economists and in particular **ACEs** themselves have a long way to go to understand complexity in terms of the Langton-Kaufmann novelty producing world with attendant problems of incompleteness and undecidability. I have made a case that 'new' evolutionary equilibrium concepts such as that of the Red Queen and the Langton (1992) equilibrium with undecidable dynamics need to be developed to further our understanding of complex market systems. To conclude, therefore, the formalist limits on algorithms and adaptive methodology of evolution and emergence (viz. columns II and III of *Table 21.1*) are two sides of the same coin that should be brought to bear on discussions of complex adaptive systems.

Acknowledgments

I'm grateful to Shu-Heng Chen for his invitation to contribute this piece. I appreciate his encouragement of the view that a powerful confluence of the deductive/rationalist and the inductive/evolutionary schools that build on remarkable 20th century intellectual achievements is necessary for the paradigm shift that is afoot in Economic analysis. I've benefited from discussions with David Easley, Ken Binmore, Ken Burdett, Edward Tsang, Nick Vriend, Seppo Honkapohja, Vela Velupillai. Gordon Kemp kindly helped in improving the presentation. I'm implicating none of them for the views expressed here or for any errors.

Notes

1. Note, a partial recursive function is number theoretic function, f: $\mathbf{N} \rightarrow \mathbf{N}$, that is not defined on the full domain of the set of all integers, **N**. In other words, on some $n \in \mathbf{N}$ which is its input, the computation being implemented by the partial recursive function will not halt. Total recursive functions are defined on the full domain on **N**. The set of all total computable functions is uncountably infinite and hence it is not recursively enumerable by any Turing machine (see, Cutland 1980).

2. Informally and in popular language, Gödel's Second Incompleteness Theorem implies that the price for logical consistency is incompleteness or the algorithmic unsolvability of a decision problem, see Binmore (1987).

3. It is increasingly being understood that the standard Savage model of choice under uncertainty is inadequate in the bigger scheme of things. See, Easley and Rustichini (1999) who develop a framework of rational choice in a complex environment and attempt to relax the assumption of a one-to-one mapping between actions and outcomes.

4. These results show that the notion of symbol sequences associated with Turing computability and formalism is not essential to computation theory. When conditions of pragmatic implementation are at stake, it is useful to consider number theoretic computable functions that have inputs and outputs that are integer encoded descriptions of finite objects. However, as there are computable continuous functions with no computable solutions in the class of ordinary differential equations (see, Pour-el and Richards 1989) what is important is the nature of the problem that brings about noncomputability rather than whether the function is continuous or discrete. I'm grateful to Ken Burdett for pressing for this clarification.

5. Bak (1996) style self-organized criticality refers to order in large systems with many interacting agents that is poised at the edge of chaos or at a phase transition. At the point of criticality, minor disturbances can cause such a system to dissipate and to reorganize new patterns.

6. It has been suggested in Markose-Cherian (1991) that the classical Kantian view that rules of just society satisfy no predetermined outcomes may formally mean that the dynamical outcomes of such rules of engagement are formally undecidable.

7. To understand that there is indeed a non-trivial mathematical problem in the modelling of complex adaptive intelligent systems, see, the account in Sugden (1989) on what is spontaneous order which makes no mention of the methodological developments in columns II and III of *Table 21.1*

8. Rust (1997) only hints at the logical necessity of the impossibility result on computation that underpins evolutionary and emergent phenomena with decentralized decision units. Lewis (1985, 1988) was the first to rigorously prove that the two most basic problems of economic theory viz. micro rational choice and Walrasian general equilibrium outcomes are *not* Turing computable. Indeed, Lewis has asserted that his results have "serious consequences for the foundations of neoclassical mathematical economics......that the foundations of neoclassical economics are hopelessly non-effective computationally and therefore must be considered *irrational* from the standpoint of *computational viability*" (Lewis, 1985, p. 46, p. 72, italics in the original). How do we come to terms with the growing number of theoretical results that state that virtually all of what is considered to be of fundamental interest to economics is *outside* the domain of Turing computability or uniform procedures ? The answer is, of course, to make a laborious intellectual journey from column I to column III of *Table 21.1.*

9. The well known Samuelsonian tenet states 'properly' anticipated prices fluctuate randomly. Here 'proper' anticipation involves using conditional expectations operators, which by the Tower/Iterated property satisfy the martingale condition. Hence, random fluctuations in properly anticipated prices follow by tautology. By the Tower property $E(E(P_{t+1}\,|H_t)|H_{t-1})$ $= E(P_{t+1}\,|H_{t-1})$ implying that $E(P_{t+1}\,|H_t) - E(P_{t+1}\,|H_{t-1}) = \varepsilon_t$, where ε_t is white noise.

10. As stated in Spear (1989) it does not follow that $g(a) = a$. It is only required that both a and $g(a)$ identify functions that produce identical outputs.

11. Note that $P(w) = 0$ for $w \leq 0$ and $\lim P(w) = 0$ as $w \to \infty$. That is, the distribution is zero if w is negative or tends to infinity.

12. There is, however, a long tradition in the macro policy literature design seminally put forward by Lucas (1972). Lucas postulated the necessity of secrecy, ambiguity and surprise strategies in policy against the possibility of a private sector that can contravene policy and render it ineffective if policy outcomes can be rationally expected. The Lucas Critique in Lucas (1976) indicates that there is a problem of predictive failure of policy outcomes by meta (econometric) models. Though not fully recognized yet (see Markose 2001b) this is the exact same logic from formalist settings that demonstrate incompleteness and undecidable dynamics.

13. Kaufmann (1993, pp. 369–404) suggests the use of random grammars to model the evolution of complexity and novelty which necessarily incorporate computationally incomplete and undecidable problems.

14. Gödel's analogue of the Liar proposition is the undecidable proposition, say A, which has the following structure: A $\leftrightarrow$ $\sim$P(A). That is, A says of itself that it is not provable ($\sim$P). However, there is no paradox here as it is indeed true that this is so. Any attempt to prove the proposition A results in a contradiction with both A and $\sim$A, its negation, being provable in the system.

15. Self-subversion is possible. In this case, the player breaks his own rule when its outcomes are desired by the other.

16. Note, the objective functions of players are computable functions Π_i, i$\in(p, g)$ defined over the partial recursive payoff/outcome functions specified in state variables in (21.15). Thus, $ArgMax_{b_i \in B_i} \Pi_i(\phi_{\sigma(b_i, \hat{b_j})}(s)), i, j \in (p, g)$, and B_i is the set of codes for i's strategies.

17. In out two place notation, the first $b_a^{\neg}$ is code for p's Liar strategy and b_a is code for g's mistaken belief of p's strategy.

18. Formally, $b_a^{\neg}$ may be viewed as the code of a refutable proposition in a formal system. A refutable proposition is one whose negation (b_a here) is provable in the system. As theoremhood is a computable relationship, the code of the refutable proposition cannot belong to the domain of any computable function. However, as b_a is provable, the set of all such refutable functions is a recursively enumerable subset of the domain of calculations such as $\phi_x(x)$, for all x, which do not terminate.

19. Assume (21.23) is computable and the **RHS** of (21.23) produces the output $q^{\sim}$ and the **LHS** by the definition of the Liar strategy produces output q. However, if (21.23) is computable then we have $q = q^{\sim}$ which is a contradiction.

20. A rigorous proof of this is omitted here and can be found in Markose (2001b).

References

Ackley, D. H. and M. Litterman (1992). "Interactions Between Learning and Evolution," in C.G Langton et al. (eds.), *Artificial Life II, Santa Fe Institute Studies in the Sciences of Complexity*, 10, MA: Addison Wesley.

Albin, P. (1998). *Barriers and Bounds to Rationality, Essays on Economic Complexity and Dynamics in Interactive Systems*, Edited and with an Introduction by D. Foley. Princeton University Press.

Anderlini, L. and H. Sabourian (1995). "Cooperation and Effective Computability," *Econometrica*, 63, 1337–1369.

Arifovic, J. (1994). "Genetic Algorithm Learning and the Cobweb Model," *Journal of Economic Dynamics and Control*, 18(1), 3–28.

Arrow, K. J. and F. H. Hahn (1971). *General Competitive Analysis*, Amsterdam: North Holland.

Arthur, W. B. (1991). "Designing Economic Agents that Act Like Human Agents: A Behavioral Approach To Bounded Rationality," *American Economic Review*, 81, 353–359.

Arthur, W. B. (1993a). "On Learning and Adaptation in the Economy," Working Paper, 92-07-038, Santa Fe, NM: Santa Fe Institute.

Arthur, W. B.(1993b). "On The Evolution of Complexity," Working Paper, 93-11-070, Santa Fe, NM: Santa Fe Institute.

Arthur, W. B. (1994). "Inductive Behaviour and Bounded Rationality," *American Economic Review*, 84, 406–411.

Arthur, W. B., J. Holland, B. LeBaron, R. Palmer, and P. Taylor (1997). "Asset Pricing Under Endogenous Expectations in an Artificial Stock Market," in Arthur W.B., Durlauf S., Lane, D. (eds.), The Economy as an Evolving Complex System II, Reading MA: Addison Wesley.

Axtell,R.(2000). "Why Agents? On the Varied Motivations For Agent Computing in the Social Sciences," Center on Social and Economic Dynamics, Working Paper No.17.

Bak, P. (1996). *How Nature Works:The Science of Self-Organized Criticality*, New York: Springer-Verlag(Copernicus).

Barry, N. (1982). "The Tradition of Spontaneous Order," *Literature of Liberty*, 7–58.

Binmore, K. (1987). "Modelling Rational Players: Part 1," *Journal of Economics and Philosophy*, 3, 179–214.

Brock, W. and C. Holmes (1997a). "A Rational Route To Randomness," *Econometrica*, 65, 1059–1095.

Burks, A. W. (1970). *Essays on Cellular Automata.* Urbana: University of Ilinois Press.

Casti, J. (1994). *Complexification: Explaining A Paradoxical World Through the Science of Surprises.* London Harper Collins.

Challet, D. and Zhang, Y.C. (1998). "On The Minority Game: Analytical And Numerical Studies," *Physica*, 256–514.

Chen, S.-H. and C.-H. Yeh (2000). *"On Emergent Properties of Artificial Stock Markets: The Efficient market Hypothesis and the Rational Expectations Hypothesis"*, Mimeo Presented at the Fourth Conference on Economics With Heterogeneous Interacting Agents (WEHIA, 1999).

Cutland, N. J. (1980). *Computability: An Introduction to Recursive Function Theory.* Cambridge University Press.

Dawid, H. (1999). *Adaptive Learning by Genetic Algorithms: Analytical Results and Applications to Economic Models*, 2nd ed. Springer.

Dosi, G. and R. Nelson (1994). "An Introduction to Evolutionary Theories in Economics," *Journal of Evolutionary Economics*, 4, 153–172.

Easley, D. and J. Ledyard (1993). "Theories of Price Formation and Exchange in Double Oral Auctions," In D. Friedman and J. Rust (eds.) *The Double Auction Market Institutions, Theories and Evidence*, Proceeding Volume XIV, Santa Fe Institute. Addison Wesley Publishing Company.

Easley, D. and A. Rustichini (1999). "Choice Without Beliefs," *Econometrica*, 69(5), 1157–1185.

Friedman D. and J. Rust (1993). *The Double Auction Market Institutions, Theories and Evidence*, Proceedings Volume XIV, Santa Fe Institute. Addison Wesley Publishing Company.

Epstein, J. and R. Axtell (1996). *Growing Artificial Societies Social Science From Bottom Up*. MIT Press.

Gode, D. K. and S. Sunder (1993). "Allocative Efficiency of Markets With Zero Intelligence Traders: Markets as a Partial Substitute for Individual Rationality," *Journal of Political Economy*, 101, 119–137.

Gödel, K. (1931). "On Formally Undecidable Propositions of Principia Mathematica and Related Systems," Translation in English in *Gödel's Theorem in Focus*, ed. S. G. Shanker, 1988, Croom Helm.

Goldberg, D. E. (1989). *Genetic Algorithms in Search, Optimization and Machine Learning*, Reading, M.A: Addison Wesley.

Goldberg, D. E. (1995). "The Existential Pleasures of Genetic Algorithms," in G. Winter, J. Perioux, M. Galan and P. Cuesta (eds.), *Genetic Algorithms in Engineering and Computer Science*, 23–31, Wiley and Sons.

Hayek, F. A. (1945). "The Use of Knowledge in Society," *American Economic Review*, 35, 519–530.

Hayek, F. A. (1953). *The Sensory Order*, University of Chicago Press, Chicago, IL.

Hayek, F. A. (1967). *Studies in Philosophy, Politics and Economics*, New York: Simon and Schuster.

Hayek, F. A. (1982). "The Sensory Order After 25 Years," in W. B. Weimar and D. S. Palermo, (eds.), *Cognition and Symbolic Processes*, 2, NJ: Lea Publishers.

Hertz, J., A. Krogh, and R. Palmer (1991). *Introduction to the Theory of Neural Computation*, Redwood City, CA: Addison Wesley.

Hillis, W. D. (1992). "Co-Evolving Parasites Improve Simulated Evolution as an Optimization Procedure," in C.G. Langton,*et al.*(eds.), *Artificial Life II*, Santa Fe Institute Studies in the Sciences of Complexity, 10. Addison-Wesley.

Hodgson, G. (1993). *Economics and Evolution: Bringing Life Back Into Economics*, Oxford: Basil Blackwell.

Holland, J. H. (1975). *Adaptation in Natural and Artificial Systems*, Ann Arbor, MI: The University of Michigan Press.

Holland, J. H.(1992). *Adaptation in Natural and Artificial Systems, An Introductory Analysis with Applications to Biology, Control and Artificial Intelligence*, 2nd ed. Cambridge, MA: MIT Press.

Hurwicz, L. (1960). "Optimality and Informational Efficiency in Resource Allocation processes," *Proceedings of the first Stanford Symposium on Mathematical Methods In Social Sciences*, Stanford, CA: Stanford University Press.

Hume, D. (1888). *A Treatise of Human Nature*, in L. A. Selby-Bigge (ed.). Oxford University Press.

Kant, I. (1965). *The Metaphysical Elements of Justice*, Translated, with an Introduction by John Ladd, Indianapolis, IN: The Bobbs-Merriel Company.

Kauffman, S. A. (1993). *The Origins of Order: Self Organization and Selection in Evolution.* Oxford University Press.

Kirman, A. and G. Teyssiere (2001). "Microeconomic Models for Long-Memory in the Volatility of Financial Time Series," *Studies in Nonlinear Dynamics and Econometrics*, forthcoming.

Koons, R. (1992). *Paradoxes of Belief and Strategic Rationality.* Cambridge University Press.

Koza, J. (1992). *Genetic Programming*, Cambridge, MA: MIT Press.

Krugman, P. (1996). *The Self-Organizing Economy.* Blackwell Publishers.

Langton, C. (1992). "Life at the Edge of Chaos," In, C. Langton *et al.* (eds.), *Artificial Life II*, Santa Fe Institute Studies in the Sciences of Complexity, 10. Addison-Wesley.

LeBaron, B. (2001). "Empirical Regularities from Interacting Long and Short Memory Investors in an Agent Based Stock Market," IEEE Transactions on Evolutionary Computation.

Lewis, A. A. (1985). "On Effectively Computable Choice Functions," *Mathematical Social Sciences*, 10.

Lewis, A. A. (1987), "On Turing Degrees of Walrasian Models and a General Impossibility Result in the Theory of Decision Making," Technical Report no. 512 (Centre for Research for Organization Efficiency, Stanford University, CA).

Lucas, R. (1972). "Expectations and the Neutrality of Money," *Journal of Economic Theory*, 4, 103–124.

Lucas, R. (1976). "Econometric Policy Evaluation: A Critique," *Carnegie-Rochester Conference Series on Public Policy*, 1, 19–46.

Lux, T. and M. Marchesi (1999). "Scaling and Criticality in Stochastic Multi-Agent Model of a Financial Market," *Nature*, 397, 498–500.

Mandelbroit, B. (1966). "When Can Price Be Arbitraged Efficiently? A Limit to The Validity of the Random Walk and Martingale Models," *The Review of Economics and Statistics*, 225–236.

Markose S. (1991). "End Independent Rules and the Political Economy of Expanding Market Societies of Europe," *European Journal of Political Economy*, 7, 579–601.

Markose, S.(2001a). Book Review, *Computable Economics* by K. Velupillai, In, *The Economic Journal*, June, 111, 468–470.

Markose, S. (2001b), "The Liar and Surprises: So What Is The Lucas Critique - A New Perspective From Computability Theory," Mimeo

University of Essex,
<http://www.essex.ac.uk/economics/staff/scher.html>.

Marschak, T. (1959). "Centralization and Decentralization in Economic Organizations," *Econometrica*, 27.

Miller, M. (1986). "Financial Innovation: The Last Twenty Years and the Next," *Journal of Financial and Quantitative Analysis*, 21, 459–471.

Nicolis, G. and I. Prigogine (1989). *Exploring Complexity: An Introduction*, New York: Freeman.

Nozick, R. (1993). *The Nature of Rationality*. Princeton University Press.

Penrose, R. (1988). "On Physics and Mathematics of Thought," in Rolf Herken (ed.), *The Universal Turing Machine: A Half Century Survey*. Oxford University Press.

Penrose, R. (1989). *The Emperor's New Mind:Concerning Computers, Minds and The Laws of Physics*. Oxford University Press.

Post, E.(1944). "Recursively Enumerable Sets of Positive Integers and Their Decision Problems," *Bulletin of American Mathematical Society*, 50, 284–316.

Pour-El, M. B. and J. I. Richards (1989). *Computability In Analysis and Physics*, Berlin: Springer-Verlag.

Ray, T. S. (1992)."An Approach to The Synthesis of Life," in C. Langton *et.al* (eds.), *Artificial Life II*, Santa Fe Institute Studies in the Sciences of Complexity, 10. Addison-Wesley.

Rust, J. (1997). "Dealing With the Complexity of Economic Calculations," Mimeo, Yale University.

Samuelson, P. A. (1965). "Proof That Properly Anticipated Prices Fluctuate Randomly," *Industrial Management Review*, VI , 41–49.

Savit, R., R. Manuca, and R. Riolo (1998). "Adaptive Competition, Market Efficiency and Phase Transition," *Physics Review Letters*, 82.

Schanze, E.(1995). "Hare and Hedgehog Revisited: The Regualtion of Markets That Have Escaped Regulated Markets," *Journal of Institutional and Theoretical Economics*, 15(1), 162–176.

Schelling, T. (1978). *Micromotives and Macrobehaviour*, New York, NY: Norton.

Shiller, R. (1981). "Do Stock Market Prices Move Too Much to be Justified by Subsequent Changes in dividends?" *American Economic Review*, 71, 421–36.

Simon, H. A. (1981). *The Sciences of the Artificial*, Cambridge, MA: MIT Press.

Solomon, S. (2000). "Generalized Lotka-Volterra (GLV) Models and Generic Emergence of Scaling Laws in Stock Markets," in G. Ballot and G. Weisbuch (eds.), Proceedings of the International Conference on *Computer Simulations and the Social Sciences*, Hermes Science Publications.

Solomon, S and M. Levy (1996). "Spontaneous Scaling Emergence in Generic Stochastic Systems," *International Journal of Modern Physics* C, 7(5).

Spear, S. (1989). "Learning Rational Expectations Under Computability Constraints," *Econometrica*, 57, 889–910.

Sugden, R. (1989). "Spontaneous Order," *Journal of Economic Perspectives*, 4, 85–87.

Suzumura, K. (1990). "Alternative Approaches to Libertarian Rights in The Theory of Social Choice," in K. J. Arrow (ed.), *Issues In Contemporary Economics*, Vol.1, Markets and Welfare, Macmillan.

Testfasion, L. 1998. ACE Website. <http://www.econ.iastate.edu/tesfatsi/ace.htm>.

Turing, A. (1936). "On Computable Numbers With an Application To the Entscheidungsproblem," in M. Davis (ed.), *The Undecidable*, New York: Raven.

Ullmann-Margalit, E. (1978). "Invisible Hand Explanations," *Synthese*, 39, 263-290.

Velupillai, K. (2000). *Computable Economics*, Arne Ryde Lectures. Oxford University Press.

Vriend, N.(2000). "Was Hayek an ACE ?" Mimeo, <http://www.qmw.qc.uk/ ugtel73/>.

Weibull, J.W.(1996). *Evolutionary Game Theory*, The MIT Press.

Weimar, W. B. (1982). "Hayek's Approach to the Problems of Complex Phenomenon: An Introduction to the Theoretical Psychology of the Sensory Order," in W.B. Weimar and D.S. Palermo (eds.), *Cognition and Symbolic Processes*, 2, NJ: Lea Publishers.

Wolfram, S. (1984). "Cellular Automata As Models of Complexity," *Physica*, 10D, 1–35.

Index